QD
305
E 7
P 413   Perst, Hartwig
     Oxonium Ions in Organic Chemistry

3 2107

DATE DUE			
DEC 17 1992			

Waubonsee Community College

Perst-Oxonium Ions
in Organic Chemistry

Hartwig Perst

Oxonium Ions in Organic Chemistry

1971
Verlag Chemie
Academic Press

Dr. Hartwig Perst
Institut für Organische Chemie
3550 Marburg/Lahn
Bahnhofstraße 7
Germany

Translated by Express Translation Service, London

With 6 figures and 12 tables

ISBN 3-527-25348-3 Verlag Chemie
ISBN 0-12-551050-0 Academic Press

LIBRARY OF CONGRESS CATALOG CARD NO. 76-159512

Copyright © 1971 by Verlag Chemie GmbH, Weinheim/Bergstr.
All rights reserved (including those of translation into foreign languages). No part of this book may be reproduced in any form — by photoprint, microfilm, or any other means — nor transmitted or translated into a machine language without written permission from the publishers.
Alle Rechte, insbesondere die der Übersetzung in fremde Sprachen, vorbehalten. Kein Teil dieses Buches darf ohne schriftliche Genehmigung des Verlages in irgendeiner Form — durch Photokopie, Mikrofilm oder irgendein anderes Verfahren — reproduziert oder in eine von Maschinen, insbesondere von Datenverarbeitungsmaschinen, verwendbare Sprache übertragen oder übersetzt werden.
Printed in The Netherlands by Mouton & Co, The Hague

PREFACE

Only during the past few years have the trialkyloxonium salts begun to attract the close attention of the chemists who perform preparative investigations, although these salts had been discovered in 1937 by H. Meerwein, and their use as alkylating agents since that time has been known. Other tertiary oxonium salts, the alkoxycarbonium and dialkoxycarbonium salts, have also recently become more important in preparative and theoretical investigations. Although pyrylium salts, the oldest class of substances among the tertiary oxonium salts, were first prepared 70 years ago, no such delay between their discovery and their general preparative use has occurred, and even continued application has not exhausted the reaction possibilities to this day.

Some aspects of oxonium salts, particularly the reactions of pyrylium salts, have been reviewed by various authors, but no comprehensive studies of the whole field of oxonium salts have been available. The present book is therefore intended to close some of the remaining gaps.

Besides the organic oxonium salts isolated as such, primary and secondary oxonium ions have been reported, which could often be detected "only" in solution by low-temperature NMR measurements; these were protonated alcohols, ethers, aldehydes, ketones, and carboxylic acid derivatives. Many of these cations observed in solution permit important conclusions to be drawn about the structure of oxonium ions.

In this book the formation, the modes of reaction, and the preparative aspects of tertiary oxonium salts are described in detail. In addition, there is a large section on oxonium ions whose existence has long been assumed in the interpretation of many reaction mechanisms in organic chemistry. Such cations are often true intermediates, which can be detected by low-temperature NMR spectroscopy and by the isolation of oxonium salts. However, the examples here are restricted to the tertiary oxonium ions occurring as intermediates.

While the field of organic oxonium ions is currently in a stage of rapid growth, it is still quite easy to review its main features. Nevertheless, no complete literature review was planned within the framework of the present book, though an attempt has been made to take into account the largest possible number of original papers, particularly when comprehensive reviews are lacking. Publications that appeared after September 1, 1969, could be included only in exceptional cases.

I am greatly indebted to Prof. K. Dimroth for numerous stimulating discussions, for reading the manuscript, and for relieving me of a substantial part of my university duties during the preparation of this book. At this point I should also like to thank particularly Prof. W. Kirmse and Dr. A. Berndt for reading the manuscript and supplying much valuable information. Dr. S. Kabuss kindly allowed me to use hitherto unpublished results.

H. Perst

Marburg, October 1969.

CONTENTS

1. Historical Introduction 1
2. Classes of Oxonium Ions 5
   2.1. Classification 5
   2.2. Nomenclature 8
3. Stability of Oxonium Salts 11
   3.1. Stabilization of the Cation 11
      3.1.1. General Considerations on Stability 11
      3.1.2. Charge Delocalization 12
   3.2. Influence of the Anion 17
4. Formation of Oxonium Salts 22
   4.1. General Methods of Preparation 22
      4.1.1. Formation of Oxonium Ions 22
         4.1.1.1. Primary and Secondary Oxonium Ions 22
         4.1.1.2. Tertiary Acyclic Oxonium Ions 22
         4.1.1.3. Tertiary Cyclic Oxonium Ions 23
      4.1.2. Alkylating Agents 26
      4.1.3. Acceptors for Anionically Leaving Groups 28
      4.1.4. Origin of the Complex Anions 32
   4.2. Possibilities of Synthesis for Individual Classes of Oxonium Salts . 34
      4.2.1. Trialkyloxonium Salts 34
      4.2.2. Dialkoxycarbonium Salts 39
         4.2.2.1. Acyclic Dialkoxycarbonium Salts 39
         4.2.2.2. 1,3-Dioxolenium Salts 39
      4.2.3. Pyrylium Salts 43
         4.2.3.1. Pyrylium Salts from Preformed Oxygen Heterocycles 45
         4.2.3.2. Pyrylium Syntheses by Ring-Closure Reactions . . 47
5. Reactions of Oxonium Salts 58
   5.1. Acidic Oxonium Salts 60
      5.1.1. Deprotonation 60
      5.1.2. Splitting Out of Carbonium Ions 63
         5.1.2.1. Acidic Saturated Oxonium Ions 63
         5.1.2.2. Acidic Carboxonium Ions 66
      5.1.3. Reactions at the Carbonium Center of Acidic Carboxonium Ions 68
   5.2. Saturated Tertiary Oxonium Salts 70
      5.2.1. Trialkyloxonium Salts 70
      5.2.2. Triaryloxonium Salts 75
      5.2.3. Acyldialkyloxonium Salts 77
   5.3. Ambident Tertiary Oxonium Salts 78
      5.3.1. General Considerations on Modes of Reaction 78

Contents

    5.3.2. Reactions with Various Nucleophiles 80
    5.3.3. Influence of the Ambident Cation on the Result of the Reaction   91
        5.3.3.1. Cations of Different Energies. 91
        5.3.3.2. Steric Effects 94

6. Tertiary Oxonium Ions as Intermediates in Chemical Reactions 100
    6.1. Neighboring-Group Participation of Ether Groups 102
    6.2. Participation of Keto Carbonyl Groups 105
    6.3. Participation of Ester Groups as Neighboring Groups 106
        6.3.1. Formation of Lactonium Ions 107
        6.3.2. Formation of 1,3-Dioxacycloalkenium Ions 110
            6.3.2.1. 1,3-Dioxolenium Ions 110
            6.3.2.2. 1,3-Dioxenium Ions 123

7. Preparative Application of Oxonium Salts 128
    7.1. Alkylation with Trialkyloxonium Salts 128
        7.1.1. Activation of C–X Multiple Bonds 128
            7.1.1.1. Activation of Ketones 129
            7.1.1.2. Conversion of Carboxylic Amide Functions 131
            7.1.1.3. Nitrile Alkylation 137
        7.1.2. Trialkyloxonium Salts as Anion–Acceptors 138
            7.1.2.1. Alkoxide Transfer 138
            7.1.2.2. Hydride Transfers 139
        7.1.3. Catalytic Action of Trialkyloxonium Salts 140
    7.2. Acyclic Alkoxycarbonium Salts 142
        7.2.1. Use as Alkylating Agents 142
        7.2.2. Use as Anion–Acceptors 143
    7.3. Preparative Use of Dialkoxycarbonium Salts 145
        7.3.1. Use of Acyclic Dialkoxycarbonium Salts as Alkylating Agents   146
        7.3.2. Use of Dialkoxycarbonium Salts as Anion Acceptors 146
    7.4. Reactions of Pyrylium Salts 149
        7.4.1. Primary Reactions 149
            7.4.1.1. Review of Reaction Possibilities 149
            7.4.1.2. Isolable Primary Products from Pyrilium Salts . . . 150
        7.4.2. Secondary Reactions of Nucleophilic Addition at C_2 . . . 156
            7.4.2.1. Review 156
            7.4.2.2. Syntheses with $C_2 \rightarrow C_6$ Linkage 158
            7.4.2.3. Syntheses with $C_2 \rightarrow C_5$ Linkage 164
        7.4.3. Secondary, Reactions of Nucleophilic Addition at C_4 . . . 166
            7.4.3.1. Review 166
            7.4.3.2. Exchange Reactions of Substituents in Position 4 . . 167
            7.4.3.3. Preparation of Naphthalene Derivatives from 4-Benzyl-4H-Pyrans 167

Appendix: The Structure of Oxonium Ions 173

A.1. Saturated Oxonium Ions	.	173
A.2. Aldehydium and Ketonium Ions	.	173
A.3. Acidium Ions	.	176
Author Index	.	180
Subject Index	.	192

1. Historical Introduction

At the turn of the century, among the compounds of the elements belonging to the sixth main group of the periodic system, trialkylsulfonium [1], trialkylselenonium [2], and trialkyltelluronium salts [2,3] were known, but the corresponding oxygen compounds had yet to be identified. The tertiary salts of the sulfur, selenium, and tellurium series were obtained by reactions of the corresponding dialkyl compounds with alkyl iodides, i.e., in a manner analogous to the preparation of quaternary ammonium halides from tertiary amines and alkyl halides:

$$R_3N + R'-Hal \longrightarrow R_3N^{\oplus}-R' \; Hal^{\ominus}$$

$$R-\underline{X}-R + R'-I \longrightarrow R-\overset{\oplus}{\underset{|}{X}}(R')-R \; I^{\ominus} \quad \begin{array}{l} X = S, Se, Te \\ X \neq O \end{array}$$

However, trialkyl*oxonium* halides (X = O) could *not* be obtained by the action of alkyl halides on dialkyl ethers [4].

Because of the failure of such attempts it was believed that oxygen generally formed no compounds with a valence higher than 2. It is true that since the middle of the nineteenth century more and more addition compounds of ethers [5,6], aldehydes [7], and ketones [8] with protonic acids had become known, but no ideas of their structure existed. Collie and Tickle in 1899 were the first to formulate the complexes of 2,6-dimethylpyrone with acids as oxonium salts *(3)* on the basis of conductivity measurements [9]. In analogy with the manner of writing ammonium salts with pentavalent nitrogen *(1)*, such oxonium salts *(2)* were then written with tetravalent oxygen [10]. The addition compounds of alcohols, ethers, aldehydes, ketones, and carboxylic acid esters with complex protonic acids that had been prepared by Baeyer and Villiger were also formulated with tetravalent oxygen [11,12].

$$R-\underset{R'}{\overset{R}{N^{V}}}\underset{Y}{\overset{H}{\diagdown}} \qquad \underset{R'}{\overset{R}{\diagdown}}\underset{Y}{\overset{IV}{O}}\underset{Y}{\overset{H}{\diagdown}} \qquad Y = halogen$$

$$(1) \qquad\qquad (2)$$

The formulation assumed by Collie and Tickle for the dimethylpyrone-HCl complex *(3)* was later replaced by the hydroxypyrylium formula *(4)* that is used today [13–15]. In this the oxygen is trivalent, since it is connected with one of the two adjacent carbon atoms by a formal double bond:

(3) *(4)*

Pyrylium salts and their derivatives with fused benzene rings [13, 14, 16] remained the only compounds in which the "trivalent" oxygen is linked only to carbon atoms —until 1937*, when the trialkyloxonium salts were discovered by H. Meerwein et al. [23, 24]. In all other oxonium salts known at that time, one of the residues bound to the oxygen is always a *proton* [17a].

However, all those oxonium compounds in which the oxygen atom is doubly bound, such as the pyrylium salts, can also be formulated as *carbonium* salts. According to the ideas of that time, the negative counter-ion was therefore regarded as being fixed to the oxygen or to the carbon atom.

In the first two decades of the twentieth century the latter theory led to vigorous arguments and, at first to a renunciation of the oxonium structures for these compounds [18–21a,b]. However, the discussion became pointless when the ionic structure of saltlike compounds was recognized and it was learned that oxonium and carbonium ions were resonance structures of the same compound [21 a,b, 22]. This is plainly seen in the examples of a pyrylium salt and a protonated ketone:

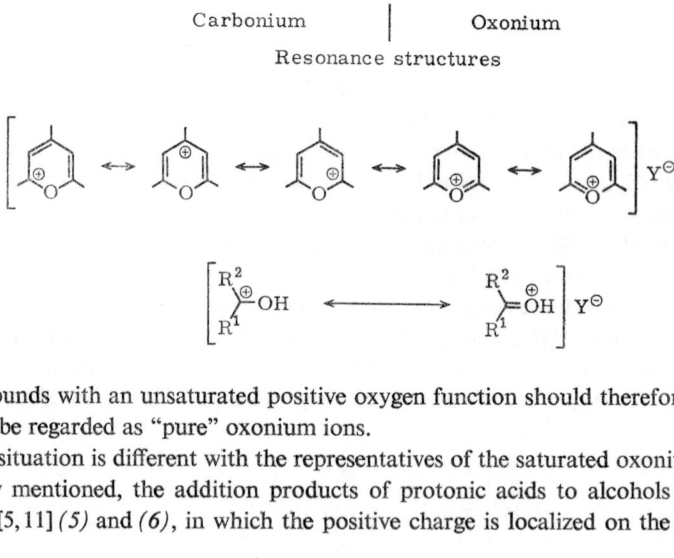

Carbonium | Oxonium

Resonance structures

Compounds with an unsaturated positive oxygen function should therefore by no means be regarded as "pure" oxonium ions.

The situation is different with the representatives of the saturated oxonium salts already mentioned, the addition products of protonic acids to alcohols [11], or ethers [5, 11] *(5)* and *(6)*, in which the positive charge is localized on the oxygen.

---

* Other tertiary oxonium salts, the dihydrofurylium salts (B), were discovered as early as 1915 by Faworskii and Venus. They thought, however, the structures to be those of "acidic" oxonium salts, formed by addition of an acid to an oxygen atom of the hydroxydihydrofuran derivatives (A) [17b]. It was not before 1956 that the correct structure (B) was confirmed [17c]:

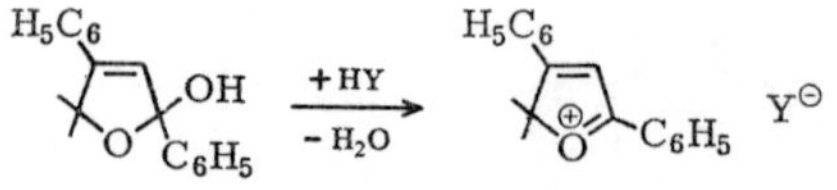

Here the final proof of the validity of the oxonium theory was supplied only by the the discovery of the trialkyloxonium salts *(7)* [23,24]:

$$R-\overset{\oplus}{O}\overset{H}{\underset{H}{\diagdown}} \quad Y^{\ominus} \qquad R-\overset{\oplus}{O}\overset{R}{\underset{H}{\diagdown}} \quad Y^{\ominus} \qquad R-\overset{\oplus}{O}\overset{R}{\underset{R}{\diagdown}} \quad Y^{\ominus}$$

$$(5) \qquad\qquad (6) \qquad\qquad (7)$$

H. Meerwein et al. [23] showed convincingly that such trialkyloxonium salts *(7)* are capable of existence only with complex anions of low polarizability (e.g., BF_4^-, $SbCl_6^-$). Being strong alkylating agents, compounds *(7)* transfer alkyl groups to halogen anions, giving alkyl halides and dialkyl ethers. This is just the converse of all previously attempted methods of preparation, which were analogous to the formations of the trialkylsulfonium, trialkylselenonium, and trialkyltelluronium halides, and it explains why trialkyloxonium halides cannot be synthesized in this way [23].

After the decisive role of the anion had been recognized, Meerwein was able to obtain on the basis of this concept the O-alkyl cations of carbonyl compounds *(8)*, carboxylic acid esters *(9)*, and other oxygen derivatives as salts with complex anions [22–25].

$$\underset{R^1}{\overset{R^2}{\diagup}}\hspace{-1em}\overset{\oplus}{=}\text{OR} \quad BF_4^{\ominus} \qquad\qquad R^1-C\overset{OR}{\underset{OR}{\diagdown}}\overset{\oplus}{}\quad BF_4^{\ominus}$$

(8a): R^1 = H, R^2 = alkyl, aryl *(9)*
(8b): R^1, R^2 = alkyl, aryl

The structure of numerous oxonium salts of this type has been confirmed by NMR spectroscopic measurements since 1963 [26–28]. Furthermore—particularly by the work of Olah and associates [29–34]—at sufficiently low temperatures the oxonium ions arising by the protonation of a wide range of oxygen-containing compounds (alcohols, ethers, aldehydes, ketones, carboxylic acids, carboxylic esters) have been detected directly in the same way. By means of low-temperature NMR measurements it has recently been proved that oxonium salts of type *(9)* with noncomplex anions (e.g., the bromide) are capable of existence at −80°C [35].

References

1. A. von Oefele, *Ann. Chem. Liebigs* **132**, 82 (1864).
2. A. Cahours, *Ann. Chem. Liebigs* **135**, 356 (1865); L. von Pieverling, *Ann. Chem. Liebigs* **185**, 333 (1877).
3. A. Cahours, *Compt. Rend.* **60**, 624 (1865).
4. H. Decker and T. von Fellenberg, *Ann. Chem. Liebigs* **364**, 1 (1909).
5. O. Wallach and W. Brass, *Ann. Chem. Liebigs* **225**, 297 (1884); K. Schäfer and F. Hein, *Z. Anorg. Allgem. Chem.* **101**, 272 (1917).
6. C. Friedel, *Ber. Deut. Chem. Ges.* **8**, 642 (1875).

7. J. Dumas and E. Peligot, *Ann. Chem. Liebigs* **14**, 50 (1835).
8. J. Kachler, *Ann. Chem. Liebigs* **159**, 281 (1871).
9. J. N. Collie and T. Tickle, *J. Chem. Soc.* **75**, 710 (1899); J. N. Collie, *J. Chem. Soc.* **85**, 973 (1904).
10. P. Walden, *Ber. Deut. Chem. Ges.* **34**, 4185 (1901); **35**, 1764 (1902); J. Schmidt, "Über die basischen Eigenschaften des Sauerstoffs und Kohlenstoffs." Verlag Gebr. Bornträger, Berlin, 1904.
11. A. Baeyer and V. Villiger, *Ber. Deut. Chem. Ges.* **34**, 2679 (1901).
12. A. Baeyer and V. Villiger, *Ber. Deut. Chem. Ges.* **34**, 3612 (1901).
13. A. Werner, *Ber. Deut. Chem. Ges.* **34**, 3300 (1901).
14. A. Baeyer, *Ber. Deut. Chem. Ges.* **43**, 2337 (1910).
15. A. Hantzsch, *Ber. Deut. Chem. Ges.* **52**, 1535 (1919); **52**, 1544 (1919).
16. H. Decker and T. von Fellenberg, *Ann. Chem. Liebigs* **356**, 281 (1907); W. Dilthey, *J. Prakt. Chem.* **94**, 53 (1916); **102**, 209 (1921); **104**, 28 (1922); C. Gastaldi, *Gazz. Chim. Ital.* **5**, 169 (1921).
17. a) P. Pfeiffer, *Ann. Chem. Liebigs* **412**, 253 (1917); b) A. Faworskii and E. Venus, *Zh. Russ. Fiz.-Khim. Obshch.* **47**, 133 (1915); c) A. Fabrycy, *Chimia (Aarau)* **15**, 552 (1961).
18. A. Baeyer, *Ber. Deut. Chem. Ges.* **38**, 569 (1905).
19. F. Kehrmann and H. de Gottrau, *Ber. Deut. Chem. Ges.* **38**, 2574 (1905); F. Kehrmann, *Ber. Deut. Chem. Ges.* **38**, 2959 (1905); F. Kehrmann and A. Duttenhöfer, *Ber. Deut. Chem. Ges.* **39**, 1299 (1906).
20. A. Hantzsch, *Ber. Deut. Chem. Ges.* **38**, 2143 (1905); **39**, 153 (1906).
21. a) D. W. Hill, *Chem. Rev.* **19**, 42 (1936); b) C. D. Nenitzescu, in "Carbonium Ions" (G. A. Olah and P. von R. Schleyer, eds.). Wiley (Interscience), p. 1. New York, 1968.
22. H. Meerwein, in "Methoden der organischen Chemie (Houben-Weyl)" (E. Müller, ed.), 4. Auflage, Bd. VI/3: Sauerstoffverbindungen I, p. 330. Thieme, Stuttgart, 1965.
23. H. Meerwein, G. Hinz, P. Hofmann, E. Kroning, and E. Pfeil, *J. Prakt. Chem.* **147**, 257 (1937).
24. H. Meerwein, E. Battenberg, H. Gold, E. Pfeil, and G. Willfang, *J. Prakt. Chem.* **154**, 83 (1939).
25. H. Meerwein, *Angew. Chem.* **67**, 374 (1955); H. Meerwein, P. Borner, O. Fuchs, H. J. Sasse, H. Schrodt, and J. Spille, *Chem. Ber.* **89**, 2071 (1956); H. Meerwein, K. Bodenbenner, P. Borner, F. Kunert, and K. Wunderlich, *Ann. Chem. Liebigs* **632**, 38 (1960).
26. C. B. Anderson, E. C. Friedrich, and S. Winstein, *Tetrahedron Letters* p. 2037 (1963).
27. H. Hart and D. A. Tomalia, *Tetrahedron Letters* p. 3383 (1966); D. A. Tomalia and H. Hart, *Tetrahedron Letters* p. 3389 (1966); B. G. Ramsey and R. W. Taft, *J. Am. Chem. Soc.* **88**, 3058 (1966); H. Hart and D. A. Tomalia, *Tetrahedron Letters* p. 1347 (1967); G. A. Olah and J. M. Bollinger, *J. Am. Chem. Soc.* **89**, 2993 (1967).
28. J. B. Lambert and D. H. Johnson, *J. Am. Chem. Soc.* **90**, 1349 (1968).
29. G. A. Olah, J. Sommer, and E. Namanworth, *J. Am. Chem. Soc.* **89**, 3576 (1967); G. A. Olah and J. Sommer, *J. Am. Chem. Soc.* **90**, 927 (1968).
30. G. A. Olah and D. H. O'Brien, *J. Am. Chem. Soc.* **89**, 1725 (1967).
31. G. A. Olah, D. H. O'Brien, and M. Calin, *J. Am. Chem. Soc.* **89**, 3582 (1967).
32. G. A. Olah, M. Calin, and D. H. O'Brien, *J. Am. Chem. Soc.* **89**, 3586 (1967); M. Brookhart, G. C. Levy, and S. Winstein, *J. Am. Chem. Soc.* **89**, 1735 (1967).
33. H. Hogeveen, A. F. Bickel, C. W. Hilbers, E. L. Mackor, and C. MacLean, *Rec. Trav. Chim.* **86**, 687 (1967); G. A. Olah and A. M. White, *J. Am. Chem. Soc.* **89**, 3591 (1967); **89**, 4752 (1967); G. A. Olah and M. Calin, *J. Am. Chem. Soc.* **90**, 405 (1968).
34. G. A. Olah, D. H. O'Brien, and A. M. White, *J. Am. Chem. Soc.* **89**, 5694 (1967); G. A. Olah and A. M. White, *J. Am. Chem. Soc.* **90**, 1884 (1968).
35. C. H. V. Dusseau, S. E. Schaafsma, H. Steinberg, and T. J. de Boer, *Tetrahedron Letters* p. 467 (1969).

2. Classes of Oxonium Ions

2.1. Classification

Oxonium ions contain positively *tri*valent oxygen with an electron octet. Cations with positively *mono*valent *sextet* oxygen, for which the name oxenium ions is customary [1], must be carefully distinguished from trivalent.

$$\overset{\oplus}{\underset{|}{\text{O}}}\diagup \qquad -\overset{\oplus}{\underline{\text{O}}}$$

Oxonium ion Oxenium ion

Our discussion will be limited to *oxonium ions*, in which the oxygen is bound to organic residues or to hydrogen. Depending on whether all the ligands are attached to the oxygen by single bonds *(7)*, or whether there is a C–O double bond *(8)* or even a C–O triple bond *(10)*, three classes of compounds can be expected:

$$\underset{\underset{R}{|}}{\overset{R\diagdown \;\oplus\; \diagup R}{\text{O}}} \qquad \underset{R^1}{\overset{R^2}{\diagdown}}\text{C}=\overset{\oplus}{\underline{\text{O}}}-\text{R} \qquad \text{R}^1-\text{C}\equiv\overset{\oplus}{\text{O}}|$$

(7)   *(8)*   *(10)*

In the first two groups the organic residues may also be linked to one another. The oxonium center is then a constituent of an *n*-membered ring; *cf.* Table 1 and *(12)* and *(14)*.

In saturated oxonium ions the positive charge is substantially localized on the oxygen, while in the cations formulated with a C–O double bond it is delocalized. These therefore behave as oxonium and carbonium ions, which will be expressed by means of the carboxonium formula [2,3]:

$$\underset{R^1}{\overset{R^2}{\diagdown}}\text{C}=\overset{\oplus}{\underline{\text{O}}}-\text{R} \quad\longleftrightarrow\quad \underset{R^1}{\overset{R^2}{\diagdown}}\overset{\oplus}{\text{C}}-\underline{\text{O}}-\text{R} \quad\longleftrightarrow\quad \underset{R^1}{\overset{R^2}{\diagdown}}\text{C}\overset{\oplus}{=}\text{O}-\text{R}$$

Oxonium ion   Alkoxycarbonium ion   Carboxonium ion

The more highly unsaturated the oxonium center, i.e., the smaller the number of ligands bound to the oxygen atom, the less pronounced will be the oxonium nature of the center. Then the unsaturated oxonium structure has a rather formal significance.

Oxonium ions with a formal C–O triple bond are the acyl cations *(10)* [4]. The term "oxinium ions" has been proposed for cations of this type [1].

However, the behavior of acyl cations can be described by the oxo*carbonium* structure; the contribution of the oxinium structure is probably small [5]:

$$\text{R}^1-\text{C}\equiv\overset{\oplus}{\text{O}}| \quad\longleftrightarrow\quad \text{R}^1-\overset{\oplus}{\text{C}}=\overline{\text{O}}|$$

Oxinium ion   Oxocarbonium ion

Consequently we shall not deal here with the acyl cations.

A further classification within the two remaining classes of compounds can be obtained according to the *number of hydrogen atoms on the oxonium center*:

Saturated oxonium ions can be regarded as derived from the hydroxonium ion $H_3O^{\oplus}$. Accordingly, if one, two, or all three hydrogen atoms are replaced by organic residues, we obtain primary *(5)*, secondary *(6)*, and tertiary *(7)* oxonium ions, which may also be regarded as alkyl derivatives of water, the alcohols, and the ethers. In the carboxonium ion series, only secondary *(13)* and tertiary *(8)* cations exist; *cf.* Table 1.

TABLE 1. Types of Oxonium Ions with One Oxygen Atom

Oxonium center	Number of ligands on the O-atom	Types of formula				
		Acyclic			Cyclic	
		Primary	Secondary	Tertiary	Secondary	Tertiary
Saturated	3	$\overset{R}{\underset{H}{\overset{\oplus}{O}}}\!-\!H$ *(5)* [6]	$\overset{R}{\underset{R}{\overset{\oplus}{O}}}\!-\!H$ *(6)* [7]	$\overset{R}{\underset{R}{\overset{\oplus}{O}}}\!-\!R$ *(7)* [8]	⊕O–H *(11)* [9]	⊕O–R *(12)* [10]
Unsaturated	2		$\overset{R^2}{\underset{R^1}{C}}\!=\!\overset{\oplus}{O}\!-\!H$ *(13)* [11]	$\overset{R^2}{\underset{R^1}{C}}\!=\!\overset{\oplus}{O}\!-\!R$ *(8)* [8,12]		$R^1\!-\!C\overset{\oplus}{O}$ *(14)* [13]
	1			$[R^1\!-\!C\!\equiv\!\overset{\oplus}{O}]$ *(10)* [4,5]		

A still further subdivision is possible in the case of the carboxonium ions because the carbonium center may bear one, two, or three oxygen atoms. Just as the cations *(8)*, *(13)*, and *(14)* with *one* oxygen atom can be obtained from aldehydes or ketones (*cf.* Table 1), carboxylic acids, carboxylic esters, or carbonic esters yield carboxonium ions with *two* or *three* O-atoms in the neighborhood of the carbonium center. Here again there are *cyclic* carbonium ions when, for example, the starting materials are the internal esters, the lactones. In these compounds, therefore, the positive charge is distributed over three *(9)*, *(17)*–*(21)* or four atoms *(22)*–*(27)*.

TABLE 2. Types of Carboxonium Ions with Two and Three Oxygen Atoms at the Carbonium Center

Number of O-atoms	Types of formula					
	Acyclic			Cyclic		
	Secondary		Tertiary		Secondary	Tertiary
2	$\overset{HO}{\underset{R^1}{C}}\!=\!\overset{\oplus}{OH}$ *(17)* [15,20]	$\overset{RO}{\underset{R^1}{C}}\!=\!\overset{\oplus}{OH}$ *(18)* [16,20]	$\overset{RO}{\underset{R^1}{C}}\!=\!\overset{\oplus}{OR}$ *(9)* [12]	$\overset{O}{C}\!=\!\overset{\oplus}{OH}$ *(19)*	$\overset{O}{C}\!=\!\overset{\oplus}{OR}$ *(20)* [17]	$\overset{O}{\underset{O}{C}}\!-\!R^1$ *(21)* [18]
3	$\overset{HO}{\underset{HO}{C}}\!=\!\overset{\oplus}{OH}$ *(22)* [19]	$\overset{HO}{\underset{RO}{C}}\!=\!\overset{\oplus}{OH}$ *(23)* [19]	$\overset{RO}{\underset{RO}{C}}\!=\!\overset{\oplus}{OH}$ *(24)* [19,20]	$\overset{RO}{\underset{RO}{C}}\!=\!\overset{\oplus}{OR}$ *(25)* [12,17]	$\overset{O}{\underset{O}{C}}\!=\!\overset{\oplus}{OH}$ *(26)* [19]	$\overset{O}{\underset{O}{C}}\!=\!\overset{\oplus}{OR}$ *(27)* [21]

Secondary and tertiary ("acidic" or "neutral") carboxonium ions can also be regarded as carbonium ions in which the carbonium center is substituted by hydroxy or alkoxy (aryloxy) groups (*cf.* Table 2).

If the carbonium center bears, besides oxygen, other electron-donating groups X (especially the heteroatoms N,S), *(15)* and *(16)*, it is often questionable whether oxonium structures make an appreciable contribution to such cations:

(15) $\overset{X}{\underset{R^1}{C}}=\overset{\oplus}{\underline{O}}R \leftrightarrow \overset{X}{\underset{R^1}{\overset{\oplus}{C}}}-\underline{\bar{O}}R \leftrightarrow \overset{\overset{\oplus}{X}}{\underset{R^1}{C}}-\underline{\bar{O}}R \qquad \overset{X}{\underset{R^1}{\overset{\cdot\cdot\oplus}{C}}}{=}\underline{O}R$

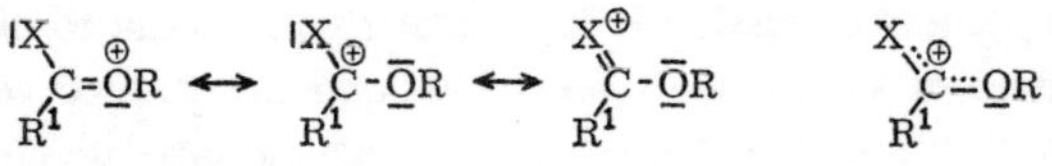

(16) $\overset{X}{\underset{X}{C}}=\overset{\oplus}{\underline{O}}R \leftrightarrow \overset{X}{\underset{X}{\overset{\oplus}{C}}}-\underline{\bar{O}}R \leftrightarrow \overset{\overset{\oplus}{X}}{\underset{X}{C}}-\underline{\bar{O}}R \leftrightarrow \overset{X}{\underset{\overset{\oplus}{X}}{C}}-\underline{\bar{O}}R \qquad \overset{X}{\underset{X}{\overset{\oplus}{C}}}{=}\underline{O}R$

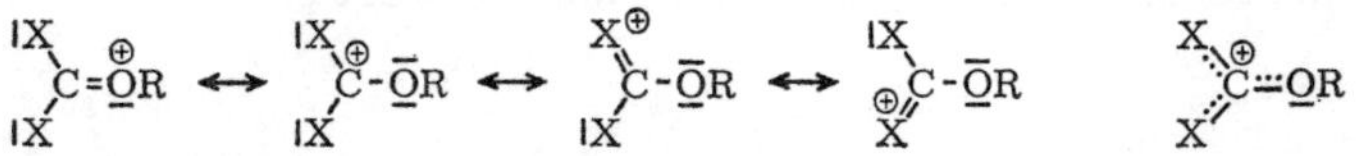

X = NR$_2$, SR

Only in exceptional cases, therefore, will cations also containing N and S atoms be dealt with below. Structures *(15)* and *(16)* belong to heteroatom-stabilized carbonium ions [14]; in the special case where X = oxygen they pass into the carboxonium ions *(9)*, *(17)–(27)*.

The electron-donating substituent X in *(15)* and *(16)* may also be a vinyl group; for such cations the participation of hydroxy- or alkoxyallyl structures *(28)* is then to be expected [22]:

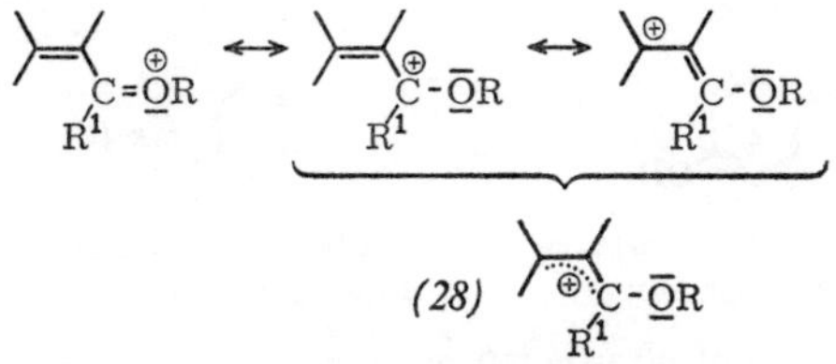

$\underset{R^1}{\searrow}C=\overset{\oplus}{\underline{O}}R \leftrightarrow \underset{R^1}{\searrow}\overset{\oplus}{C}-\underline{\bar{O}}R \leftrightarrow \underset{R^1}{\overset{\oplus}{\searrow}}C-\underline{\bar{O}}R$

(28) $\underset{R^1}{\searrow}\overset{\oplus}{C}-\underline{\bar{O}}R$

It is worth noting their relationships with enamines.

Cyclic carboxonium ions having double bonds in conjugation with the C–O double bond are particularly numerous. Cyclic conjugation can then lead to *aromatic* systems as in the pyrylium ions or their derivatives with fused benzene rings *(29)–(31)*.

(29) Pyrylium ion 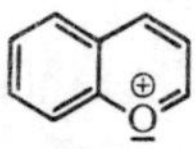   *(30)* Benzopyrylium ion 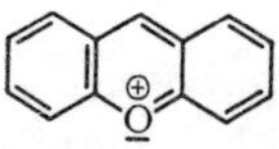   *(31)* Xanthylium ion

Similarly, 1,3-dioxolium ions *(32)*, as 6-π-electron heterocycles, may exhibit aromatic characteristics [19, 21]:

Classes of Oxonium Ions

(32)

2.2. Nomenclature [2]

In the subsequent discussion it is particularly important to distinguish between the two classes of saturated *oxonium* ions and unsaturated *carboxonium* ions*. Only the *acyclic* primary, secondary, and tertiary cations with an oxygen coordination number of 3 are uniformly named oxonium ions; e.g.:

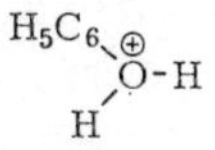

Phenyloxonium ion 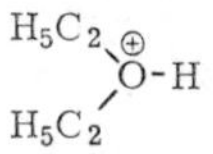   Diethyloxonium ion 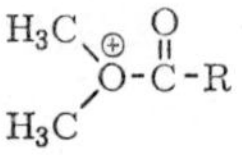   Dimethylacyloxonium ion

In general, however, the suffix "-ium" is added to the parent compound from which the cation is derived; e.g.:

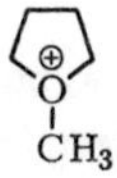

1-Methyltetrahydro-   Benzaldehydium ion 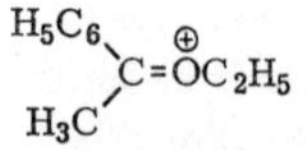   O-Ethyraceto-
furanium ion   phenonium ion

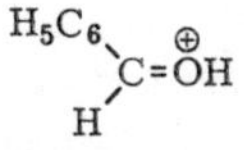

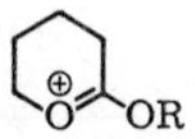

O-Alkyl-δ-valero-   2-Alkyl-1,3-benzo
lactonium ion   dioxolium ion

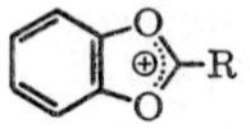

Apart from this type of nomenclature there is that of the acyclic carboxonium ions derived from carboxylic acids or carboxylic esters, which are called acylate acidium ions; e.g.:

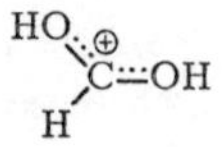

Formate   O-Methylacetate   O,O-Diethylcarbonate
acidium ion   acidium ion   acidium ion

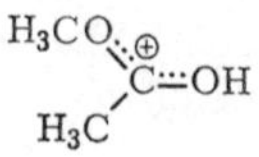

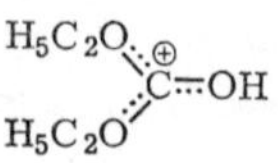

All these carboxonium ions may also be called "substituted" carbonium ions. This method is frequently used in the case of acyclic cations. Consequently, the following names are obtained for the carboxonium ions named differently above:

---

* Acyl cations were originally termed "carboxonium ions" [23]. In the meantime, in accordance with Meerwein's proposal, the term "carboxonium ions" has come into use for oxonium ions having an oxygen coordination number of 2 [2].

| Hydroxy phenyl carbonium ion | Ethoxy methyl phenyl carbonium ion | Dihydroxy carbonium ion | Hydroxy methoxy methyl carbonium ion | Diethoxy hydroxy carbonium ion |

(structures shown above: H_5C_6–$\overset{\oplus}{C}$(H)–OH ; H_5C_6–$\overset{\oplus}{C}$(H_3C)–OC_2H_5 ; H–$\overset{\oplus}{C}$(HO)–OH ; H_3CO–$\overset{\oplus}{C}$(H_3C)–OH ; H_5C_2O–$\overset{\oplus}{C}$(H_5C_2O)–OH)

Here the sequences of prefixes in German is prescribed by the Beilstein nomenclature [24]. In English-speaking countries, the substituents are given in alphabetical order.

The -a- nomenclature is generally applicable to cyclic carboxonium ions*; e.g.:

  2-Methyl-1,3-dioxacyclohex-1-enium ion

  2-Methyl-*cis*-4,5-tetramethylene-1,3-dioxacyclopent-1-enium ion

In addition to this, these compounds are also called 2-methyl-*1,3-dioxenium* and 2-methyl-*cis*-4,5-tetramethylene-*1,3-dioxolenium* ions [2] in order to show their relationship to 1,3-dioxane and 1,3-dioxolane, respectively; the new international nomenclature uses the terms 1,3-dioxanylium and 1,3-dioxolanylium ions [26]. Furthermore, for the two 2-methyl-substituted systems the names 1,3- and 1,2-acetoxonium ions have become established, being intended to show their method of formation from acetates by a ring-closure reaction [27]; (correspondingly, the 2-phenyl derivatives are called "benzoxonium ions").

For the aromatic oxygen heterocycle the term "pyrylium ion" has become established, and trivial names are also customary for its derivatives with fused benzene rings; *cf.* Section 2.1.

References

1. J. J. Jennen, *Chimia (Aarau)* **20**, 309 (1966).
2. H. Meerwein, *in* "Methoden der organischen Chemie (Houben-Weyl)" (E. Müller, ed.), 4. Auflage, Bd. VI/3: Sauerstoffverbindungen I, p. 330. Thieme, Stuttgart, 1965.
3. S. Hünig, *Angew. Chem.* **76**, 400 (1964); *Angew. Chem. Intern. Ed. English* **3**, 548 (1964).
4. G. A. Olah, S. J. Kuhn, W. S. Tolgyesi, and E. B. Baker, *J. Am. Chem. Soc.* **84**, 2733 (1962); G. A. Olah, W. S. Tolgyesi, S. J. Kuhn, M. E. Moffatt, I. J. Bastien, and E. B. Baker, *J. Am. Chem. Soc.* **85**, 1328 (1963); G. A. Olah and M. B. Comisarow, *J. Am.*

---

* According to the IUPAC nomenclature, the oxonium center in cyclic compounds is denoted by "oxonia" [25].

Chem. Soc. **88**, 3313, 4442 (1966); N. C. Deno, C. U. Pittman, Jr., and M. J. Wisotsky, *J. Am. Chem. Soc.* **86**, 4370 (1964); D. Cook, in "Friedel-Crafts and Related Reactions" (G. A. Olah, ed.), Vol. 1, p. 790. Wiley (Interscience), New York, 1963.
5. F. P. Boer, *J. Am. Chem. Soc.* **88**, 1572 (1966).
6. G. A. Olah, J. Sommer, and E. Namanworth, *J. Am. Chem. Soc.* **89**, 3576 (1967); G. A. Olah and J. Sommer, *J. Am. Chem. Soc.* **90**, 927 (1968).
7. F. Klages, H. Meuresch, and W. Steppich, *Ann. Chem. Liebigs* **592**, 116 (1955); G. A. Olah and D. H. O'Brien, *J. Am. Chem. Soc.* **89**, 1725 (1967).
8. H. Meerwein, G. Hinz, P. Hofmann, E. Kroning, and E. Pfeil, *J. Prakt. Chem.* **147**, 257 (1937).
9. O. Wallach and W. Brass, *Ann. Chem. Liebigs* **225**, 297 (1884); O. Wallach and E. Gildemeister, *Ann. Chem. Liebigs* **246**, 280 (1888); H. Meerwein, see Reference 2, p. 334.
10. H. Meerwein, see Reference 2, p. 338; A. Kirrmann and L. Wartski, *Bull. Soc. Chim. France* p. 3825 (1966).
11. F. Klages, H. Träger, and E. Mühlbauer, *Chem. Ber.* **92**, 1819 (1959); G. A. Olah, M. Calin, and D. H. O'Brien, *J. Am. Chem. Soc.* **89**, 3586 (1967).
12. H. Meerwein, *Angew. Chem.* **67**, 374 (1955).
13. K. Dimroth and W. Mach, *Angew. Chem.* **80**, 489 (1968); *Angew. Chem. Intern. Ed. English* **7**, 460 (1968).
14. G. A. Olah, *Chem. Eng. News* **45**, 77 (1967).
15. G. A. Olah and A. M. White, *J. Am. Chem. Soc.* **89**, 3591, 4752 (1967).
16. G. A. Olah, D. H. O'Brien, and A. M. White, *J. Am. Chem. Soc.* **89**, 5694 (1967).
17. H. Meerwein, P. Borner, O. Fuchs, H. J. Sasse, H. Schrodt, and J. Spille, *Chem. Ber.* **89**, 2071 (1956).
18. H. Meerwein, K. Bodenbenner, P. Borner, F. Kunert, and K. Wunderlich, *Ann. Chem. Liebigs* **632**, 38 (1960).
19. G. A. Olah and A. M. White, *J. Am. Chem. Soc.* **90**, 1884 (1968).
20. F. Klages and E. Zange, *Chem. Ber.* **92**, 1828 (1959).
21. C. P. Heinrich, Ph.D. Thesis, University of Marburg, 1966.
22. D. M. Brouwer, *Tetrahedron Letters* p. 453 (1968).
23. F. Seel, *Z. Anorg. Allgem. Chem.* **250**, 331 (1943).
24. "Beilstein's Handbuch der Organischen Chemie" (Beilstein-Institut für Literatur der Organischen Chemie, ed.), 4. Auflage, 3. Ergänzungswerk Bd. 5/1, p. XXXVII. Springer, Berlin, 1963.
25. "IUPAC Nomenclature of Organic Chemistry," 2nd Ed., p. 68. Butterworths, London and Washington, D.C., 1966.
26. S. Kabuss, *Angew. Chem.* **78**, 940 (1966); *Angew. Chem. Intern. Ed. English* **5**, 896 (1966).
27. C. B. Anderson, E. C. Friedrich, and S. Winstein, *Tetrahedron Letters* p. 2037 (1963).

3. Stability of Oxonium Salts

Whether oxonium salts can be isolated as such depends both on the stabilization of the oxonium cation and on the nature of the anion.

3.1. Stabilization of the Cation

3.1.1. General Considerations on Stability

Onium compounds of oxygen with hydrogen as ligands are thermodynamically more stable than the corresponding sulfonium complexes [1]. In contrast, the stability of the trialkyloxonium salts is much lower than that of the analogous compounds of the sulfonium series:

$$\text{Stability} \quad \underset{R}{\overset{R}{>}}\overset{\oplus}{O}\text{-H} \;\; Y^{\ominus} \;\;>\;\; \underset{R}{\overset{R}{>}}\overset{\oplus}{S}\text{-H} \;\; Y^{\ominus} \qquad \underset{R}{\overset{R}{>}}\overset{\oplus}{O}\text{-R} \;\; Y^{\ominus} \;\;<\;\; \underset{R}{\overset{R}{>}}\overset{\oplus}{S}\text{-R} \;\; Y^{\ominus}$$

where R = alkyl

This behavior can be explained by R. G. Pearson's concept of "hard" and "soft" acids and bases [2]: Oxonium and sulfonium ions are connected with the con-conjugate bases R_2X by the relationships

$$H^{\oplus} + R_2X \rightleftharpoons R_2XH^{\oplus}$$

$$R^{\oplus} + R_2X \rightleftharpoons R_3X^{\oplus}$$

where X = O, S; R = alkyl.

According to the Lewis definition of an acid as an electron-pair acceptor (A), and of a base as an electron-pair donor (:B), the general acid–base reaction that satisfies the above two relationships is to be regarded as the formation or dissociation of the acceptor–donor complex A:B.

$$\underset{\substack{\text{Lewis}\\\text{acid}}}{A} \;+\; \underset{\text{Base}}{:B} \;\rightleftharpoons\; \underset{\substack{\text{Lewis acid-base}\\\text{complex}}}{A:B}$$

Pearson's classification into "hard" and "soft" acids and bases gives the following general reaction principle: Hard acids react preferentially with hard bases, and soft acids with soft bases.

According to this idea, RSH and R_2S (the valence electrons of which can easily be polarized or removed) are regarded as *soft*, and ROH and R_2O (with their much more firmly fixed valence electrons) as *hard* bases. On the other hand, the proton, as a Lewis acid, is extremely *hard* (because of the concentration of the positive charge into a small space). A proton will therefore add more readily to an alcohol or ether than to a thioalcohol (thiol) or thioether (sulfide). Alkyl cations $CR_3^{\oplus}$ do indeed rate as "hard," but their acceptor atom is large compared with the proton;

therefore they are much softer acceptors than the proton and add to thioethers more readily than to ethers.

The base strength of a donor therefore depends on the nature of the acceptor (acid). The acceptor–donor compound of the (relatively "soft") alkyl cation with ("soft") dialkyl sulfide may be stable with respect to ("soft) iodide, while the ("soft") alkyl cation is transferred from the corresponding complex with ("hard") dialkyl ether to the ("soft") iodide.

According to Pearson's classification the phenyl cation is a very much harder Lewis acid than alkyl cations. In harmony with this is the striking stability of the triphenyloxonium ion, even with respect to halide ions [3].

The concept of stability is far from simple. In general, stability denotes a lack of reactivity with respect to other agents. In order to compare the "stability" of different oxonium ions, it would in principle be possible to allow an oxonium ion to act on another oxygen base and to determine the position of the equilibrium. Thus the O-alkylation of ketones with trialkyloxonium salts leads to ketonium salts; e.g. [4]:

$$(CH_3)_3C-C(=O)-CH_3 + (C_2H_5)_3O^\oplus BF_4^\ominus \rightarrow (CH_3)_3C-C(OC_2H_5)^\oplus-CH_3 + (C_2H_5)_2O$$

However, most reactions of this type take place heterogeneously. Either one of the components (here the ether formed) is very much more volatile, or one of the two salts is sparingly soluble, so that no homogenous equilibrium of all the components is established under the reaction conditions. Accurate determinations of the equilibrium position of such reactions in homogeneous solution are indeed possible by spectroscopy, but have scarcely been performed at the present time [6]. Another complicating factor is that many oxonium ions tend to undergo subsequent reactions. Finally, cations of very different structural types with different activation energies must be taken into account. It is therefore quite possible that cations possessing a considerable stabilization energy may be very unstable from the chemical point of view because of a low activation energy. A meaningful measure of "stabilization" is found in the degree of charge delocalization, and we shall use this to compare carboxonium ions.

3.1.2. Charge Delocalization

Oxonium ions should be lower in energy the more highly the positive charge is delocalized. Increasing charge delocalization, and thus decreasing energy, are to be expected in the following sequence:

$$\text{Charge delocalization} \quad \underset{R^1}{\overset{R^2}{>}}C\overset{\oplus}{=}OR \quad < \quad \underset{R^1}{\overset{RO}{>}}C\overset{\oplus}{=}OR \quad < \quad \underset{RO}{\overset{RO}{>}}C\overset{\oplus}{=}OR$$

$$(8) \qquad\qquad (9) \qquad\qquad (25)$$

Stabilization of the Cation   13

For the gaseous state it can be shown by mass spectroscopic measurements that this sequence is valid for carboxonium ions of types *(8)*, *(9)*, and *(25)*. Stabilization energies (determined from the difference in the appearance potentials of $CH_3^{\oplus}$ and the corresponding cation) increase in the direction from $CH_3^{\oplus}$ to $C(OCH_3)_3^{\oplus}$ [5a,b]:

Cation	Stabilization energy [kcal·mole^{-1}]
$CH_3^{\oplus}$	Ref. (0)
$CH_3CH_2^{\oplus}$	37 ± 3
$(CH_3O)CH_2^{\oplus}$	66 ± 3
$(CH_3O)_2CH^{\oplus}$	85 ± 3
$(CH_3O)_3C^{\oplus}$	90 ± 3

The numerical values found for the gaseous state cannot in general be transferred to the situation in solution, but here the same change in the stabilization of the cations can be shown by NMR measurements in H_2SO_4–SO_3 solution: With cations of a similar structural type, the resonance signal of a given type of proton at the positive centers is to be expected at a field that becomes higher the more the charge is delocalized. Thus, for example, the chemical shift of the O-methyl protons changes fully in the sense of the sequence given above [6]:

δ_{OCH_3} = 5.3   5.2   4.95   4.9   ppm*

The marked stabilization of the carbonium ion by methoxy groups can also be recognized in the sequence of cations *(33)*–*(35)* by the chemical shift of the fluorine atom. With increasing shielding of the fluorine atom (increasing π-electron density), this shifts in the direction of higher fields from *(33)* to *(35)* [6]:

(33)   *(34)*   *(35)*

Finally, the particularly well-stabilized trimethoxycarbonium salt *(36)* can be actually obtained quantitatively in homogeneous solution from equivalent amounts of methyl orthocarbonate and triphenylcarbonium tetrafluoroborate [6,7]:

---

* Measurements at room temperature: The corresponding δ figure for the trimethoxycarbonium ion *(36)* departs from this sequence with a value of 5.0–5.1 ppm and is not (as was to be expected) at the highest field [6]. No satisfactory explanation has yet been found for this observation [6,8].

Stability of Oxonium Salts

$$C(OCH_3)_4 + C(C_6H_5)_3^{\oplus} BF_4^{\ominus} \longrightarrow \underset{H_3CO}{\overset{H_3CO}{>}}C{\cdots}OCH_3\ BF_4^{\ominus} + C(C_6H_5)_3OCH_3$$

(36)

In the sequence of the cations *(8)*, *(9)*, *(25)* according to decreasing energy given above, an influence of the substituents R, R^1, and R^2 is also to be expected. Systematic NMR investigations on this point have been carried out, in particular for dialkoxycarbonium salts: In the case of the 2-aryl-1,3-dioxolenium ions *(37)* [8], the resonance signals of the ring methylene protons were observed as a function of the substitution of the aryl nucleus. A linear correlation was in fact obtained as indicated by the increase in the δ figures with increasing Hammett σ-constants; however, the effects are very small (Fig. 1). This is ascribed in part to a high weight of the oxonium structures and a relatively small contribution from the carbonium structure [8, 12]:

$$R^1{-}C\underset{\bar{O}-}{\overset{\overset{\oplus}{O}-}{\diagdown}} \longleftrightarrow R^1{-}\overset{\oplus}{C}\underset{\bar{O}-}{\overset{\bar{O}-}{\diagdown}} \longleftrightarrow R^1{-}C\underset{\overset{\oplus}{O}-}{\overset{\bar{O}-}{\diagdown}}$$

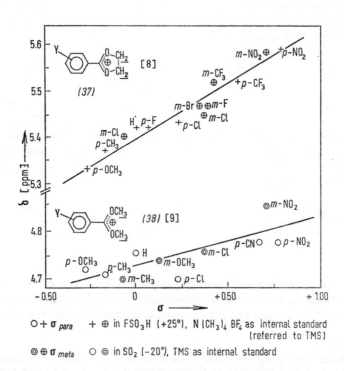

Fig. 1. Chemical shifts δCH_2 of *(37)* and δCH_3 of *(38)* as functions of the Hammett σ-constants.

The analogous investigation of the NMR signals for the O-methyl protons of aryldialkoxycarbonium ions *(38)* showed scarcely significant changes with different substitutions of the aryl nucleus [9]. This is probably due essentially to a more pronounced twisting out of the benzene nucleus in relation to the O–C–O plane.

In the case of protonated aldehydes and ketones *(8)*, the chemical shift of the O–H proton proved to be very sensitive to different substituents R^1 and R^2 [10a,b, 11a,b,c,d]. On the basis of the remarkably pronounced shift of the O–H resonance in the low-field direction (in general, 13.5–15 ppm), a high weight of the oxonium structure was assumed for *(8)* [11b].

$$(8) \quad \begin{matrix} R^2 \\ \\ R^1 \end{matrix}\!\!\!=\!\overset{\oplus}{\underline{O}}H \quad \longleftrightarrow \quad \begin{matrix} R^2 \\ \\ R^1 \end{matrix}\!\!\!-\!\overset{\oplus}{\underline{O}}H$$

If O–H resonance signals are found at a higher field than these "normal" values, the deviation is ascribed to a delocalization of the positive charge [11c]. Thus, for example, the chemical shift for the O–H proton in the series of cations *(39)–(41)* changes in the high-field direction [11c]:

	(39)	*(40)*	*(41)*
structure	$H_3C\!\!>\!\!=\!\!\overset{\oplus}{O}H$ / H_3C	$H_5C_6\!\!>\!\!=\!\!\overset{\oplus}{O}H$ / H_3C	$H_5C_6\!\!>\!\!=\!\!\overset{\oplus}{O}H$ / H_5C_6
δ_{O-H} =	14.24	13.03	12.23 ppm

(measured in SO_2–FSO_3H–SbF_5 in a molar ratio of 9:7:2 at -65 to $-50\,°C$).

Alkyl groups also cause a stabilization of the positive charge [11c]. As shown by the following series, the O–H resonance signal shifts increasingly toward higher fields when the carbonium center bears alkyl groups with an increasing number of carbon atoms [11a,b]:

		(39)		
structure	$H\!\!>\!\!=\!\!\overset{\oplus}{O}H$ / H	$H_3C\!\!>\!\!=\!\!\overset{\oplus}{O}H$ / H_3C	$H_5C_2\!\!>\!\!=\!\!\overset{\oplus}{O}H$ / H_5C_2	$(CH_3)_2CH\!\!>\!\!=\!\!\overset{\oplus}{O}H$ / $(CH_3)_2CH$
δ_{O-H} = 16.73		14.93	14.06	13.86 ppm

(measured in an equimolar mixture of FSO_3H and SbF_5, diluted with the same volume of SO_2, at $-60\,°C$).

Because of the different conditions of measurements, the two sequences do not give the same values for the same substance; cf. *(39)*.

The cations obtained by protonation from benzylideneacetophenone, dibenzylideneacetone, and biscinnamylideneacetone, and those from the corresponding methoxy derivatives *(42)–(44)* have been known for a very long time [13]. Their ease of formation can be explained by the marked charge delocalization of the cations. Correspondingly, in the electronic spectrum the long-wave absorption moves increasingly into the regions of longer wavelengths in the visible region from *(42)* to *(44)*:

(42) Ph–CH=CH–C(Ph)=ŌH⁺

(43) (4-H₃CO-C₆H₄)–CH=CH–C(2,4-(H₃CO)₂C₆H₃)=ŌH⁺

(44) Ph–(CH=CH)₂–C(Ph-(CH=CH)₂)=ŌH⁺

Finally, mention may also be made of cations that are stabilized in a special manner by the formation of aromatic systems. This is the case with the alkylation products of diphenylcyclopropenone *(45)* [14] and tropone *(46)* [15]:

(45) diphenylcyclopropenyl–OR cation (with resonance structures)

(46) tropylium–OR cation (with resonance structures)

The analogy of these cations with alkylated pyrones, the alkoxypyrylium ions *(47)*, is close [18a]. Repeated reference has already been made to the particular stability of the pyrylium system:

(47) alkoxypyrylium ion (with resonance structures)

S. Winstein *et al.* have drawn attention to the fact that charge delocalization might also arise by homoallyl resonance, which is indicated in the case of protonated norbornanone and norbornenone, for example, by a considerable difference of the NMR signals for the OH groups [11c]:

δ_{O-H} = 14.30   12.75 ppm

In general, vinyl substitution at the *carbonium* center of a carboxonium ion leads to charge delocalization and therefore to a lowering of the energy of the cation. If, on the other hand, vinyl groups enter the *oxonium* center, cations of *higher* energy, e.g. *(48)*, are to be expected. As a result of O-vinyl mesomerism [16], here the free-electron pair of the oxygen is less available for delocalization of the positive charge:

(48) R¹–C⁺(Ō–CH=CH₂)(Ō–CH=CH₂)

Substitution of the oxonium center by phenyl residue has a similar effect. Cations of type *(48)* have not so far been isolated in the form of their salts, but they must be assumed as intermediates in certain reactions [16].

3.2. Influence of the Anion

If stable oxonium salts are to be formed, the anions must satisfy the condition that they neither *dealkylate* or *deprotonate* the oxonium ion

$$\begin{matrix} R \\ \diagdown \\ R \diagup \end{matrix} \!\!\!O\text{-}R' + |Y^{\ominus} \longrightarrow \begin{matrix} R \\ \diagdown \\ R \diagup \end{matrix} \!\!\!O| + R'\text{-}Y$$

$$\text{or:} \quad {=}\overset{\oplus}{O}\text{-}R' + |Y^{\ominus} \longrightarrow {=}O| + R'\text{-}Y$$

where R = org. residue, H, nor, in the case of carboxonium ions, that they be capable of *addition* to the carbonium center:

$$={\overset{\oplus}{O}}\text{-}R' + |Y^{\ominus} \longrightarrow {>}\!\!<\!\!{\overset{O\text{-}R'}{}_{Y}}$$

Stable salts of oxonium ions are obtained with strongly nucleophilic anions only in exceptional cases. As examples we may mention the halides of the triphenyloxonium [3], dibenzylideneacetonium [17], and pyrylium ions [18b]:

$$\underset{C_6H_5}{\overset{C_6H_5 \diagup \overset{\oplus}{O} \diagdown C_6H_5}{|}} \quad Br^{\ominus} \qquad (C_6H_5\text{-}CH{=}CH)_2 C{=}\overset{\oplus}{O}\text{-}H \;\; Cl^{\ominus} \qquad \underset{C_6H_5 \;\; O \;\; C_6H_5}{\overset{CH_3}{\bigcirc}} \;\; Br^{\ominus}$$

However, all types of oxonium ions can combine with complex anions of low polarizability and weak nucleophilic properties to form salts. Table 3 gives only the most important anions of this kind and also the types of salts (except pyrylium salts) that have been isolated as such.

The hexachloroantimonate (V) ion is generally used with all secondary and most tertiary oxonium salts, and the tetrafluoroborate ion with all tertiary types. Most of the anions of Table 3 are halogen complexes of the type $Z(Hal)^-_{n+1}$. They are the more satisfactory as the following decompositions proceed with more difficulty:

$$\begin{matrix} R \\ \diagdown \\ R \diagup \end{matrix} \!\!\!\overset{\oplus}{O}\text{-}R' \;\; Z(Hal)^{\ominus}_{n+1} \longrightarrow \begin{matrix} R \\ \diagdown \\ R \diagup \end{matrix} \!\!\!O| + R'\text{-}Hal + Z(Hal)_n$$

$$={\overset{\oplus}{O}}\text{-}R' \;\; Z(Hal)^{\ominus}_{n+1} \Bigg\langle \begin{matrix} \longrightarrow {=}O| + R'\text{-}Hal + Z(Hal)_n \\ \\ \longrightarrow {>}\!\!<\!\!{\overset{O\text{-}R'}{}_{Hal}} + Z(Hal)_n \end{matrix}$$

where R′ = H, org. residue.

TABLE 3. ISOLABLE OXONIUM SALTS

	Secondary					Oxonium Ions					Tertiary			
Complex anion	$R_2OH^\oplus$ (6)	$R^1R^2\!\!=\!\!OH$ (13)	$HO\!\!=\!\!R\!\!=\!\!OH$ (17)	$RO\!\!=\!\!R\!\!=\!\!OH$ (18)	$RO\!\!=\!\!R\!\!=\!\!OH$ / $RO\!\!=\!\!R\!\!=\!\!OH$ (24)	$R_3O^\oplus$ (7)	(11)	$R^2R^1\!\!=\!\!OR$ (8)	(14)	$R^1\!\!=\!\!OR$ (9)	(20)	(21)	$RO\!\!=\!\!OR$ (25)	(27)
$BF_4^\ominus$	[19b]					[27]	[29, 31]	[27, 33]	[35a]	[33, 36]	[33, 34]	[36]	[33, 34]	[39]
$B(C_6H_5)_4^\ominus$						[28]		[4]						
$AlCl_4^\ominus$	[20]					[29]								
$GaCl_4^\ominus$	[20]													
$InCl_4^\ominus$	[20]													
$TlCl_4^\ominus$	[20]													
$SnCl_6^{2\ominus}$	[19b]					[29]								
$NO_3^\ominus$		[22]												
$SbF_6^\ominus$						[30a]	[30b]							
$SbCl_6^\ominus$	[19a,b]	[23]	[26]	[26]	[26]	[29]	[29, 31]	[33]	[35b]	[33]	[38]	[37]	[26]	
$BiCl_4^\ominus$	[21]											[36]		
$HSO_4^\ominus$		[24]										[36]		
$ClO_4^\ominus$		[13, 25a]							[35c]			[25b,c]		
$FeCl_4^\ominus$	[19b]	[17]				[29]								
$PtCl_6^{2\ominus}$		[17]				[29]								
$AuCl_4^\ominus$						[29]	[29]							
$ZnCl_3^\ominus$	[19b]													
$HgI_3^\ominus$						[29]	[29]							
Picrate$^\ominus$						[29]	[29]							
2,4,6-Trinitro- benzenesul- fonate$^\ominus$									[32]					

For secondary oxonium salts (R' = H), the following sequence of anions, in decreasing stability, is obtained from their hydrogen halide dissociation pressures [19b]:

$$SbCl_6^- > FeCl_4^- > AlCl_4^- > BF_4^- > SnCl_6^{2-} > ZnCl_3^-$$

Meerwein et al. [29] give another sequence for trialkyloxonium salts (R' = alkyl), which differs with respect to the position of the tetrafluoroborate ion:

$$SbCl_6^- > BF_4^- > FeCl_4^- > AlCl_4^- > SnCl_6^{2-}$$

Acidic oxonium salts are capable of forming hydrogen bonds with the anion*, which can be regarded as transitional stages to the deprotonation of the oxonium ion [29]. As can be shown with a dialkyloxonium salt as an example, the following main types of increasing proton transfer to the anion with its increasing basicity must be distinguished:

$$\underset{(6)}{\overset{R}{\underset{R'}{\diagdown}}\!\!\overset{\oplus}{O}\!\!-\!H\ |Y^\ominus} \qquad \underset{(49)}{\overset{R}{\underset{R'}{\diagdown}}\!\!\overset{\oplus}{O}\!\!-\!H\cdots|Y^\ominus} \qquad \underset{(50)}{\overset{R}{\underset{R'}{\diagdown}}\!\!O|\cdots H\!-\!Y}$$

According to this, in addition to (6) we shall regard (49) as an oxonium salt, and (50) will be considered as a "molecular compound."

In the case of the isolable addition compounds of alcohols with protonic acids [40, 41] (always 2 moles of alcohol to 1 mole of acid*) it is often uncertain whether these are oxonium salts or hydrogen-bond complexes of type (50); this is even questionable when complex anions such as hexacyanoferrate(II) and hexacyanoferrate(III) [42] or hexacyanocobaltate(III) [43] are involved.

However, G. A. Olah et al. [11e] have satisfactorily detected primary oxonium ions in solution (namely, in FSO_3H–SbF_5–SO_2) by means of low-temperature NMR measurements. The anion arising here, $SbF_5(FSO_3)^-$, is also suitable under the same conditions for the formation of all types of secondary oxonium ions. e.g. [11a,b,c,d] (see Tables 1 and 2 in Chapter 2). Likewise, it has been possible to show by means of low-temperature NMR spectroscopy that secondary oxonium ions with the anions BF_4^- (in HF–BF_3) [44a, b], FSO_3^- (in FSOH) [45], SbF_6^- (in SbF_5 (in SbF_5-HF) [10], and HSO_4^- (in H_2SO_4 or H_2SO_4–SO_3) [6, 46] exist in solution.

---

* Such hydrogen bonds are also probably responsible for the fact that all crystalline secondary oxonium salts are capable of binding a second molecule of the conjugate base, which can frequently be removed only with difficulty and always promotes stabilization [19a,b, 20, 23, 26]; e.g.:

$$\underset{SbCl_6^\ominus}{\overset{R}{\underset{R'}{\diagdown}}\!\!\overset{\oplus}{O}\!\!-\!H\cdots|O\!\!\overset{R}{\underset{R}{\diagup}}} \qquad \underset{SbCl_6^\ominus}{\overset{R^2}{\underset{R^1}{\diagdown}}\!\!\overset{\oplus}{C}\!\!=\!\!O\!\!-\!H\cdots|O\!\!=\!\!C\!\!\overset{R^2}{\underset{R^1}{\diagup}}}$$

References

1. W. Hütz, Ph.D. Thesis, University of Marburg, 1938.
2. R. G. Pearson, *Chem. Britain* **3**, 103 (1967); *J. Am. Chem. Soc.* **85**, 3533 (1963); *J. Am. Chem. Soc.* **89**, 1827 (1967).
3. A. N. Nesmeyanov, L. G. Makarova, and T. P. Tolstaya, *Tetrahedron* **1**, 145 (1957).
4. H. Meerwein, in "Methoden der organischen Chemie (Houben-Weyl)" (E. Müller, ed.), 4. Auflage, Bd. VI/3: Sauerstoffverbindungen I, p. 346. Thieme, Stuttgart, 1965.
5. a) R. W. Taft, R. H. Martin, and F. W. Lampe, *J. Am. Chem. Soc.* **87**, 2490 (1965); b) R. H. Martin, F. W. Lampe, and R. W. Taft, *J. Am. Chem. Soc.* **88**, 1353 (1966).
6. B. G. Ramsey and R. W. Taft, *J. Am. Chem. Soc.* **88**, 3058 (1966).
7. H. Meerwein, V. Hederich, H. Morschel, and K. Wunderlich, *Ann. Chem. Liebigs* **635**, 1 (1960).
8. D. A. Tomalia and H. Hart, *Tetrahedron Letters* p. 3389 (1966).
9. A. Gerlach, Ph.D. Thesis, University of Marburg, 1969.
10. a) D. M. Brouwer, *Tetrahedron Letters*, p. 453 (1968); b) D. M. Brouwer, *Rec. Trav. Chim.* **86**, 879 (1967).
11. a) G. A. Olah, D. H. O'Brien, and M. Calin, *J. Am. Chem. Soc.* **89**, 3582, 3586 (1967); b) G. A. Olah and M. Calin, *J. Am. Chem. Soc.* **90**, 938 (1968); c) M. Brookhart, G. C. Levy, and S. Winstein, *J. Am. Chem. Soc.* **89**, 1735 (1967); d) H. Hogeveen, *Rec. Trav. Chim.* **86**, 696 (1967); e) G. A. Olah, J. Sommer, and E. Namanworth, *J. Am. Chem. Soc.* **89**, 3576 (1967).
12. H. Hart and D. A. Tomalia, *Tetrahedron Letters* p. 3383 (1966); see also J. W. Larsen, S. Ewing, and M. Wynn, *Tetrahedron Letters* p. 539 (1970).
13. P. Pfeiffer, *Ann. Chem. Liebigs* **412**, 275 (1917).
14. R. Breslow, T. Eicher, A. Krebs, R. A. Peterson, and J. Posner, *J. Am. Chem. Soc.* **87**, 1320 (1965).
15. K. Hafner, H. W. Riedel, and M. Danielisz, *Angew. Chem.* **75**, 344 (1963).
16. S. Hünig, *Angew. Chem.* **76**, 400 (1964); *Angew. Chem. Intern. Ed. English* **3**, 584 (1964).
17. F. Straus, *Ber. Deut. Chem. Ges.* **37**, 3277 (1904).
18. a) A. Baeyer, *Ber. Deut. Chem. Ges.* **43**, 2337 (1910); b) W. Schneider and A. Ross, *Ber. Deut. Chem. Ges.* **55**, 2775 (1922).
19. a) F. Klages and H. Meuresch, *Chem. Ber.* **85**, 863 (1952); **86**, 1322 (1953); b) F. Klages, H. Meuresch, and W. Steppich, *Ann. Chem. Liebigs* **592**, 81 (1955).
20. E. Wiberg, M. Schmidt, and A. Galinos, *Angew. Chem.* **66**, 443 (1954).
21. K. Schäfer and F. Hein, *Z. Anorg. Allgem. Chem.* **101**, 268 (1917).
22. J. Dumas and E. Peligot, *Ann. Chem. Liebigs* **14**, 50 (1835); J. Kachler, *Ann. Chem. Liebigs* **159**, 281 (1871); E. Wedekind and A. Koch, *Ber. Deut. Chem. Ges.* **38**, 421 (1905); A. Zhukov and F. Kassatkin, *Zh. Russ. Fiz.-Khim. Obshch.* **4J**, 157 (1909); *Chem. Zentr.* **80/I**, 1760 (1909); G. Reddelien, *J. Prakt. Chem.* **91**, 213 (1915).
23. F. Klages, H. Träger, and E. Mühlbauer, *Chem. Ber.* **92**, 1819 (1959).
24. D. Vorländer and E. Mumme, *Ber. Deut. Chem. Ges.* **36**, 1481 (1903); S. Hoogewerff and W. A. van Dorp, *Rec. Trav. Chim.* **21**, 339 (1902).
25. a) K. A. Hofmann and H. Kirmreuther, *Ber. Deut. Chem. Ges.* **42**, 4864 (1909); b) J. W. Blunt, M. P. Hartshorn, and D. N. Kirk, *Chem. Ind. (London)* p. 1955 (1963); c) G. N. Dorofeenko and L. V. Mezheritskaya, *Zh. Obshch. Khim.* **38**, 1192 (1968).
26. F. Klages and E. Zange, *Chem. Ber.* **92**, 1828 (1959).
27. H. Meerwein, G. Hinz, P. Hofmann, E. Kroning, and E. Pfeil, *J. Prakt. Chem.* **147**, 257 (1937).
28. A. N. Nesmeyanov, V. A. Sazanova, G. S. Liberman, and L. I. Emel'yanova, *Izv. Akad. Nauk SSSR, Otd. Khim. Nauk* p. 48 (1955); *Chem. Abstr.* **50**, 1644 (1956).
29. H. Meerwein, E. Battenberg, H. Gold, E. Pfeil, and G. Willfang, *J. Prakt. Chem.* **154**, 83 (1939).

30. a) G. A. Olah, J. R. DeMember, and R. H. Schlosberg, *J. Am. Chem. Soc.* **91,** 2112 (1969); b) F. Klages and H. A. Jung, *Chem. Ber.* **98,** 3757 (1965).
31. H. Meerwein, see Reference 2, p. 341; A. Kirrmann and L. Wartski, *Bull. Soc. Chim. France* p. 3825 (1966).
32. D. J. Pettitt and G. K. Helmkamp, *J. Org. Chem.* **28,** 2932 (1963).
33. H. Meerwein, *Angew. Chem.* **67,** 374 (1955).
34. H. Meerwein, P. Borner, O. Fuchs, H. J. Sasse, H. Schrodt, and J. Spille, *Chem. Ber.* **89,** 2060 (1956).
35. a) K. Dimroth and W. Mach, *Angew. Chem.* **80,** 489 (1968); *Angew. Chem. Intern. Ed. English* **7,** 760 (1968); b) H. R. Ward and P. D. Sherman, Jr., *J. Am. Chem. Soc.* **90,** 3812 (1968); c) W. Rundel and K. Besserer, *Tetrahedron Letters* p. 4333 (1968).
36. H. Meerwein, K. Bodenbenner, P. Borner, F. Kunert, and K. Wunderlich, *Ann. Chem. Liebigs* **632,** 38 (1960).
37. J. F. King and D. A. Allbutt, *Chem. Commun.* p. 14 (1966); *Tetrahedron Letters* p. 49 (1967).
38. K. Wunderlich, Ph.D. Thesis, University of Marburg, 1957.
39. C. P. Heinrich, Ph.D. Thesis, University of Marburg, 1966.
40. O. Wallach, *Ann. Chem. Liebigs* **230,** 225 (1885).
41. A. Faworsky, *J. Prakt. Chem.* **88,** 480 (1913).
42. A. Baeyer and V. Villiger, *Ber. Deut. Chem. Ges.* **35,** 1203 (1902).
43. F. Hölzl, T. Meier-Mohar, and F. Viditz, *Monatsh. Chem.* **52,** 73 (1929).
44. a) C. MacLean and E. L. Mackor, *J. Chem. Phys.* **34,** 2207 (1961); b) H. Hogeveen, A. F. Bickel, E. L. Mackor, and C. MacLean, *Chem. Commun.* p. 898 (1966).
45. T. Birchall and R. J. Gillespie, *Can. J. Chem.* **43,** 1045 (1965).
46. N. C. Deno, H. G. Richey, Jr., N. Friedmann, J. D. Hodge, J. J. Houser, and C. U. Pittman, Jr., *J. Am. Chem. Soc.* **85,** 2991 (1963).

4. Formation of Oxonium Salts

4.1. General Methods of Preparation

4.1.1. Formation of Oxonium Ions

In relation to their general preparation, we shall divide all the various types of oxonium ions into three groups, namely:
Primary and secondary (= acidic oxonium ions)
Tertiary acyclic
Tertiary cyclic oxonium ions.

4.1.1.1. Primary and Secondary Oxonium Ions

All acidic oxonium ions (see Tables 1 and 2 in Chapter 2) can be obtained exclusively by the *protonation* of the conjugate oxygen bases, regardless of whether the cations are saturated or unsaturated. The second conceivable method of obtaining primary oxonium ions by the *alkylation* of water and the secondary ions by the alkylation of carbon-substituted hydroxy compounds has *not* been successful for the preparation of isolable salts; e.g.:

$$\underset{H}{\overset{R}{>}}\ddot{O}| + H^{\oplus} \longrightarrow \underset{H}{\overset{R}{>}}\overset{\oplus}{O}\text{-H} \quad \xleftarrow{\quad/\!\!/\quad} \quad R^{\oplus} + \underset{H}{\overset{}{\ddot{O}}}\text{-H}$$

(6)

$$\underset{R^1}{\overset{R\bar{O}}{>}}=\bar{O} + H^{\oplus} \longrightarrow \underset{R^1}{\overset{RO}{>}}\overset{\oplus}{\cdots}\text{OH} \quad \xleftarrow{\quad/\!\!/\quad} \quad R^{\oplus} + \underset{R^1}{\overset{|O|}{>}}\text{-OH}$$

(18)

However this second method is important in the *intermediate* formation of acidic oxonium ions in reaction sequences.

4.1.1.2. Tertiary Acyclic Oxonium Ions

The saturated tertiary oxonium ions can be obtained by O-alkylation, i.e., formally by the addition of an alkyl cation to an ethereal oxygen atom:

$$\underset{R}{\overset{R}{>}}\ddot{O}| + R^{\oplus} \longrightarrow \underset{R}{\overset{R}{>}}\overset{\oplus}{O}\text{-R}$$

(7)

Correspondingly, the unsaturated tertiary oxonium ions, the carboxonium ions, are obtained by the O-alkylation of compounds with carbonyl functions (*route A*). However, carboxonium ions also arise as mesomeric carbonium–oxonium ions by the production of a carbonium ion in the α-position to an ether function. In this case, the known methods of preparation for carbonium ions can be used:

(a) Anionic splitting out of a suitable group Y from the α-position of an ether (*route B*), or

(b) The addition of a cation (generally a proton or a alkyl cation) to the β-position (of the double bond) of a vinyl ether (*route C*):

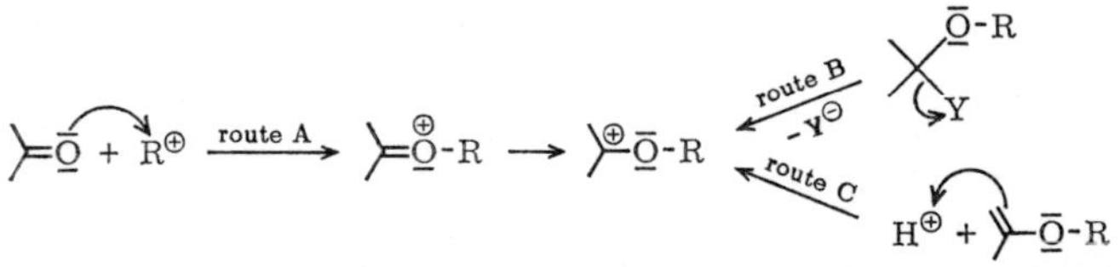

However, route C is hardly of preparative importance, although reactions with vinyl ethers [1b] or ketene acetals [1a] frequently take place with the intermediate formation of oxonium ions.

Syntheses by routes A and B can be illustrated on the basis of the formation of the triethoxycarbonium ion:

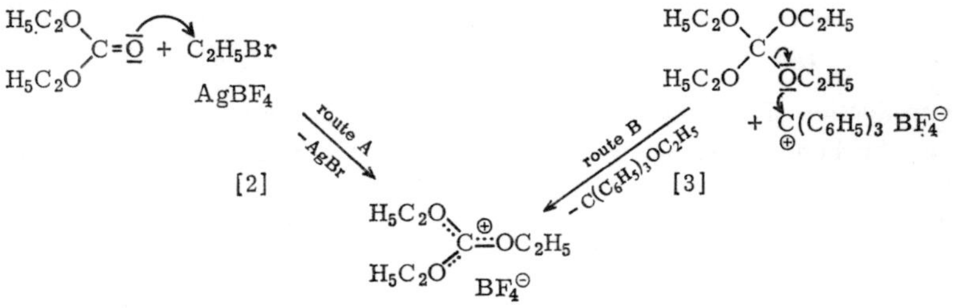

(51)

Ethyl bromide–AgBF$_4$ acts as a donor for the alkyl groups (route A) and the triphenylcarbonium salt as an acceptor for the leaving group (Y = OC$_2$H$_5$, route B). The selection of suitable alkylating agents and acceptors for different leaving groups Y will be discussed in detail in Sections 4.1.2 and 4.1.3.

4.1.1.3. Tertiary Cyclic Oxonium Ions

The tertiary cyclic oxonium ions can generally be obtained:

(*a*) By intramolecular O-alkylation with *synthesis* of the oxygen heterocycle, or,

(*b*) From *preformed* oxygen heterocycles to which the methods of formation of acyclic oxonium ions can be transferred.

Intramolecular O-Alkylations

The intramolecular addition of a carbonium ion to an ethereal oxygen or a carbonyl oxygen (of a ketone or an ester) leads to the formation of cyclic oxonium ions; schematically,

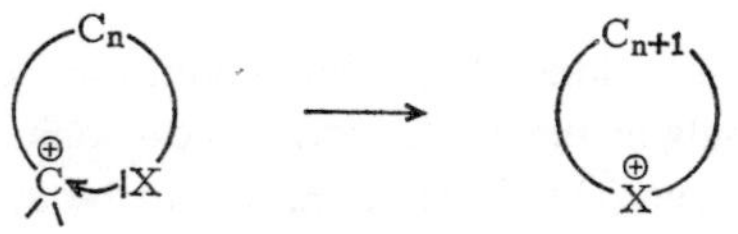

where X = ethereal or carbonyl oxygen.

Formation of Oxonium Salts

In general the carbonium ion does not occur in the free state, since the alkylation usually takes place in the nature of an internal S_N2 reaction with anionic elimination of a suitable group Y. By this principle it is possible to obtain five- or six-membered saturated tertiary cyclic oxonium ions [4a,b]:

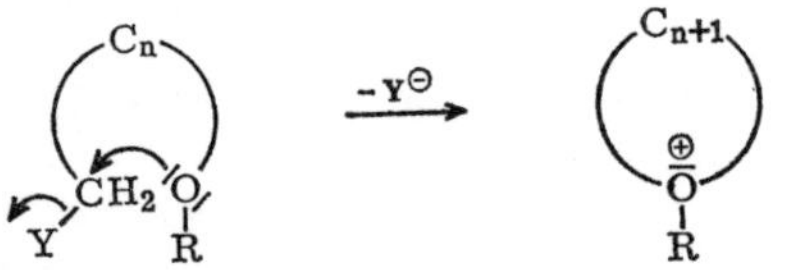

or cyclic carboxonium ions in which this method of preparation corresponds to the method A mentioned above [5a,b,c, 6a,b,c]:

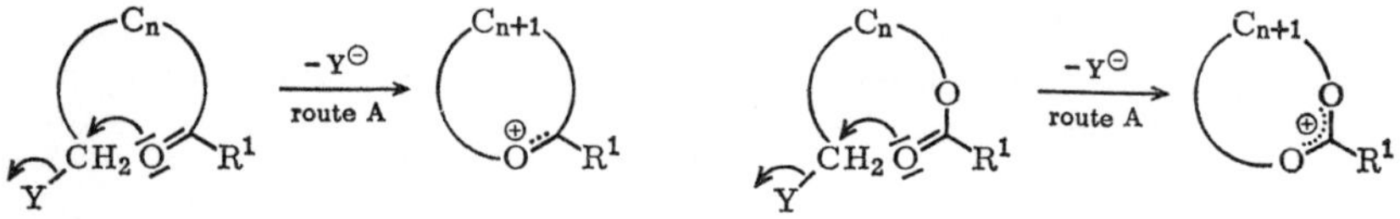

The production of the alkylating carbonium ion by the protonation (or alkylation) of a double bond in a sterically favorable position with respect to the oxygen function

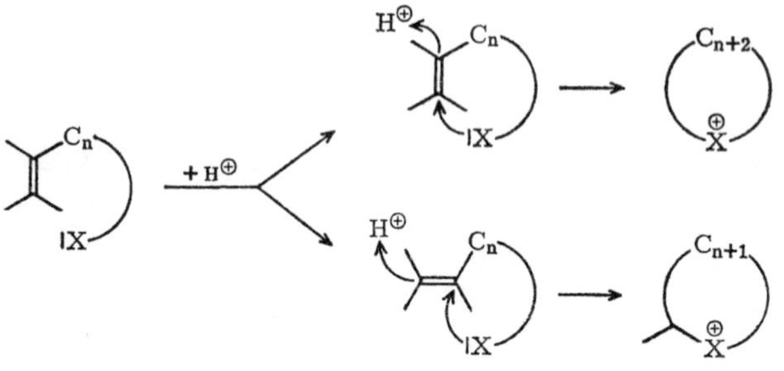

is less important preparatively [7a]. Intramolecular alkylations are also described as "the neighboring-group participation" of oxygen functions [7b].

Syntheses from Preformed Oxygen Heterocycles

Saturated cyclic oxonium ions can be obtained by the O-alkylation of the ethereal oxygen of a cyclic ether [8]:

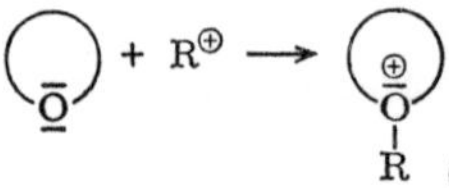

(11)

Unsaturated cyclic oxonium ions, the carboxonium ions, can be formed from a preformed heterocycle only by the production of a carbonium ion in the α-position to a cyclic ethereal oxygen. This is achieved, as in the synthesis of acyclic carboxonium ions, by the (*a*) Anionic elimination of groups Y on the carbon atom in the α-position (method B), or

(*b*) Addition of cations (protons, alkyl cations) to cyclic vinyl ethers (ketene acetals) in the β-position to the ethereal oxygen (method C).

The double bond may be in an endocyclic *(52)* or semicyclic *(53)* position:

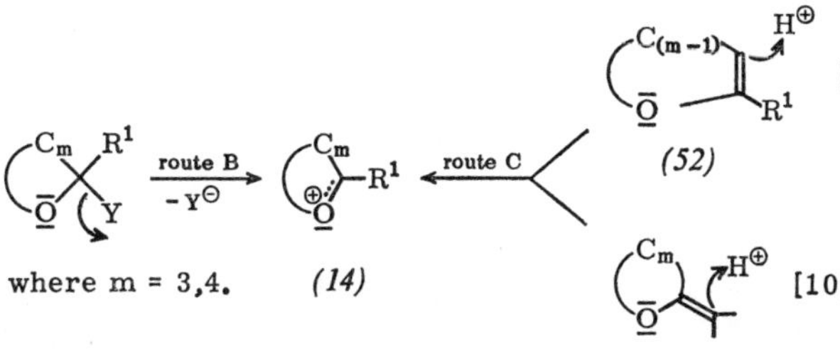

where m = 3,4. *(14)* *(52)* *(53)*

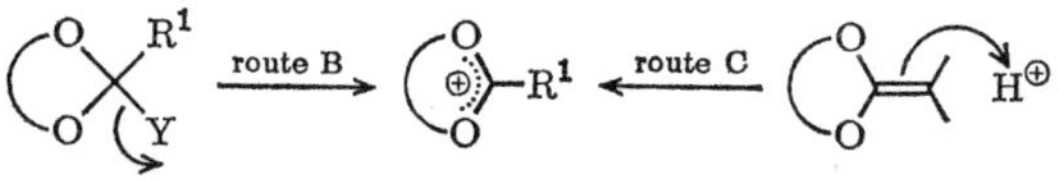

[6a, 12b] *(21)* [6b, 12à,b]

Here route C is of preparative importance, too. The addition of cations to semicyclic double bonds in fact corresponds to O-alkylation of the carbonyl groups of lactones and cyclic carbonates *(20)* and *(27)*. Indeed this reaction with respect to the cyclic ethereal oxygen atoms takes place by route C, but since an acyclic oxonium center arises simultaneously, the analogy with route A is also close.

We refer to this special case as route A':

[11a] 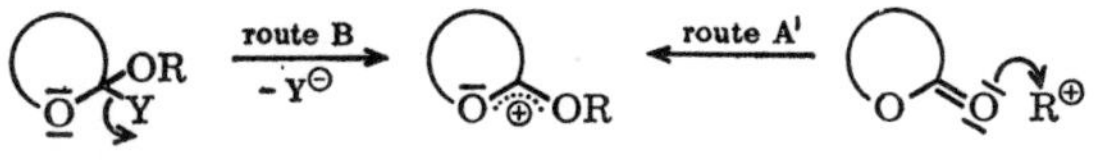 [13]

(20)

[14] 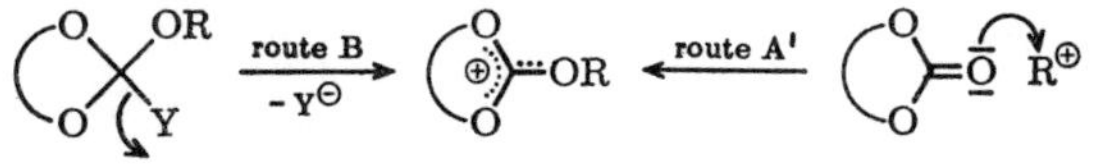 [14]

(27)

The various routes of formation for cyclic tertiary carboxonium ions will be illustrated with the O-ethylbutyrolactonium ion *(54)*:

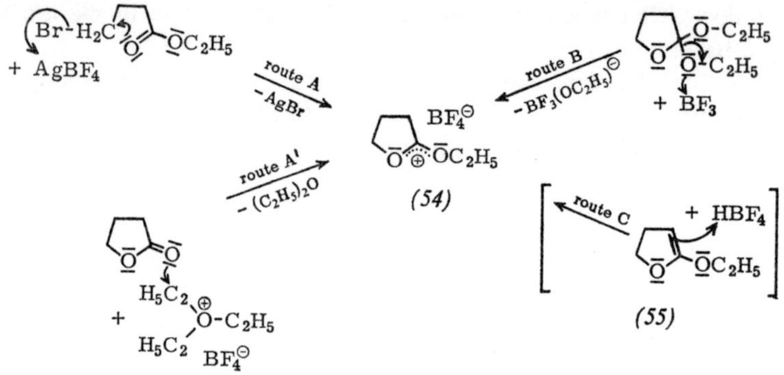

(54) (55)

With the exception of the addition of protons to the ketene acetal *(55)*, all these reactions have been realized [15]. The addition of protons to 2,3-dihydrofuran derivatives *(56)* [9] and to methylene-2,5-dihydrofuran derivatives *(58)* [10] may serve as examples for route C:

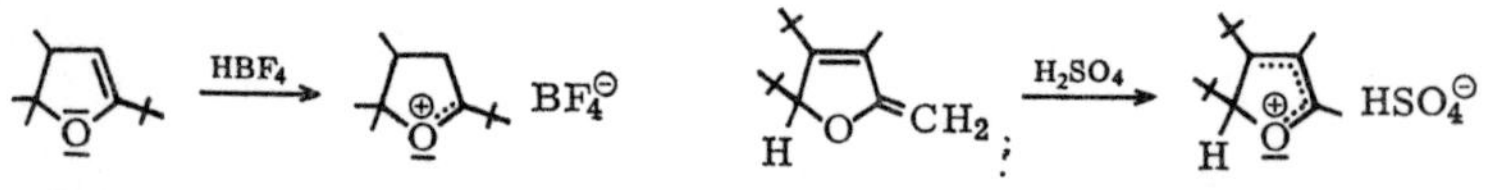

(56) *(57)* *(58)* *(59)*

where +— = *tert*-butyl.

4.1.2. Alkylating Agents

The formation of tertiary oxonium ions by the alkylation of a compound with an ethereal or carbonyl oxygen atom has been described schematically by the action of free alkyl cations:

However, the added alkyl cations arise from alkylating agents R–Y. In the simplest case the anion Y^- remaining after the transfer of the alkyl group can form a salt with the oxonium ion produced:

However, this scheme can be realized only rarely, as in the alkylation of a γ-pyrone with a dialkyl sulfate [18a]

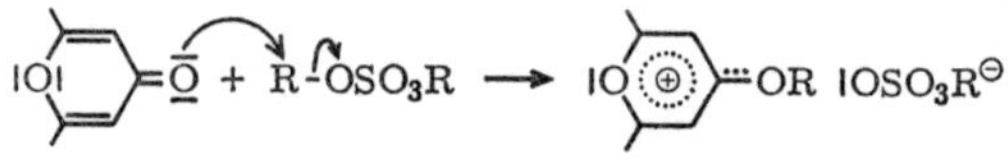

or—to present a newer example—in the alkylation of dimethyl ether with methylfluorosulfonate [18c]:

General Methods of Preparation

$$\mathrm{R_2\ddot{O}} + \mathrm{R{-}OSO_2F} \longrightarrow \mathrm{R_2\overset{\oplus}{O}{-}R} \ \mathrm{|OSO_2F^{\ominus}}$$

The reaction does not take place in the desired direction if the anion Y^- is more strongly nucleophilic than the oxygen base (with respect to the alkyl cation). Consequently, the synthesis of oxonium salts with alkyl halides normally does not take place. Nevertheless, alkyl halides can be used as alkylating agents for oxygen compounds if the halide ions are trapped by a suitable acceptor in a coupled reaction. This is possible, for example, with the addition of $AgBF_4$ [2] or $AgSbF_6$ [19a,b], halide ions being precipitated as silver halides:

$$\mathrm{(RO)_2C{=}\ddot{O}} + \mathrm{R{-}Br} + \mathrm{AgBF_4} \longrightarrow \mathrm{(RO)_2\overset{\oplus}{C}{-}OR} \ \mathrm{BF_4^{\ominus}} + \mathrm{AgBr} \quad [2]$$

An elegant possibility of coupling the removal of the halide ion with the production of the complex anion consists in the reaction of alkyl halides with the Lewis acid–base complexes from nonmetal (or metal) halides and oxygen compounds; e.g. [8]:

$$\mathrm{R_2O{\rightarrow}BF_3} + \mathrm{R{-}F} \longrightarrow \mathrm{R_2\overset{\oplus}{O}{-}R} \ \mathrm{BF_4^{\ominus}}$$

Such coupled reactions with alkyl halides are also used for internal alkylation reactions:

[cyclic ether with $-CH_2I$ + $AgSbF_6$ → cyclic oxonium $SbF_6^{\ominus}$ + AgI] [19b]

[tetrahydrofuran with $CH_2{-}Cl$, C_2H_5 + $SbCl_5$ → intermediate with $O{\rightarrow}SbCl_5$ → cyclic oxonium $SbCl_6^{\ominus}$, C_2H_5] [4a]

[dioxolane with $-CH_2F$, CH_3 + BF_3 → intermediate with $O{\rightarrow}BF_3$ → cyclic dioxonium $BF_4^{\ominus}$, CH_3] [6a]

In some cases, cationic alkylating agents (tertiary oxonium ions, diazonium ions), as salts with complex anions, can be used advantageously for the alkylation of oxygen bases. Here the group leaving the alkylating agent is an uncharged molecule, and the complex anion is added in the preformed state.

Thus, tertiary oxonium salts can in their turn alkylate other oxygen compounds [11a]:

Formation of Oxonium Salts

$$\text{RO} \diagup \!\!\!\!\!\! =\overset{\ominus}{\underset{}{O}} + R\text{-}\overset{\oplus}{\underset{BF_4^\ominus}{O}}\diagdown\!\!{}^R_R \longrightarrow \text{RO}\diagup\!\!\!\!\!\!\overset{\oplus}{=}\!\!\underset{BF_4^\ominus}{OR} + |\overset{}{\underset{}{O}}|\diagdown\!\!{}^R_R$$

The arylation of diphenyl ether with an aryldiazonium salt [17] also belongs here:

$$\begin{array}{c} H_5C_6 \\ H_5C_6 \end{array}\!\!\!\!\!\!\!\!>\!|\overset{}{O}| + C_6H_5\text{-}\overset{\oplus}{N}\!\!\equiv\!\!N|\; BF_4^\ominus \longrightarrow \begin{array}{c} H_5C_6 \\ H_5C_6 \end{array}\!\!\!\!\!\!\!\!>\!\overset{\oplus}{|O|}\text{-}C_6H_5\; BF_4^\ominus + N_2$$

The reaction can be compared formally with the alkylation of secondary oxonium salts by diazoalkanes [16]:

$$\begin{array}{c} R \\ R \end{array}\!\!\!\!\!\!>\!\!\overset{\oplus}{|O|}\text{-H} + \begin{array}{c} H \\ R' \end{array}\!\!\!\!\!\!>\!\overset{\ominus}{C}\text{-}\overset{\oplus}{N}\!\!\equiv\!\!N| \longrightarrow \left[\begin{array}{c} R \\ R \end{array}\!\!\!\!\!\!\!\!\overset{\oplus}{>|O|} + H\text{-}\!\!\overset{H}{\underset{R'}{C}}\!\!\text{-}\overset{\oplus}{N}\!\!\equiv\!\!N| \right]_{SbCl_6^\ominus} \!\!\!\!\longrightarrow \begin{array}{c} R \\ R \end{array}\!\!\!\!\!\!\!\!\overset{\oplus}{>|O|}\text{-CH}_2R' + N_2$$
$$\text{SbCl}_6^\ominus \text{SbCl}_6^\ominus$$

Special methods of alkylation in which the active alkylating agent is formed only during the reaction will be described in connection with the trialkyloxonium salts.

4.1.3. Acceptors for Anionically Leaving Groups

In the synthesis of oxonium ions by the action of alkyl halides on oxygen compounds, as mentioned above, the halide ion must be taken up by suitable acceptors (e.g., the Lewis acids Ag^+, BF_3, $SbCl_5$). Besides halide ions, other Y groups may be eliminated anionically in *internal* alkylation, for which again acceptors A^m are necessary; e.g.:

$$A^m + Y \diagdown\!\!\!\!\overset{H_2C}{\underset{O}{\diagup\!\!\!\!\!\!\bigcirc\!\!\!\!\!\!\diagdown}}\!\!\!\!\!O\!\!\text{-}R^1 \longrightarrow [A\text{-}Y]^{m-1} + \bigcirc\!\!\!\!\!\!\!\overset{\oplus}{\diagdown}\!\!\!\text{-}R^1 \qquad (21)$$

The most important leaving groups are hydroxy, acyloxy, and alkoxy groups, while protons, BF_5, $SbCl_3$, and alkyl cations act as their acceptors A^m. Details will be found in Table 4.

The application of suitable acceptors A^m is also important in the preparation of carboxonium ions by route B, namely, the splitting out of anions Y^- in the α-position to an ethereal oxygen (cf. Sections 4.1.1.2 and 4.1.1.3):

$$\underset{Y + A^m}{\diagup\!\!\!\!\!\!\times\!\!\!\!\!\diagdown}^{OR} \longrightarrow \diagup\!\!\!\!\!\!\overset{\oplus}{=}\!\!OR + [A\text{-}Y]^{m-1}$$

The splitting out of halide ions is limited to only a few cases in this synthetic route [22a, 23], and the preferred anionically eliminated groups Y are alkoxide ions [2, 3, 6a, 11a, 13–15, 20, 21] and hydride ions [3, 26a, 27a, b] and also, more rarely, carbeniate [27a,b] or cyanide ions [3].

Halides can be used as leaving groups only in the preparation of *mono*alkoxy-

TABLE 4. Leaving Groups and Splitting-out Reagents in the Synthesis of Cyclic Oxonium Ions by Internal Alkylation

$$A^m + Y{-}CH_2X \longrightarrow [A{-}Y]^{m-1} + \overset{\oplus}{O}X$$

Leaving groups Y	Splitting-out reagents for Y										Reactions			
	HCO_2H	CF_3CO_2H	FSO_3H	$HClO_4$	$AgBF_4$	$AgSbF_6$	BF_3	SbF_5	$SbCl_5$	$R_3O^{\oplus}BF_4^{\ominus}$	(11)	(14)	(20)	(21)
F							[6a]							[6a]
Cl					[30]			[31]ᵃ	[4a, 5b, 6a, 15, 32]		[4a]	[5b]	[15]	[4a, 6a, 15, 30–32]
Br					[2, 6a,b, 19a, 28a,b,c,d]			[31]ᵃ	[4b]		[4b]		[2]	[2, 6a,b, 19a, 28a,b,c,d, 31]
I						[19b]		[31]ᵃ			[19b]			[31]
OH			[28a]ᵃ											
O-Alkyl			[28a]ᵃ				[6a]		[6a, 15]	[6a]			[15]	[6a, 28a]
$OCOCH_3$		[28a]ᵃ	[28a]ᵃ	[29]						[32]				[28a, 29, 32]
OSO_2-⟨C₆H₄⟩-Br	[5a]ᵃ	[5a]ᵃ			[6c]							[5a]ᵃ		[6c]

ᵃ Only in solution.

carbonium ions *(8)*; for this purpose α-chloroethers are treated with SbF_5 [22a] or $SbCl_5$ [23] as the acceptors:

$$\underset{R^1}{\overset{R^2}{>}}\!\!\!\underset{Cl}{\overset{OR}{<}} + SbCl_5 \text{ (or } SbF_5\text{)} \longrightarrow \underset{R^1}{\overset{R^2}{>}}\!\!\overset{\oplus}{=}OR\ SbCl_6^{\ominus} \text{ (or } SbF_5Cl^{\ominus}\text{)}$$
$$(8)$$

In contrast to the α-haloethers, dialkoxy- and trialkoxyhalomethane derivatives are usually incapable of existence [12a,b, 24, 25] and therefore they do not come into question for the synthesis of di- and trialkoxycarbonium salts.

On the other hand, all types of carboxonium salts are available by the splitting out of alkoxy groups from acetals and ketals *(60)*, orthocarboxylic esters *(61)*, *(63)*, *(64)*, or orthocarbonic esters *(62)*, *(65)*. Carbonium ions (dialkoxycarbonium salts and triphenylcarbonium salts) and also compounds that produce alkyl cations (such as alkyl halides–$AgBF_4$ or trialkyloxonium salts) can be used as alkoxide acceptors in addition to protons BF_3 and $SbCl_5$; *cf.* Table 5.

Occasionally, as in the formation of 1,3-dioxolenium ions *(66)* from 1,3-dioxolane derivatives, the splitting out of hydride ions is also preparatively important. Carbonium ions such as triphenylcarbonium salts act as hydride acceptors [3, 6b,c]:

$$\begin{bmatrix}\underset{O}{\overset{O}{>}}\!\!\!\underset{H}{\overset{R^1}{<}}\end{bmatrix} + R^{\oplus} \longrightarrow \begin{bmatrix}\underset{O}{\overset{O}{>}}\!\!\overset{\oplus}{-}R^1\end{bmatrix} + H\text{-}R$$
$$(66)$$

We have also learned to recognize triphenylcarbonium salts as acceptors for alkoxy groups (Table 5). In cases where the splitting out of both hydride and alkoxide groups is possible, only the alkoxy group is transferred [3]. Indeed, in the present example, the trialkoxycarbonium ions formed by hydride cleavage should be more stable than the dialkoxycarbonium ion, but the alkoxide ion is a better leaving group than is a hydride ion:

$$\begin{bmatrix}\underset{O}{\overset{O}{>}}\!\!\!\underset{OC_2H_5}{\overset{H}{<}}\end{bmatrix} + (C_6H_5)_3C^{\oplus}\ BF_4^{\ominus} \nrightarrow \begin{bmatrix}\underset{O}{\overset{O}{>}}\!\!\overset{\oplus}{-}OC_2H_5\end{bmatrix} + (C_6H_5)_3C\text{-}H\ \ BF_4^{\ominus}$$
$$\searrow \begin{bmatrix}\underset{O}{\overset{O}{>}}\!\!\overset{\oplus}{-}H\end{bmatrix} + (C_6H_5)_3C(OC_2H_5)\ BF_4^{\ominus}$$

With compounds yielding alkyl cations, such as alkyl halides–$AgBF_4$ [2] and trialkyloxonium salts [2], 1,3-dioxolane derivatives also yield 1,3-dioxolenium salts. In this case a carboxonium ion, formed as an intermediate by the addition of an alkyl cation to a 1,3-dioxolane molecule, probably acts as hydride acceptor. The overall reaction therefore represents a disproportionation of a cyclic acetal [27b]:

TABLE 5. ACCEPTORS FOR THE ALKOXY GROUP IN THE SYNTHESIS OF CARBOXONIUM IONS

$$\!$

Splitting-out reagent	Effective alkoxide acceptor A^m	$[A-OR]^{m-1}$	Reactions (60) → (8)	(61) → (9)	(62) → (25)	(63) → (20)	(64) → (21)	(65) → (27)
HBF_4	$H^\oplus$	HOR			[20]			
H_2SO_4	$H^\oplus$	HOR			[21]^a			
BF_3	BF_3	$BF_3(OR)^\ominus$	[11a]	[11a]	[11a]	[11a, 15]	[6a,b,c, 11a]	[14]
$SbCl_5$	$SbCl_5$	$SbCl_5(OR)^\ominus$	[11a]	[11a]	[11a]	[15]	[6a, 11a]	
R'-Hal/$AgBF_4$	$R'^\oplus$	$R'OR$	[13]		[2]			
$R'_3O^\oplus BF_4^\ominus$	$R'^\oplus$	$R'OR$	[13]		[3, 13]	[13]		
HC(OR')₂$^\oplus$ BF₄$^\ominus$	HC(OR')₂$^\oplus$	HC(OR')₂-OR			[3]		[3]	
$(C_6H_5)_3 C^\oplus BF_4^\ominus$ or $SbCl_6^\ominus$	$(C_6H_5)_3 C^\oplus$	$(C_6H_5)_3 C(OR)$	[3]	[3]	[3]		[3]	

^{a} In solution.

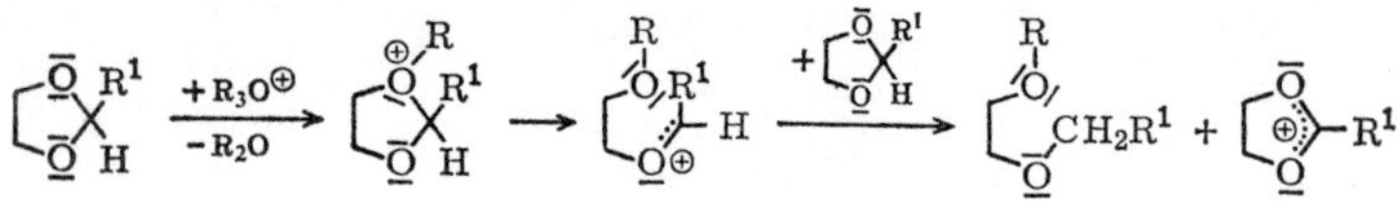

Formation of Oxonium Salts

$$(66)$$

Aryldiazonium salts are also suitable as hydride acceptors: Under certain circumstances not only the aryl cations but also the diazonium ions themselves may act as acceptors [26a]:

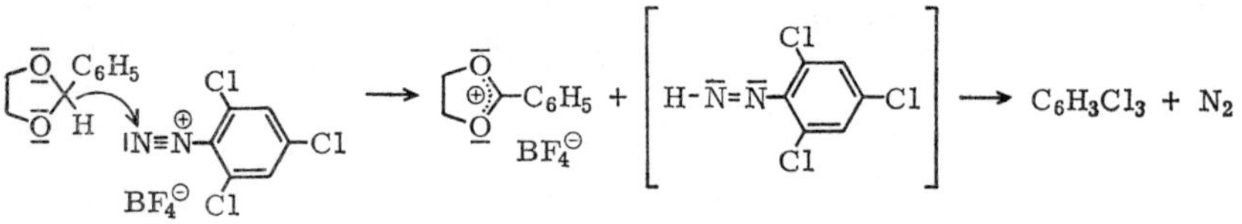

The mechanism has not been elucidated in all cases. A radical reaction is also possible [26a].*

It may be mentioned that in the case of 1,3-dioxolane and 1,3-dioxane derivatives the splitting out of carbeniate groups may also be observed.** Alkoxycarbonium ions in particular act as acceptors [27a]:

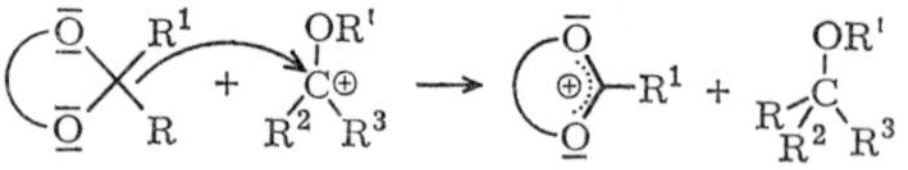

$$(21)$$

4.1.4. Origin of the Complex Anions

As already mentioned, in many cases the complex anions necessary for the isolation of stable oxonium salts arise in a coupled reaction with the formation of the oxonium ion—for example, in the action of an alkyl halide or a hydrogen halide

---

* *Note added in proof (Jan. 1970)*: Recently the mechanism was reinvestigated [26b]. There is *no* evidence of intermediate aryldiimine formed by hydride transfer. The results support a radical mechanism, as was suggested by *Meerwein et al.* [26a] in the case of the corresponding reactions with copper catalyst.

** A special example of the splitting out of an alkyl group with simultaneous elimination of a tosylate group was reported in the case of ring cleavage reactions of cycloalkanone ketals. Here the dialkoxycarbonium ion is formed as a reactive intermediate; e.g. [27c]:

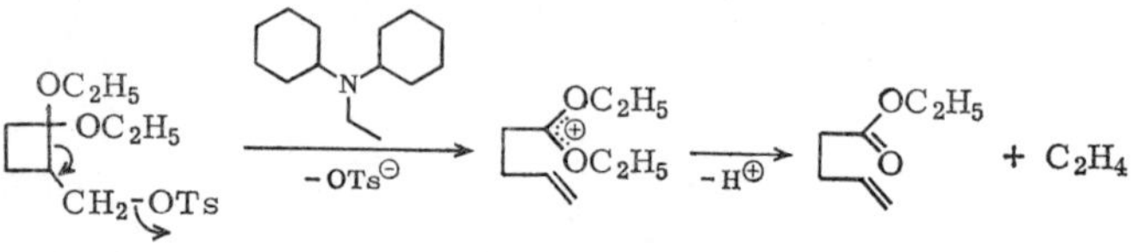

where $Ts = p\text{-}CH_3\text{-}C_6H_4\text{-}SO_2$. For a similar fragmentation with formation of 1,3-dioxolenium ion, cf. [27d].

on the Lewis acid–base complexes of nonmental (or metal) halides with oxygen compounds:

$$\underset{R}{\overset{R}{>}}\!\!O\rightarrow BF_3 + RF \longrightarrow \underset{R}{\overset{R}{>}}\!\!\overset{\oplus}{O}\!-\!R\ BF_4^{\ominus} \qquad [8]$$

$$\underset{R^1}{\overset{R^2}{>}}\!\!=\!O\rightarrow SbCl_5 + HCl \longrightarrow \underset{R^1}{\overset{R^2}{>}}\!\!\overset{\oplus}{=}\!OH\ SbCl_6^{\ominus} \qquad [23]$$

The same reaction principle is utilized in intramolecular alkylations or in the formation of carboxonium salts from α-haloethers:

$$\underset{OR}{\overset{CH_2Cl}{\bigcirc}} \xrightarrow{SbCl_5} \underset{OR}{\overset{\oplus}{\bigcirc}}\ SbCl_6^{\ominus} \qquad [15]$$

$$\underset{R^1}{\overset{R^2}{>}}\!\!\underset{Cl}{\overset{OR}{<}} + SbCl_5 \longrightarrow \underset{R^1}{\overset{R^2}{>}}\!\!\overset{\oplus}{=}\!OR\ SbCl_6^{\ominus} \qquad [22a, 23]$$

In all the present reactions, the complex anion arises *directly*. In the synthesis of carboxonium ions by the action of BF_3 or $SbCl_5$ on acetals, ketals, or orthocarboxylic esters, however, alkoxyhalogeno complex ions are formed first. From these the anions BF_4^- and $SbCl_6^-$ are produced in a rapid subsequent reaction because of their high formation tendency, by a kind of disproportionation [11a]; e.g., cleavage:

$$\underset{R^1}{\overset{R^2}{>}}\!\!\underset{OR}{\overset{OR}{<}} + BF_3 \longrightarrow \underset{R^1}{\overset{R^2}{>}}\!\!\overset{\oplus}{=}\!OR\ [BF_3(OR)]^{\ominus}$$

(8)

$$\underset{O}{\overset{O}{\bigcirc}}\!\!\!\underset{R^1}{\overset{OR}{<}} + SbCl_5 \longrightarrow \underset{O}{\overset{O}{\bigcirc}}\!\!\!\overset{\oplus}{-}R^1\ [SbCl_5(OR)]^{\ominus}$$

(66)

disproportionation:

$$3\,[BF_3(OR)]^{\ominus} + BF_3 \rightarrow 3\,BF_4^{\ominus} + B(OR)_3$$
$$[SbCl_5(OR)]^{\ominus} + SbCl_5 \rightarrow \underline{SbCl_6^{\ominus}} + SbCl_4(OR)$$

However, the complex anion may also be added *preformed* to the reaction mixture, as by the action of complex protonic acids:

$$\underset{R^1}{\overset{R^2}{>}}\!\!=\!\bar{O} + HClO_4 \longrightarrow \underset{R^1}{\overset{R^2}{>}}\!\!\overset{\oplus}{=}\!OH\ ClO_4^{\ominus} \qquad [18b]$$

(13)

$$\underset{R^1}{\overset{RO}{>}}\!\!\underset{OR}{\overset{OR}{<}} + HBF_4 \xrightarrow{-ROH} \underset{R^1}{\overset{RO}{>}}\!\!\overset{\oplus}{=}\!OR\ BF_4^{\ominus} \qquad [14]$$

(18)

34 Formation of Oxonium Salts

Preformed complex anions are also added when, as already mentioned, alkyl halides–AgBF$_4$ [2] (or AgSbF$_6$ [19a,b]), trialkyloxonium [13] and dialkoxycarbonium tetrafluoroborates or hexachloroantimonates [3], triphenylcarbonium [3] and aryldiazonium tetrafluoroborates [14,26a,b] are added in the synthesis of oxonium salts as alkylating agents or as anion–acceptors.

For trialkyloxonium salts, Meerwein *et al.* [4a, 8] also investigated the conversion of the tetrafluoroborates into sparingly soluble salts with other anions. Thus, HgI$_3$ salts can be precipitated by the addition of the trialkyloxonium tetrafluoroborates to an aqueous solution of K[HgI$_3$]; trialkyloxonium salts with the anions SbCl$_6^-$, AuCl$_4^-$, SnCl$_6^{2-}$, and picrate are produced in exactly the same way from solutions of the corresponding alkali–metal salts as sparingly soluble precipitates. This insolubility is the only reason why the oxonium salts, normally readily decomposed by water, can be isolated. Precipitation reactions with other anions in nonaqueous solvents are also known [4a, 33].

4.2. Possibilities of Synthesis for Individual Classes of Oxonium Salts

In this section we describe only methods of preparation for those types of oxonium salts that have acquired special preparative importance, namely, the trialkyloxonium, dialkoxycarbonium, and pyrylium salts.

4.2.1. Trialkyloxonium Salts

The principle of all syntheses of trialkyloxonium salts, the most important class of saturated oxonium salts, is based on the alkylation of dialkyl ethers. In the following review, detailed treatment will be given in particular to those methods of alkylation that differ from the general methods (Section 4.1).

Alkyl halides can be used for alkylation only in the presence of halide acceptors; as already mentioned, AgBF$_4$ [2,34] and AgSbF$_6$ [19b] are suitable for the preparation of acyclic or cyclic trialkyloxonium salts; e.g.:

$$\triangleright\!\!\!O + Cl\text{-}\underset{CH_3}{\overset{CH_3}{C}}\text{-}H + AgBF_4 \longrightarrow \triangleright\!\!\!\overset{\oplus}{O}\text{-}\underset{CH_3}{\overset{CH_3}{C}}\text{-}H\; BF_4^{\ominus} + AgCl \quad [34]$$

Again, as already mentioned, nonmetal or metal halides (BF$_3$, SbCl$_5$) also act as halide acceptors. However, acyclic trialkyloxonium salts can be obtained in this way with alkyl halides only after very long reaction times (several weeks) [8],* while the cyclic salts are formed readily by internal alkylation (Table 4). In many cases it is desirable to use the adducts of BF$_3$ and SbCl$_5$ with ether.

Diazoalkanes can alkylate sufficiently acidic compounds and therefore also the conjugate acids of the ethers (dialkyloxonium ions) [16, 35a, 36a,b]; e.g.:

---

* Methyl fluoride in SbF$_5$ or SbF$_5$–SO$_2$, however, a new powerful methylating agent, reacts readily with dimethyl ether to yield trimethyl oxonium hexafluoroantimonate [22b].

Possibilities of Synthesis for Individual Classes of Oxonium Salts

$$\mathrm{(H_3C)_2\overset{\oplus}{O}\text{-}H\ AlCl_4^{\ominus} + CH_2N_2 \longrightarrow (H_3C)_2\overset{\oplus}{O}\text{-}CH_3\ AlCl_4^{\ominus} + N_2}$$

A variant of this synthesis is the simultaneous reaction of complex protonic acids (HBF_4 [36a], $HSbCl_6$ [36b], 2,4,6-trinitrobenzenesulfonic acid [33]) and diazomethane on a dialkyl ether.

Dialkyloxonium salts normally crystallize as etherates with one mole of ether, which can be removed only with difficulty. Toward diazoalkanes these etherates behave like the ether-free salts but toward diazoacetic ester they behave differently. The oxonium salt formed first is not stable; it acts on a second molecule of ether as an alkylating agent so that trialkyloxonium salts are isolated [35b]:

$$\mathrm{R_2\overset{\oplus}{O}\text{-}H\cdots\!\!IO R_2\ SbCl_6^{\ominus} + N_2CHCO_2C_2H_5 \longrightarrow R_2\overset{\oplus}{O}\text{-}CH_2CO_2C_2H_5\ SbCl_6^{\ominus} + IOR_2}$$

$$\downarrow$$

$$\mathrm{R\text{-}O\text{-}CH_2CO_2C_2H_5 + R\text{-}\overset{\oplus}{O}R_2\ SbCl_6^{\ominus}}$$

Adducts of Trialkyl Phosphates and Trialkyl Phosphorothionates with $SbCl_5$

These adducts are strong alkylating agents. Their action on $SbCl_5$ etherates affords, *inter alia*, mixed trialkyloxonium salts [37a]:

$$\mathrm{R_2O\!\rightarrow\!SbCl_5 + R'O\text{-}\overset{\oplus}{P}(OR')_2\text{-}X\text{-}\overset{\ominus}{S}bCl_5 \longrightarrow R_2\overset{\oplus}{O}\text{-}R'\ SbCl_6^{\ominus} + O\!=\!P(OR')_2\text{-}X\text{-}SbCl_4}$$

where $X = O, S$.

The same alkylating action is shown by the $SbCl_5$ adducts of dialkylphosphorochloridates or dialkylphosphorochloridothioates. In these cases the active alkylating agent perhaps may be an $SbCl_6^{\ominus}$ salt [37a]:

$$\mathrm{R'O\text{-}\overset{\oplus}{P}(OR')(Cl)\text{-}X\text{-}\overset{\ominus}{S}bCl_5 \rightleftharpoons \overset{\oplus}{P}(OR')_2(X)\ SbCl_6^{\ominus}}$$

where $X = O, S$.

The disproportionation of an ether–phosphorus pentafluoride adduct into a trialkyloxonium hexafluorophosphate and phosphoryl fluoride should also be considered in connection with this alkylating action of alkoxyphosphonium compounds [37b]:

$$\mathrm{3\ R_2O\!\rightarrow\!PF_5 \longrightarrow 2\ R_2\overset{\oplus}{O}\text{-}R\ PF_6^{\ominus} + O\!=\!PF_3}$$

Tertiary Oxonium Salts

Ethers can be alkylated with *trialkyloxonium salts*. Here an equilibrium is established whose position is often determined by the different solubilities of the salts [4a, 8]:

$$R-\overset{\oplus}{O}\!\!\diagdown_{R}^{R}\ BF_4^{\ominus} + IO\!\!\diagdown_{R'}^{R'} \rightleftharpoons R-\overset{\oplus}{O}\!\!\diagdown_{R'}^{R'}\ BF_4^{\ominus} + IO\!\!\diagdown_{R}^{R}$$

$$R-\overset{\oplus}{O}\!\!\diagdown_{R'}^{R'}\ BF_4^{\ominus} + IO\!\!\diagdown_{R'}^{R'} \rightleftharpoons R'-\overset{\oplus}{O}\!\!\diagdown_{R'}^{R'}\ BF_4^{\ominus} + IO\!\!\diagdown_{R'}^{R}$$

Thus, for example, the trimethyloxonium salts can be prepared from the readily available triethyloxonium salts. Cyclic ethers are alkylated analogously [8]:

$$\bigcirc\!\!\diagup OI + R-\overset{\oplus}{O}\!\!\diagdown_{R}^{R}\ BF_4^{\ominus} \longrightarrow \bigcirc\!\!\diagup\overset{\oplus}{O}\!-R\ BF_4^{\ominus} + IO\!\!\diagdown_{R}^{R}$$

Dialkoxycarbonium salts can also be used as alkylating agents for ethers [38, 39]:

$$R^1-\overset{\oplus}{C}\!\!\diagdown_{OC_2H_5}^{OC_2H_5}\ SbCl_6^{\ominus} + IO\!\!\diagdown_{C_2H_5}^{C_2H_5} \longrightarrow R^1-C\!\!\diagdown_{OC_2H_5}^{O} + H_5C_2-\overset{\oplus}{O}\!\!\diagdown_{C_2H_5}^{C_2H_5}\ SbCl_6^{\ominus}$$

(67)

Trialkyloxonium salts are likewise obtained by the simultaneous action of orthocarboxylic esters and SbCl$_5$ on ethers; here the formation of dialkoxycarbonium ions should probably be assumed again. Nevertheless, it is not only the dealkylation mechanism of *(67)* that appears to take place. In fact, 1 mole of triethyl orthoformate gives 1.4 mole of triethyloxonium salt; similarly, 1 mole of an orthocarbonic ester gives up to 3.3 mole [4a]. The scheme proposed by Meerwein, in which the ortho ester is re-formed, should certainly also be taken into account [4a, 11a]; e.g.:

$$\begin{array}{c} OC_2H_5 \\ H-\overset{|}{C}-OC_2H_5 \\ \overset{|}{O}C_2H_5 \end{array} \xrightarrow{+2\ SbCl_5} H-\overset{\oplus}{C}\!\!\diagdown_{OC_2H_5}^{OC_2H_5}\ SbCl_6^{\ominus} + SbCl_4(OC_2H_5)$$

$$\Big\downarrow +IO\!\!\diagdown_{C_2H_5}^{C_2H_5}$$

$$\begin{array}{c} OC_2H_5 \\ H-\overset{|}{C}-O\overset{\oplus}{\diagdown}_{C_2H_5}^{C_2H_5} \\ \overset{|}{O}C_2H_5 \\ SbCl_6^{\ominus} \end{array} \xrightarrow{+IO\diagdown_{C_2H_5}^{C_2H_5}} \begin{array}{c} OC_2H_5 \\ H-\overset{|}{C}-OC_2H_5 \\ \overset{|}{O}C_2H_5 \end{array} + H_5C_2-\overset{\oplus}{O}\!\!\diagdown_{C_2H_5}^{C_2H_5}\ SbCl_6^{\ominus}$$

Tertiary Oxonium Ions Formed as Intermediates

In the alkylation of dialkyloxonium salts with diazoacetic ester [35a, b] or that of ethers with SbCl$_5$ and orthocarboxylic esters [4a, 11a], the true alkylating agents are

Possibilities of Synthesis for Individual Classes of Oxonium Salts

probably particularly reactive tertiary oxonium salts occurring as *intermediates*. Meerwein's discovery of the trialkyloxonium salts must be ascribed to the intermediate formation of an oxonium ion [8,40a,b]. Here, epichlorohydrin and BF_3 etherate first give an internal tertiary oxonium salt *(68)*, which acts as the alkylating agent for another molecule of ether [40a,b]:

$$ClH_2C-\underset{H_2C}{\overset{H}{\underset{|}{C}}}\!\!\diagdown_{\!\!O}\!\!\diagup + \underset{H_5C_2}{\overset{H_5C_2}{\diagdown}}\!\!O\rightarrow BF_3 \longrightarrow ClH_2C-\underset{H_2C-\overset{\oplus}{O}\diagdown_{C_2H_5}^{C_2H_5}}{\overset{H}{\underset{|}{C}}}-\overset{\ominus}{O}BF_3$$

$$(68)$$

$$\xrightarrow{+10\diagdown_{C_2H_5}^{C_2H_5}} ClH_2C-\underset{H_2C-OC_2H_5}{\overset{H}{\underset{|}{C}}}-\overset{\ominus}{O}BF_3 + H_5C_2-\overset{\oplus}{O}\diagdown_{C_2H_5}^{C_2H_5}$$

$$(69)$$

This is followed by disproportionation of the alkoxyborofluoride *(69)*:

$$3\ \ ClH_2C-\underset{H_2C-OC_2H_5}{\overset{H}{\underset{|}{C}}}-\overset{\ominus}{O}BF_3 \longrightarrow 2\ BF_4^{\ominus} + \left[ClH_2C-\underset{H_2C-OC_2H_5}{\overset{H}{\underset{|}{C}}}-O-\right]_3 B$$

$$(69)$$

The overall reaction therefore satisfies the equation

$$3\ ClH_2C\!\!\diagdown\!\!\diagup\!\!O + 4\ \underset{R'}{\overset{R}{\diagdown}}O\rightarrow BF_3 + 2\ \underset{R'}{\overset{R}{\diagdown}}O \longrightarrow 3\ \underset{R'}{\overset{R}{\diagdown}}\overset{\oplus}{O}\text{-R}\ BF_4^{\ominus} + \left[ClH_2C\!\!-\!\!\underset{H_2C-OR}{\overset{H}{\underset{|}{C}}}\!\!-O-\right]_3 B$$

The etherates of $FeCl_3$, $AlCl_3$, and $SbCl_5$ react with epichlorohydrin in the same way to give the corresponding trialkyloxonium salts. In the case of $SbCl_5$ etherate, the internal oxonium salt *(70)* formed first can be isolated at a low temperature. It decomposes rapidly at room temperature with liberation of alkyl chloride; however, with $SbCl_5$ etherate it reacts quantitatively to give the trialkyloxonium salt [8]:

$$\underset{(70)}{R^1-\underset{H_2C-\overset{\oplus}{O}\diagdown_R^R}{\overset{H}{\underset{|}{C}}}-O\overset{\ominus}{S}bCl_5} \begin{array}{c} \xrightarrow{R-O\rightarrow SbCl_5} \\ \\ \xrightarrow{\Delta} \end{array} \begin{array}{l} R^1-\underset{H_2C-OR}{\overset{H}{\underset{|}{C}}}-OSbCl_4 + \underset{R}{\overset{R}{\diagdown}}\overset{\oplus}{O}\text{-R}\ SbCl_6^{\ominus} \quad R^1 = H,\ CH_2Cl \\ \\ R^1-\underset{H_2C-OR}{\overset{H}{\underset{|}{C}}}-OSbCl_4 + R\text{-Cl} \end{array}$$

Acyclic trialkyloxonium salts are most easily obtained by the epoxide method, but the other methods of formation are also of preparative importance; *cf.* Table 6.

TABLE 6. METHODS OF PREPARATION OF TRIALKYLOXONIUM SALTS

Trialkyloxonium Salt			Methods of Preparation (yields)			
			Dialkyloxonium salt +		SbCl$_5$ etherate + trialkylphosphorothionate– SbCl$_5$ adduct, %	Nonmetal (metal) halide etherate + epichlorohydrin, %
Oxonium ion	Anion	Mp, °C	Diazoacetic ester, %	Diazoalkane, %		
Trimethyl	BF$_4^\ominus$	143 [4a]	80 [35a]	68 [16]	—	98.2 [40a]
	SbCl$_6^\ominus$	159 [35b]	97.5 [35b]	65 [16]	—	95.5 [8]
	FeCl$_4^\ominus$	81 [35a]	80 [35a]	—	—	—
Triethyl	BF$_4^\ominus$	92 [8]	—	53 [35a]	—	100 [8]
	SbCl$_6^\ominus$	135–137 [8]	85 [35b]	—	95–99 [37a]	95 [8]
	FeCl$_4^\ominus$	74 [35a]	80 [35a]	—	—	100 [8]
Tri-n-propyl	BF$_4^\ominus$	73–74 [8]	—	—	—	30 [8]
	SbCl$_6^\ominus$	105 [35b]	73 [35b]	—	—	—
Tri-n-butyl	SbCl$_6^\ominus$	118–119 [35b] (123–124 [37a])	33 [35b]	—	47.3 [37a]	74.5 [4a]
Ethyldimethyl	SbCl$_6^\ominus$	119 [16]	—	37 [16]	—	—
Diethylmethyl	SbCl$_6^\ominus$	114 [16]	—	83 [16]	87.2 [37a]	—
Diethyl-n-propyl	SbCl$_6^\ominus$	112–114 [37a]	—	—	71.9 [37a]	—

Possibilities of Synthesis for Individual Classes of Oxonium Salts 39

In general the yields of oxonium salts decrease with increasing size of the alkyl groups; nevertheless, whereas the tri-*n*-butyloxonium salt can still be obtained satisfactorily, all attempts at the synthesis of triisopropyloxonium or tri-*tert*-butyloxonium salts have so far been unsuccessful. This has been ascribed to steric effects, but in any case *tert*-butyloxonium ions should be very unstable, since they can readily split out the stable trimethylcarbonium ion; *cf.* Section 5.2.

4.2.2. Dialkoxycarbonium Salts

The two main methods of preparing dialkoxycarbonium salts (O-alkylation of an ester carbonyl by route A or anionic splitting out of suitable substituents in the α-position to two ethereal oxygen atoms by route B), namely,

$$R^1\underset{OR}{\overset{O}{\diagup}} + R\text{-}Y \xrightarrow{\text{route A}} R^1\underset{OR}{\overset{OR}{\diagup}}\oplus\ Y^\ominus \xleftarrow{\text{route B}} R^1\underset{OR'}{\overset{OR}{\diagup}}Y$$

(9)

are capable of wide variation only in the synthesis of the *cyclic* representatives.

4.2.2.1. Acyclic Dialkoxycarbonium Salts

In the formation of acyclic dialkoxycarbonium salts, only route B has achieved preparative importance; it is *exclusively* the orthocarboxylic esters (Y = OR) that can be used as starting materials for this; therefore *(9)* is also occasionally called an "ortho-ester cation." Acceptors that can be employed for the alkoxy groups are BF_3, $SbCl_5$ [11a], ethereal HBF_4 [20], and (more rarely) triphenylcarbonium salts [3] (*cf.* Table 5); e.g.;

$$3\ \underset{R^1}{\overset{RO}{\diagdown}}\!\!\underset{OR}{\overset{OR}{\diagup}} + 4\ BF_3 \longrightarrow 3\ R^1\!\!\underset{OR}{\overset{OR}{\diagup}}\oplus\ BF_4^\ominus + B(OR)_3$$

(9)

Alkylation of the ester carbonyl (route A) to *(9)* is *not* possible even with an alkyl halide–$AgBF_4$ or a trialkyloxonium salt; on the other hand, the transfer of alkyl groups to the ester carbonyl does take place with diethoxycarbonium hexachloroantimonate [39], which in turn can be obtained from the ortho ester (route B):

$$R^1\underset{OC_2H_5}{\overset{O}{\diagup}} + H\!\!\underset{OC_2H_5}{\overset{OC_2H_5}{\diagup}}\!\!\oplus\ SbCl_6^\ominus \longrightarrow R^1\!\!\underset{OC_2H_5}{\overset{OC_2H_5}{\diagup}}\!\!\oplus\ SbCl_6^\ominus + H\!\!\underset{OC_2H_5}{\overset{O}{\diagup}}$$

where $R^1 = (CH_3)_2CH$, $C_6H_5CH_2$. (67)

The yields of alkylated ester are good when the ester is used in excess and the arising formate is removed from the reaction mixture.

4.2.2.2. 1,3-Dioxolenium Salts

In the series of cyclic dialkoxycarbonium salts, only methods for the synthesis of

1,3-dioxolenium salts will be treated. These are also readily obtained by route A (internal alkylation) and route B:

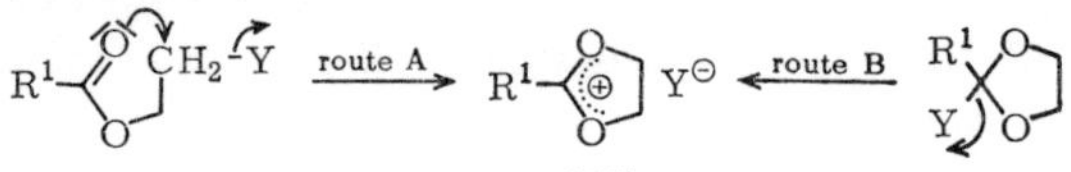

$$R^1 \overset{O}{\underset{O}{\diagup}} CH_2 - Y \xrightarrow{\text{route A}} R^1 \overset{O}{\underset{O}{\diagup}} Y^\ominus \xleftarrow{\text{route B}} R^1 \overset{O}{\underset{O}{\diagup}} Y$$

(66)

Very diverse substituents may be used as the leaving groups Y.

Syntheses by Route A

The leaving substituents in internal alkylation are halide ions [2, 4a, 6a,b, 15, 19a, 28a,b,c,d, 30–32], hydroxy groups [28a], alkoxy groups [6a, 28a], and acyloxy groups [28a, 29, 32]; *cf.* Table 4 in Chapter 3. Syntheses of *(66)* from β-haloethyl esters with silver ions as halide acceptors in the form of $AgBF_4$ [2, 6a,b, 28a,b,c,d, 30] or $AgSbF_6$ [19a] are particularly widely used.

Using the preparation of the 4-vinyl-1,3-dioxolenium salt *(72)* as an example, it may be shown that in addition to the usual method of synthesis from the corresponding substituted β-chloroethyl ester *(71)*, the vinylogous reaction from the allyl halide *(73)* is also possible [30]:

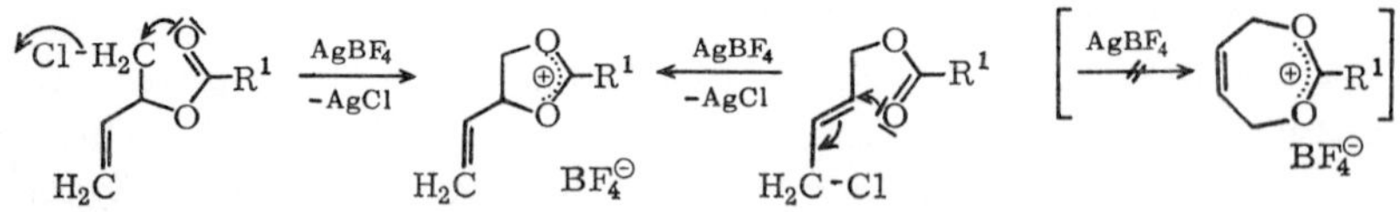

(71)   (72)   (73)   (74)

This is simultaneously an example of the formation of an oxonium ion by the interaction of an oxygen function with a C–C double bond. Because of the particular stability of the 1,3-dioxolenium system, only the five-membered heterocycle *(72)* and not the seven-membered heterocycle *(74)* arises.

The $AgBF_4$ method has also been extended to the preparation of bisdioxolenium ions *(75)* from bis(β-bromoethyl ester)s [28b]:

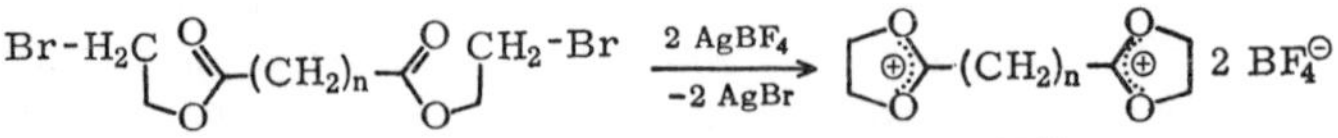

(75)

Here the reaction times increase with decreasing number n of methylene groups (2 hours for $n = 3$–5, and 8 hours for $n = 1, 2$; the compound with $n = 0$ has not yet been prepared) [28b]. The arylbis- and -tris(1,3-dioxolenium) ions *(76)–(78)* can be synthesized by the same process [28c].

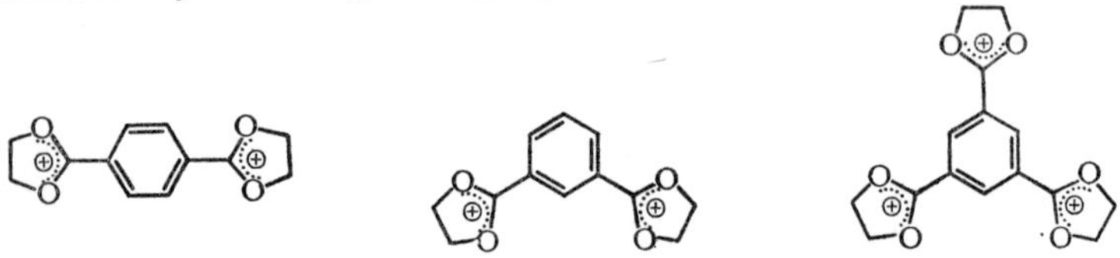

(76)   (77)   (78)

Internal alkylations of β-haloethyl esters with BF_3 or $SbCl_5$ [4a, 6a, 15, 32] have been mentioned above (Sections 4.1.2 and 4.1.3).

In the splitting out of alkoxy groups (Y = OR) from β-alkoxyethyl esters, the rate of formation of the 1,3-dioxolenium ion depends on the nature of the acceptor. The process of formation takes place extraordinarily slowly with BF_3 (it requires 6–12 months) [6a]:

$$RO-H_2C\overset{O}{\underset{O}{\bigcirc}}-R^1 + BF_3 \rightleftharpoons RO\overset{BF_3}{\underset{}{\stackrel{\downarrow}{-}}}H_2C\overset{O}{\underset{O}{\bigcirc}}-R^1 \rightleftharpoons \overset{O}{\underset{O}{\bigcirc}}\stackrel{\oplus}{-}R^1 \;\; [BF_3(OR)]^{\ominus}$$

$$\xrightarrow{+\tfrac{1}{3}BF_3} \overset{O}{\underset{O}{\bigcirc}}\stackrel{\oplus}{-}R^1 \; BF_4^{\ominus} + \tfrac{1}{3} B(OR)_3$$

(66)

It is very much faster with $SbCl_5$, and proceeds particularly well with trialkyl-oxonium tetrafluoroborates [6a] (which provide one alkyl group as an acceptor). For β-alkoxy-, β-hydroxy-, and β-acetoxyethyl esters, FSO_3H is also suitable as a cleavage reagent: The rate of formation of *(66)* decreases in the following sequence [28a]:

$$Y\overset{O}{\underset{O}{\bigcirc}}-R^1 \text{ with Y: } HO- > H_3C-\overset{O}{\underset{\|}{C}}-O- > H_3C-O-$$

Syntheses by Route B

1,3-Dioxolane derivatives substituted in position 2 are used as starting materials for syntheses of *(66)* by route B. We have already given an exhaustive account of the splitting out of hydride ions (Y = H) by carbonium salts [2, 3, 6b] or diazonium salts [26a], Section 4.1.3. Cyanide can also act as the leaving group Y [3], and here triphenylcarbonium or dialkoxycarbonium salts act as acceptors. However, this reaction has no preparative importance, since 2-cyano-1,3-dioxolanes are best prepared by the reverse reaction from 1,3-dioxolenium salts and cyanides.

The splitting of alkoxides out of 2-alkoxy-1,3-dioxolanes *(79)* with BF_3, $SbCl_5$, or carbonium salts is of great preparative value; Table 7. Addition products with the Lewis acid, e.g. *(80)*, are formed first, and the alkoxy groups are split out from these [11a]:

$$\overset{O}{\underset{O}{\bigcirc}}\!\!\!\overset{OR}{\underset{R^1}{\diagdown}} \underset{-BF_3}{\overset{+BF_3}{\rightleftharpoons}} \overset{O}{\underset{O}{\bigcirc}}\!\!\!\overset{\overset{BF_3}{\stackrel{\downarrow}{O}-R}}{\underset{R^1}{\diagdown}} \rightleftharpoons \overset{O}{\underset{O}{\bigcirc}}\stackrel{\oplus}{-}R^1 + [BF_3(OR)]^{\ominus}$$

(79)   *(80)*   *(66)*

The alkoxytrifluoroborate ion so produced disproportionates as usual with BF_3 to give BF_4^- and $B(OR)_3$. Only this disproportionation appears to be irreversible: The addition of the Lewis acid and the splitting out of the alkoxytrifluoroborate ion are regarded as reversible steps. In fact, by adding 2,6-dimethyl-γ-pyrone as a BF_3 acceptor, the 2-alkoxy-1,3-dioxolane *(79)* is re-formed from the primary products [6a]:

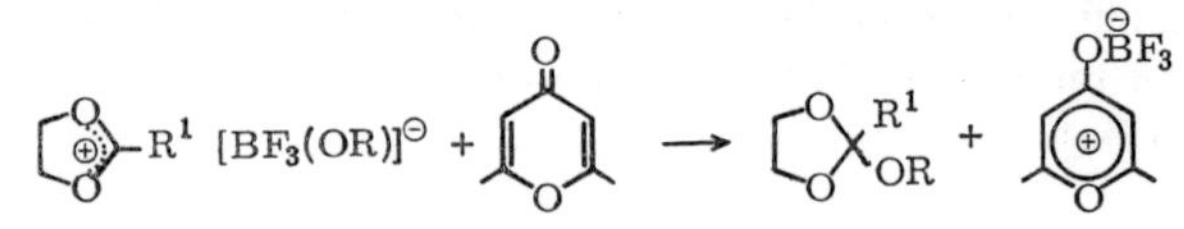

2-Alkoxy-1,3-dioxolanes are readily available by transesterification of acyclic orthocarboxylic esters with 1,2-glycols in the presence of catalytic amounts of a strong acid [11a]. Another route to *(79)* involves the reaction of carboxylic esters with epoxides, using catalytic amounts of BF_3 [11a]; this reaction probably taking the following course*:

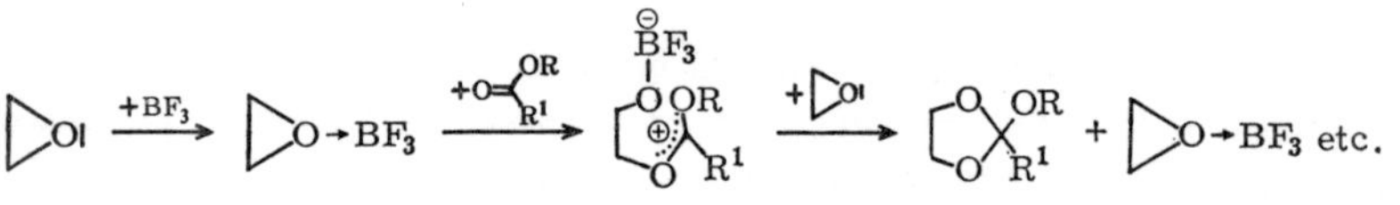

If *molar* (instead of catalytic) amounts of BF_3 are allowed to act on the epoxide and the carboxylic ester, no 2-alkoxy-1,3-dioxolane *(79)* can be isolated, but the 1,3-dioxolenium salt is obtained (it is usual to use the BF_3 adduct of the carboxylic ester instead of BF_3 itself). In this dioxolenium synthesis the reaction again probably passes through the complex *(80)* as an intermediate [11a].

2,2-Dialkyl- and 2-alkyl-2-aryl-1,3-dioxolanes can also be used for the synthesis of 1,3-dioxolenium ions. Here the leaving group is a carbeniate ion [27a,b]; *cf.* Section 4.1.3. The reaction is particularly clear when alkoxycarbonium or diphenyl-carbonium salts are used:

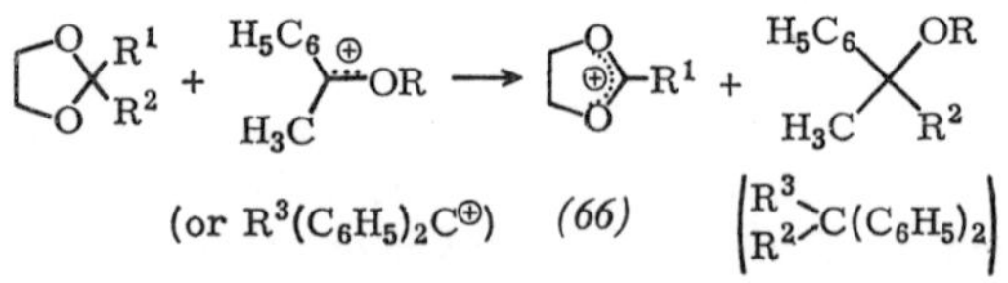

The yields of *(66)* are low when R^1 and R^2 are alkyl groups and better when R^1 is phenyl and R^2 is methyl, i.e., with cyclic ketals *(81)* of acetophenone. The electrophilic agent attacking the alkyl group may also be a 1,3-dioxolane molecule activated by the addition of a Lewis acid (BF_3, $SbCl_5$), or by alkylating agents, and which probably forms a ketonium ion—e.g., *(82)* or *(83)*—as an intermediate. The

---

* *Note added in proof (June 1970)*: Recently this mechanism was proved for the formation of 1,3-dioxolanes from the BF_3-catalyzed reaction of epoxides with carbonyl compounds [11b].

alkyl group transfer then takes place by a *disproportionation* reaction similar to that in the case of the cyclic acetals (Section 4.1.3): One dioxolane molecule is converted into the dioxolenium ion and the second splits to a glycol ether with a tertiary alkyl residue. Under the reaction conditions, however, the tertiary ether undergoes cleavage into alkene (dimers have been detected) and glycol derivatives (esters of antimonic or boric acid on action of $SbCl_5$ and BF_3, respectively, and glycol dialkyl ethers when alkylating agents are used) [27a]: Disproportionation scheme is

Table 7 shows the various possibilities for the synthesis of 2-methyl-1,3-dioxolenium tetrafluoroborate.

4.2.3. Pyrylium Salts

Being cyclic conjugated carboxonium ions (or monoalkoxycarbonium ions) with a π-electron sextet, pyrylium ions *(29)* exhibit particular stability when the reactive positions 2, 4, and 6 are protected by substituents.

Most pyrylium syntheses have therefore been developed for 2,4,6-trisubstituted derivatives. In general, *no* complex anion is in this case necessary for the isolation

TABLE 7. METHODS OF PREPARING 2-METHYL-1,3-DIOXOLENIUM TETRAFLUOROBORATE

[2-methyl-1,3-dioxolenium cation with BF_4^-]

Type of reaction	Initial compound	Method	Effective acceptor for Y^-	Yield, %	Reference
Intramolecular alkylation (route A)	$Br-H_2C-C(O)(CH_3)-O-CH_3$	$AgBF_4$	$Ag^{\oplus}$	84.5	[2]
	$F-H_2C-C(O)(CH_3)-O-CH_3$	BF_3	BF_3	61	[6a]
	$H_5C_2-O-H_2C-C(O)(CH_3)-O-CH_3$	BF_3	BF_3	49	[6a]
		$(C_2H_5)_3O^{\oplus}BF_4^{\ominus}$	$C_2H_5^{\oplus}$	75	[6a]
Splitting out of Y^- from dioxolane derivatives (route B)	[1,3-dioxolane with CH_3, H]	$AgBF_4/C_2H_5Br$	$H_3C-C^{\oplus}(H)-O-(CH_2)_2-OC_2H_5$ [a]	41.5	[2]
		$(C_6H_5)_3C^{\oplus}BF_4^{\ominus}$	$(C_6H_5)_3C^{\oplus}$	69	[3]
		[3,5-dichlorophenyl-$N_2^{\oplus}$ $BF_4^{\ominus}$ (Cu)]	b	71	[26a]
	[1,3-dioxolane with 2 CH_3]	BF_3	$H_3C-C^{\oplus}(CH_3)-O-(CH_2)_2-O\overset{\oplus}{B}F_3$ [a]	20	[27a]
		$(C_2H_5)_3O^{\oplus}BF_4^{\ominus}$	$H_3C-C^{\oplus}(CH_3)-O-(CH_2)_2-OC_2H_5$ [a]	33	[27a]
	[1,3-dioxolane with CH_3, CN]	$(C_6H_5)_3C^{\oplus}BF_4^{\ominus}$	$(C_6H_5)_3C^{\oplus}$	59	[3]
		$HC(OC_2H_5)_2^{\oplus}BF_4^{\ominus}$	$HC(OC_2H_5)_2^{\oplus}$	88	[3]
	[1,3-dioxolane with CH_3, OC_2H_5]	BF_3	BF_3	93	[11a]
		$(C_6H_5)_3C^{\oplus}BF_4^{\ominus}$	$(C_6H_5)_3C^{\oplus}$	86	[3]
		$HC(OC_2H_5)_2^{\oplus}BF_4^{\ominus}$	$HC(OC_2H_5)_2^{\oplus}$	73	[3]
	[cyclopropane + $CH_3-C(=O)-OC_2H_5$]	BF_3	BF_3	80.5	[4a, 11a]

[a] Assumed as acceptor [27b].

Possibilities of Synthesis for Individual Classes of Oxonium Salts

of the salts; nevertheless, salts with the anions BF_4^-, $FeCl_4^-$, and ClO_4^- are frequently prepared.

While it is true that the synthetic routes A (internal O-alkylation), B (splitting out of Y^- from position 2, 6, or 4 of pyran derivatives *(85)*, *(90)*), and C (addition of a cation to exocyclic double bonds of *(87)*, *(88)*) can also be found here, a clear distinction is only possible if in these syntheses we start from preformed heterocycles.

More frequently syntheses of the pyrylium nucleus are carried out, in which, before the final stage of the reaction to give the pyrylium ion, compounds of types *(84)*, *(85)*, or *(90)* are formed without being isolated. Consequently, the classification under route A or route B is uncertain in the case of reactions in which ring closure *and* the anionic detachment of a group Y take place, particularly because ring-chain tautomerism can exist between the pentadienone *(84)* and the pyran derivative *(85)* [41a,b]. However the interaction of the carbonyl function of the pentenynone *(89)* with the C–C triple bond (with proton addition) probably follows route A. The following scheme gives a review of the final stages of the formation of pyrylium ions:

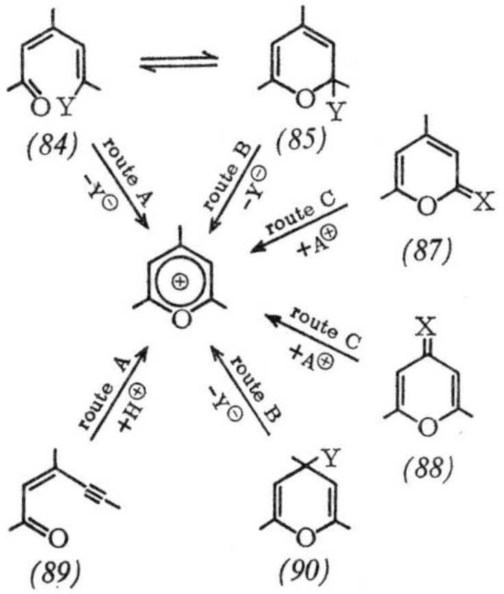

where Y = H, OH, OR, NR_2, Cl
X = O, S, Se, CR_2
A = H, alkyl, acyl

We shall follow the usual subdivision of pyrylium syntheses, namely: Methods of formation starting from *preformed heterocycles* and *syntheses with ring closure* [4a, 42, 43].

4.2.3.1. Pyrylium Salts from Preformed Oxygen Heterocycles

The starting materials here are pyran derivatives *(85)*, *(90)*, or pyrones *(87)*, *(88)* (X = O). Pyran derivatives react by route B with the anionic detachment of Y from *(85)* or *(90)* to form the pyrylium salt: Thus, for example, 4*H*-pyrans *(86)* react with the splitting out of a hydride ion (Y = H):

Formation of Oxonium Salts

$$\text{(86)} \quad \xrightarrow{-H^{\ominus}}$$

At the same time, this is the best synthesis for unsubstituted pyrylium salts *(29)* [44a,b]. The following compounds can be used as hydride acceptors: triphenylcarbonium salts (*inter alia*, in the preparation of *(29)*) [44a,b], PCl_5 [44a], potassium permanganate [45a,b], potassium hexacyanoferrate(III) [45a,b], $FeCl_3$ [45b], or even $HClO_4$ [45b]. The 4-hydroxybenzopyrylium salts *(92)* that can be obtained from chromanones by the elimination of hydride can be imagined as arising from the 2H-pyran *(91)* [46]:

$$\text{(91)} \quad \xrightarrow{+(C_6H_5)_3C^{\oplus}ClO_4^{\ominus}} \quad \text{(92)} \quad ClO_4^{\ominus} + (C_6H_5)_3CH$$

The synthesis of pyrylium salts by the action of nucleophilic agents on pyrones and subsequent treatment with acid is based on the anionic elimination of OH groups from the (generally not isolated) pyranols *(94)* or *(96)*: the reaction of γ-pyrones *(93)* with Grignard compounds [18b, 47] takes place in this way:

$$\text{(93)} \xrightarrow{R^1MgX} \text{(94)} \xrightarrow[-H_2O]{+H^{\oplus}}$$

and it can be extended to coumarins [48a,b] and isocoumarins [49], compounds of the series with fused benzene rings. α-Pyrones can react with diarylethylenes or N,N-dialkylaniline (in an acid medium) in a similar manner [50]:

$$\text{(95)} \xrightarrow{+R^1-H} \text{(96)} \xrightarrow[-H_2O]{+H^{\oplus}}$$

$$R^1 = \underset{R'}{\overset{R'}{\diagdown}}N - \bigcirc -$$

$$\underset{H}{\overset{}{\diagdown}}C = C\underset{Ar}{\overset{Ar}{\diagup}}$$

These reactions of the pyrones with *nucleophilic* agents are based on the intermediate conversion of *(87)* into *(85)* or of *(88)* into *(90)*, as the case may be. However, pyrones are also converted into pyrylium derivatives under the attack of *electrophilic* agents. A special case of the addition of cations $A^{\oplus}$ to the semicyclic double bonds of *(87)* or *(88)*, route C:

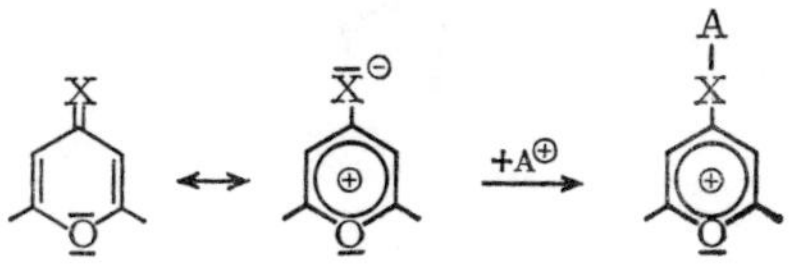

(88)

is the addition of protons (from mineral or organic acids, $A^\oplus = H^\oplus$) to the carbonyl oxygen (X = O) of γ-pyrones *(93)* to form 4-hydroxypyrylium salts [51].

Similarly, with alkylating agents (dialkyl sulfates [18a, 52a,b], nitrobenzenesulfonic esters [53], trialkyloxonium salts [4a, 40a], $A^\oplus = $ alkyl$^\oplus$) 4-alkoxypyrylium salts, and with acylating agents ($A^\oplus = $ acyl$^\oplus$) [54] 4-acyloxypyrylium salts are obtained. α-Pyrones *(95)* can be converted into 2-alkoxypyrylium salts only with strong alkylating agents (trialkyloxonium salts) [42, 55] (this reaction is also possible with coumarin [40a]).

With 4-thiopyrone (X = S) [56], alkylation takes place, as in the case of the γ-pyrones *(93)*, to give 4-(alkylthio)pyrylium salts and with 4-selenopyrone (X = Se) to give 4-(alkylseleno)pyrylium salts [57]. The addition of protons to 4-methylenepyrans ("4-dehydropyrans") *(97)* is basically the same type of reaction [43, 45a,b]:

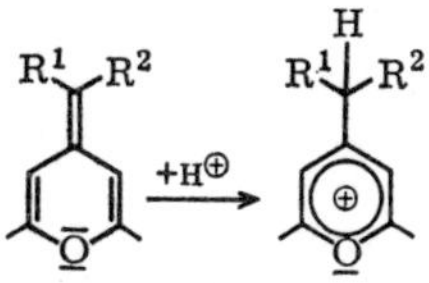

(97)

The methylene derivatives *(97)* are readily available from γ-pyrones with active methylene components [58].

4.2.3.2. Pyrylium Syntheses by Ring-Closure Reactions

All the ring-closure reactions to form pyrylium ions known hitherto take place finally through cyclization of a C_5 structural component. According to Schroth and Fischer, even the multicomponent syntheses lead to this final phase [43]: Their scheme distinguishes C_5, C_1C_4, C_2C_3, $C_1C_2C_2$, $C_2C_1C_2$, and $C_1C_3C_1$ syntheses according to the size of the carbon components that are used in a single procedure to yield pyrylium ions directly. It is assumed that the smaller components form themselves into C_5 chains before ring closure. The scheme below gives the relationships between the possible combinations [43]:

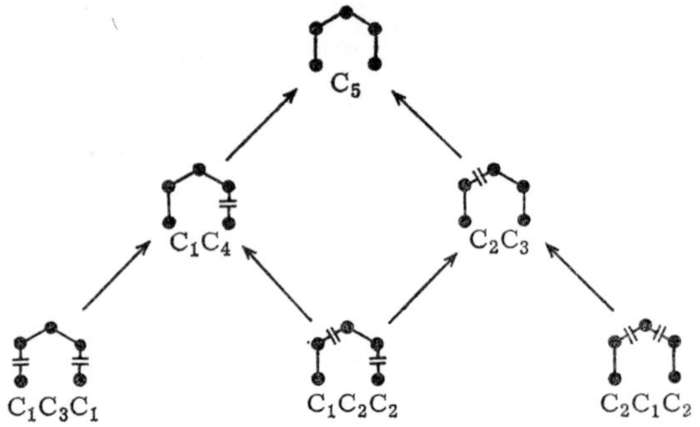

A synthesis with C_5 ring closure is carried out in the case of a pent-2-en-4-yn-1-one *(89)* under the action of protonic acids; e.g. [59a,b]:

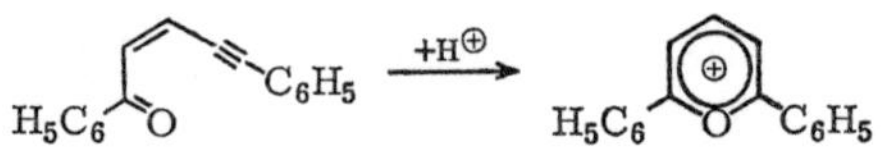

without the anionic elimination of suitable groups Y. However, C_5 cyclizations usually take place by the general reaction:

Here it is not clear whether the splitting out of Y takes place synchronously with the ring closure or whether pyran derivatives of type *(85)* arise as intermediate products [43]. Different cations *(100)* [60] and *(101)* [61] are also assumed as intermediates in protonation. However, in accordance with the overall reaction, the following processes must be formulated in this way:

Cyclization of unsaturated 1,5-dicarbonyl compounds *(102)* with strong acids: Ring closure obviously takes place via the enolic form, i.e., the doubly unsaturated 1,5-hydroxyketone *(103)* [47,62] (Y=OH), provided the *cis* configuration is present [63]:

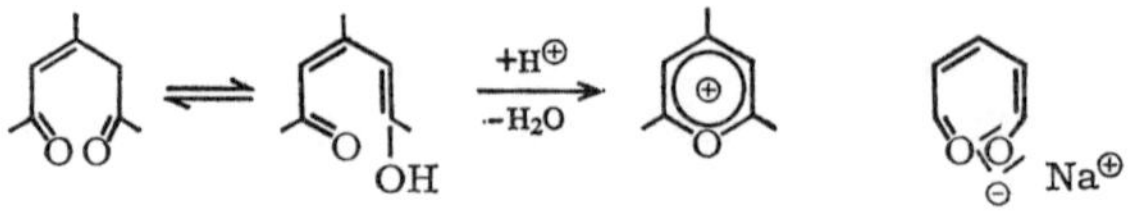

(The unsubstituted pyrylium ion *(29)* has also been synthesized by this method from the sodium salt of glutacondialdehyde *(104)* [60].)

Quite similarly, doubly unsaturated 1,5-aminoketones *(105)*, (Y = NR$_2$) can be converted with protonic acids [64], or 1,5-chloroketones, e.g. *(106)* (Y = Cl), can be converted with Lewis acids (SbCl$_5$, SnCl$_4$, FeCl$_3$) [65], into pyrylium salts; e.g.:

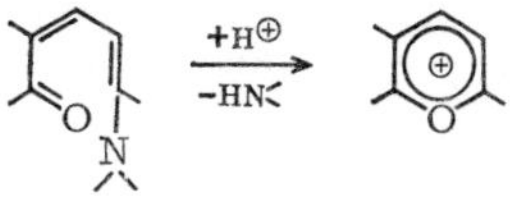

(105) 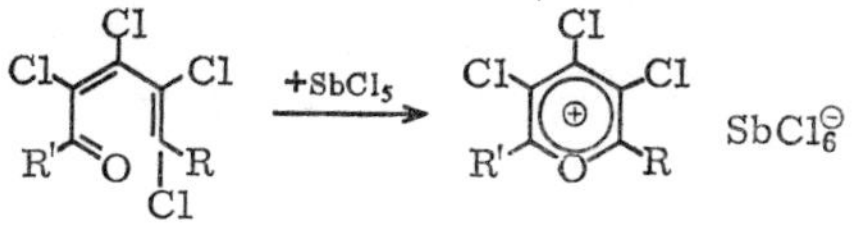   *(106)*

Finally, the cyclodehydrogenation of the pentadienones *(107)* (Y = H) also belongs here [66]:

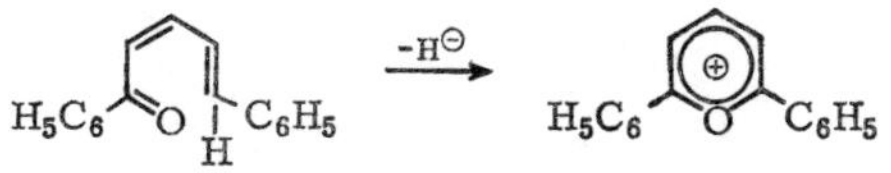

(107)   *(99)*

FeCl$_3$ [42, 66, 67] or triphenylcarbonium salts [61] act as hydride acceptors.

The formation of pyrylium salts by the cyclodehydrogenation of 1,5-diketones *(108)* with FeCl$_3$ and acetic anhydride [68a,b,c,d] takes place with splitting out of a hydride ion *and* a molecule of water; if the detachment of the hydride ion is regarded as the primary step, the synthesis reduces finally to cyclization of an enol of type *(103)*:

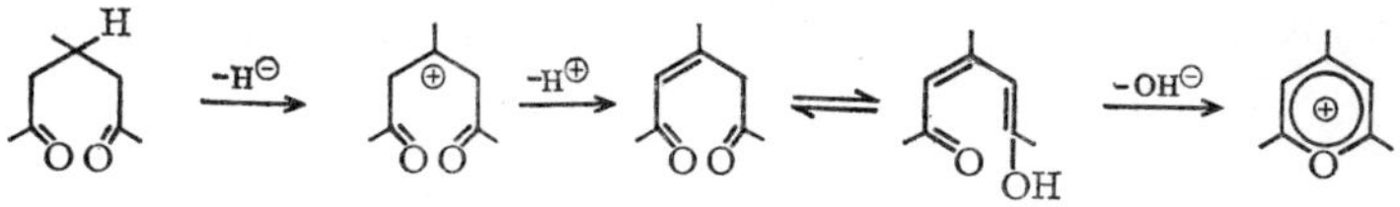

(108)   *(109)*   *(102)*   *(103)*

Other hydride acceptors that can be used besides FeCl$_3$ are triphenylcarbonium ions [69, 70a], *tert*-butyl [69, 70a], hydroxyallyl or acetyl cations [69], and in some cases even atmospheric oxygen [71]. The use of the *tert*-butyl cation (obtained from *tert*-butyl chloride–AlCl$_3$) leads in the case of aryl-substituted 1,5-diketones to the *tert*-butyl substitution of the phenyl nucleus in position 4 of the pyrylium ring [70a]; e.g.:

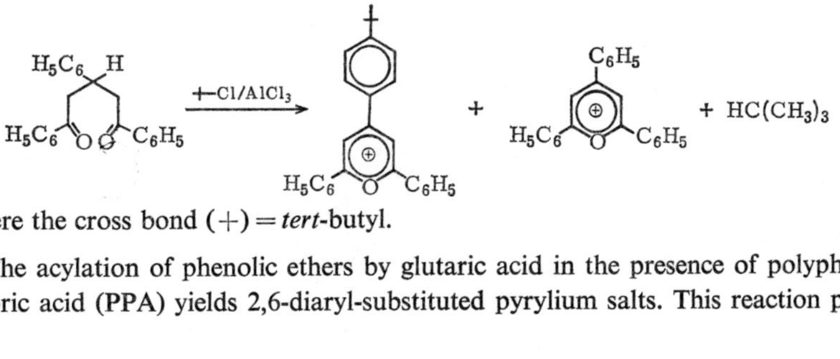

where the cross bond (+) = *tert*-butyl.

The acylation of phenolic ethers by glutaric acid in the presence of polyphosphoric acid (PPA) yields 2,6-diaryl-substituted pyrylium salts. This reaction proceeds probably via a 1,5-diketone, too, which is formed in situ and cyclizes in subsequent reaction steps [70b]:

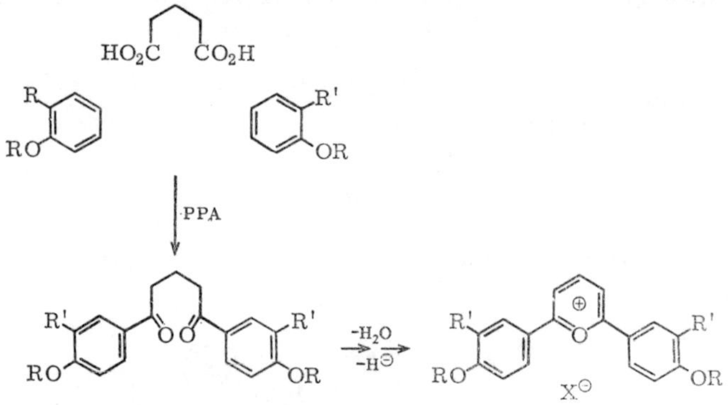

where $R^1 = H, OCH_3; R = CH_3; X = H_2PO_4$.

Multicomponent Syntheses

Processes that construct one of the aforementioned C_5 chains in several steps and then convert the isolated C_5 component in a subsequent definite reaction step into the pyrylium salt (e.g. [72]) will not be regarded as multicomponent syntheses in the sense of our classification. These always lead in a single procedure to the pyrylium salt. Here we shall indicate only the most important methods; for details, reference may be made to Schroth and Fischer's summary [43]. The following scheme gives a review of the possibilities of multicomponent syntheses:

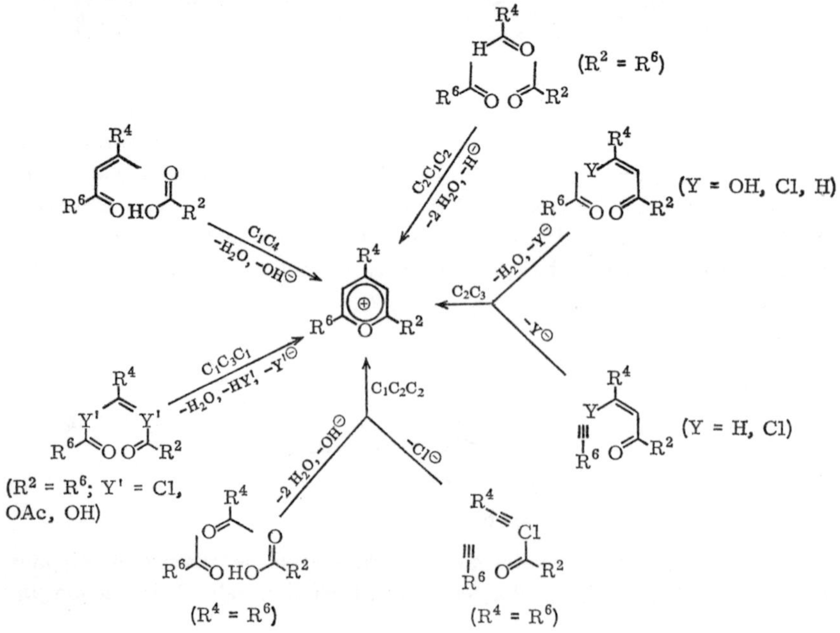

For syntheses of types C_1C_4, $C_1C_3C_1$, and $C_1C_2C_2$, carboxylic acids or their derivatives are necessary in the condensation. The reactions consequently reduce to the acylation of a double or triple bond. Here the intermediate formation of acyl cations is attained by the action of protonic acids ($HClO_4$, H_2SO_4) or Lewis acids (e.g., $AlCl_3$) on carboxylic acid anhydrides or chlorides (more rarely the carboxylic acids themselves). The addition step in C_1C_4 syntheses probably takes place as follows [52b, 68b, 73, 79a,b,c]:

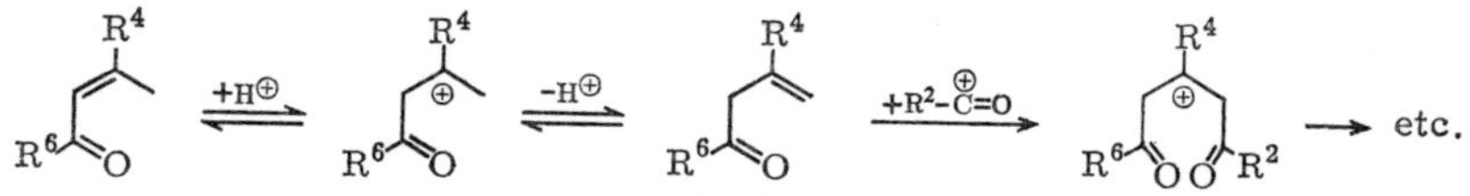

to give the carbonium ion *(109)* already mentioned via the β,γ-unsaturated ketone *(111)*, which arises by proton catalysis from *(110)* [41b, 74a,b]. This is indicated in any case by the fact that the yield of trimethylpyrylium salts ($R^2 = R^4 = R^6 = CH_3$) is very much higher when, instead of mesityl oxide (*(110)*, $R_4 = R^6 = CH_3$), isomesityl oxide (*(111)*, $R^4 = R^6 = CH_3$) is condensed with acetic anhydride–$HClO_4$ [74b, 75]. A variant of the synthesis starts from β-hydroxy ketals *(112)* and acetic anhydride [76]; here again the reaction can probably be ascribed to the formation of *(111)*:

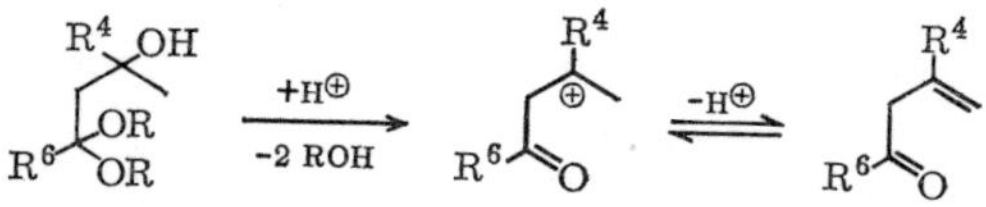

The diacylation of propenes ($C_1C_3C_1$ synthesis) takes place very similarly [41b, 63, 77a,b,c,d,e]; the 1,5-dicarbonyl carbonium ion formed as an intermediate corresponds to *(109)*:

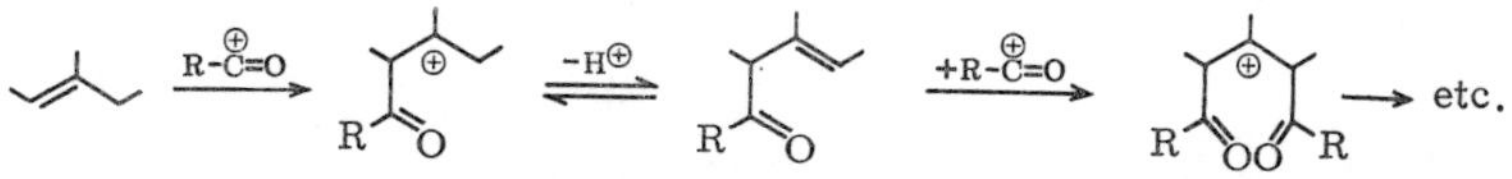

Nevertheless, the propene derivative need not be added as such but may also be obtained by the action of protonic or Lewis acids on a tertiary alcohol; *cf.* *(112)* [41b, 74b]. The synthesis of pyrylium salts with fused rings, e.g. *(113)*, is also possible by this procedure [78]:

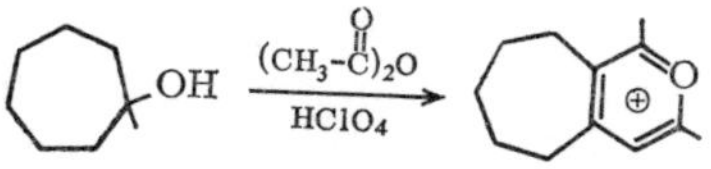

The $C_1C_2C_2$ synthesis from one mole of carboxylic acid (or carboxylic acid derivative) and 2 moles of methyl ketone (*cf.* scheme on page 50) offers no basic

Formation of Oxonium Salts

differences from the C_1C_4 synthesis; however, only pyrylium salts with the same substituents in positions 4 and 6 are accessible in this way [66, 68b, 79a,b,c]. In the case of the other possibility, of $C_1C_2C_2$ synthesis with 1 mole of aromatic carboxylic acid chloride and 2 moles of phenylacetylene in the presence of $SnCl_4$, perhaps a chlorovinyl ketone *(114)* is formed as an intermediate [80a,b]:

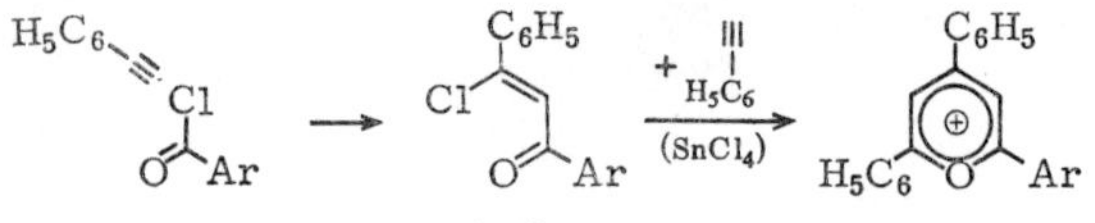

(114)

In fact, C_2C_3 syntheses also take place with β-chlorovinyl ketones and acetylene derivatives [80b]. An analog of this reaction is that of chalcones with substituted acetylenes [81]:

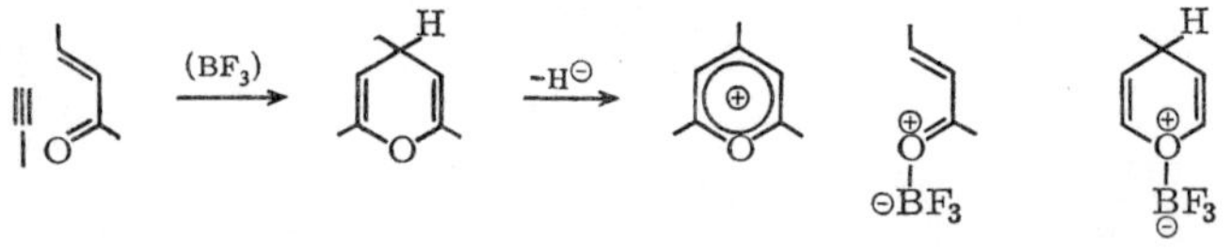

(115) *(116)*

The BF_3 adducts *(115)* and *(116)* are assumed to be intermediate products; *(115)* probably acts as a hydride acceptor.

If β-chlorovinyl ketones are treated with monoketones, in addition to the splitting out of chloride, the elimination of a molecule of water is also necessary [82a,b]:

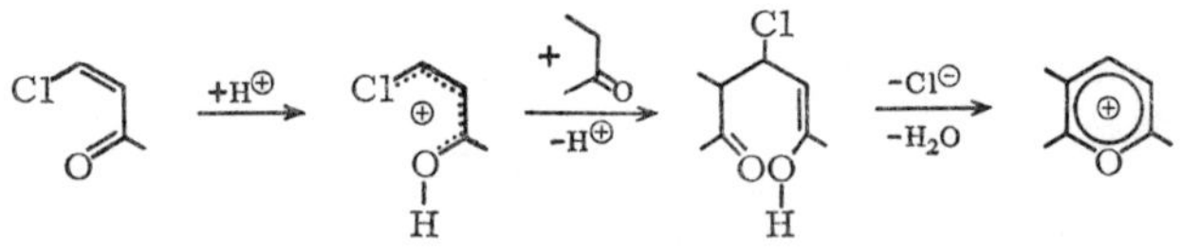

Phenols and naphthols can react similarly with β-chlorovinyl ketones to give pyrylium salts with fused benzene rings [83]. The condensation of β-dicarbonyl compounds with monoketones containing methylene groups takes place similarly, with the splitting off of a hydroxy group *and* a molecule of water [84a,b]; e.g.:

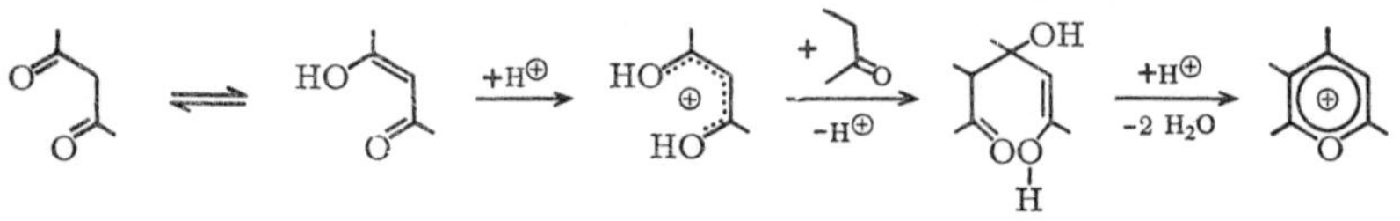

The condensation of salicylaldehyde with a ketone [85] to give the flavylium salt *(117)* and the reaction of phenols with β-diketones [48a, 86] to give *(118)* also belong here:

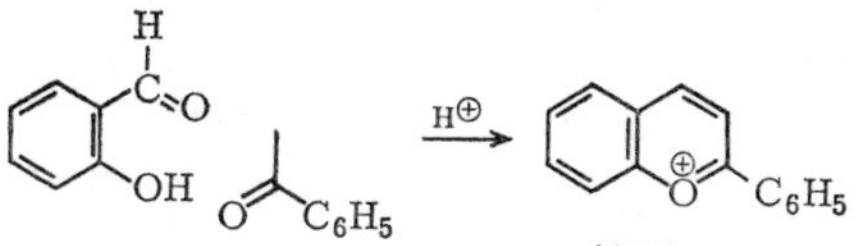

(117) 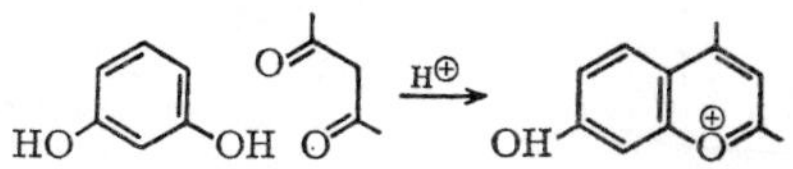 (118)

Finally, the chalcones themselves can also react with methyl ketones to give pyrylium salts. The overall reaction takes place with the detachment of a hydride ion *and* elimination of a molecule of water [68a,c]:

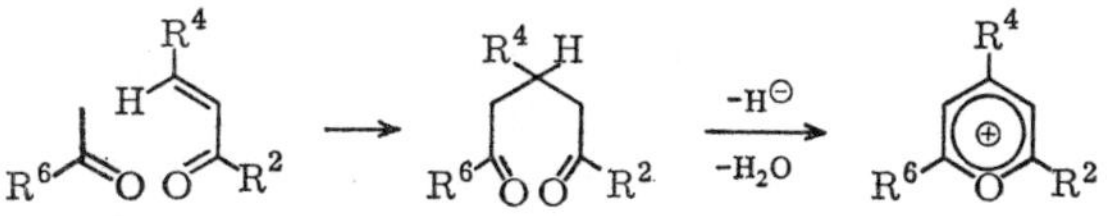

(108)

The reaction probably takes place in the manner already mentioned, via the 1,5-diketone *(108)*. FeCl$_3$ [68a,c], BF$_3$ [87], POCl$_3$ [88], H$_2$SO$_4$ [88] can be used as condensing agents. FeCl$_3$ or triphenylcarbonium perchlorate [91] can act as hydride acceptors, but it is primarily the chalcone or the hydroxyallyl cation *(119)* arising from it by protonation that is capable of taking up the hydride ion [69, 89]:

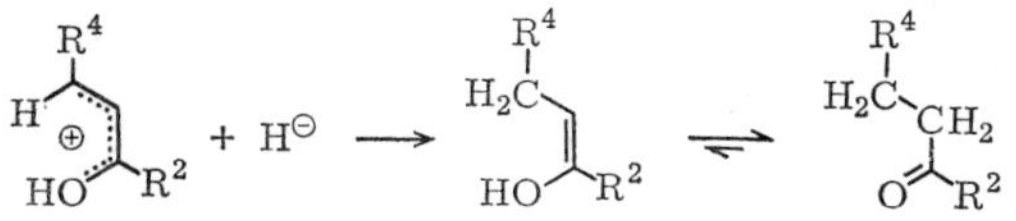

(119)

see also *(115)*. When no additional hydride acceptor is present, 2 moles of chalcone are required for 1 mole of methyl ketone.

The C$_2$C$_1$C$_2$ synthesis from 2 moles of aryl methyl ketone and 1 mole of aryl aldehyde also reduces to the formation of the chalcone and its condensation with the methyl ketone [68a,b,c, 88, 90] (*cf.* the scheme of multicomponent syntheses on p. 50).

References

1. a) S. M. McElvain and A. N. Bolstad, *J. Am. Chem. Soc.* **73**, 1988 (1951); b) F. Effenberger, *Angew. Chem.* **81**, 374 (1969); *Angew. Chem. Intern. Ed. English* **8**, 295 (1969).
2. H. Meerwein, V. Hederich, and K. Wunderlich, *Arch. Pharm.* **291/63**, 541 (1958).
3. H. Meerwein, V. Hederich, H. Morschel, and K. Wunderlich, *Ann. Chem. Liebigs* **635**, 1 (1960).
4. a) H. Meerwein, *in* "Methoden der organischen Chemie (Houben-Weyl)" (E. Müller, ed.), 4. Auflage, Bd. VI/3: Sauerstoffverbindungen I, p. 338. Thieme, Stuttgart, 1965; b) A. Kirrmann and L. Wartski, *Bull. Soc. Chim. France* p. 3825 (1966).
5. a) H. R. Ward and P. D. Sherman, Jr., *J. Am. Chem. Soc.* **89**, 4222 (1967); b) *J. Am. Chem. Soc.* **90**, 3812 (1968); c) A. Fabrycy, *Chimia (Aarau)* **15**, 552 (1961).

6. a) H. Meerwein, K. Bodenbenner, P. Borner, F. Kunert, and K. Wunderlich, *Ann. Chem. Liebigs* **632**, 38 (1960); b) C. B. Anderson, E. C. Friedrich, and S. Winstein *Tetrahedron Letters* p. 2037 (1963); c) Ö. K. J. Kovács, G. Schneider, L. K. Láng, and J. Apjok, *Tetrahedron* **23**, 4181 (1967).
7. a) C. U. Pittman, Jr. and S. P. McManus, *Chem. Commun.* p. 1479 (1968); *Tetrahedron Letters* p. 339 (1969); S. P. McManus, *Chem. Commun.* p. 235 (1969); J. W. Larsen, S. Ewing, and M. Wynn, *Tetrahedron Letters* p. 539 (1970); b) B. Capon, *Quart. Rev. (London)* **14**, 45 (1964); W. Lwoski, *Angew. Chem.* **70**, 483 (1958).
8. H. Meerwein, E. Battenberg, H. Gold, E. Pfeil, and G. Willfang, *J. Prakt. Chem.* **154**, 83 (1939); H. Meerwein, see Reference 4a, p. 341.
9. K. Dimroth and W. Mach, *Angew. Chem.* **80**, 490 (1968); *Angew. Chem. Intern. Ed. English* **7**, 461 (1968).
10. V. E. Wiersum and H. Wynberg, *Tetrahedron Letters* p. 2951 (1967).
11. a) H. Meerwein, *Angew. Chem.* **67**, 374 (1955); b) B. N. Blackett, J. M. Coxon, M. P. Hartshorn, A. J. Lewis, G. R. Little, and G. J. Wright, *Tetrahedron* **26**, 1311 (1970).
12. a) S. M. McElvain and M. J. Curry, *J. Am. Chem. Soc.* **70**, 3781 (1948); S. M. McElvain, *Chem. Rev.* **45**, 453 (1949); b) H. Gross, J. Freiberg, and B. Costisella, *Chem. Ber.* **101**, 1250 (1968).
13. H. Meerwein, P. Borner, H. J. Sasse, H. Schrodt, and J. Spille, *Chem. Ber.* **89**, 2060 (1956).
14. C. P. Heinrich, Ph.D. Thesis, University of Marburg, 1966.
15. K. Wunderlich, Ph.D. Thesis, University of Marburg, 1957.
16. F. Klages and H. Meuresch, *Chem. Ber.* **85**, 863 (1952).
17. A. N. Nesmeyanov, L. G. Makarova, and T. P. Tolstaya, *Tetrahedron* **1**, 145 (1957).
18. a) F. Kehrmann and A. Duttenhöfer, *Ber. Deut. Chem. Ges.* **39**, 1299 (1906); A. Baeyer, *Ber. Deut. Chem. Ges.* **43**, 2337 (1910); b) K. A. Hofmann, A. Metzler, and H. Lecher, *Ber. Deut. Chem. Ges.* **43**, 178 (1910); c) M. G. Ahmed, R. W. Alder, G. H. James, M. L. Sinnott, and M. C. Whiting, *Chem. Commun.* p. 1533 (1968).
19. a) J. F. King and D. A. Allbutt, *Chem. Commun.* p. 147 (1966); b) F. Klages and H. A. Jung, *Chem. Ber.* **92**, 3757 (1965).
20. A. Gerlach, Ph.D. Thesis, University of Marburg, 1969.
21. B. G. Ramsey and R. W. Taft, *J. Am. Chem. Soc.* **88**, 3058 (1966).
22. a) G. A. Olah and J. M. Bollinger, *J. Am. Chem. Soc.* **89**, 2993 (1967); b) G. A. Olah, J. R. DeMember, and R. H. Schlosberg, *J. Am. Chem. Soc.* **91**, 2112 (1969).
23. F. Klages, H. Träger, and E. Mühlbauer, *Chem. Ber.* **92**, 1819 (1959).
24. S. Hünig, *Angew. Chem.* **76**, 400 (1964); *Angew. Chem. Intern. Ed. English* **3**, 548 (1964).
25. H. Gross and J. Freiberg, *Chem. Ber.* **100**, 3777 (1967); H. Gross and E. Höft, *Z. Chem.* **4**, 417 (1964).
26. a) H. Meerwein, H. Allendörfer, P. Beekmann, F. Kunert, H. Morschel, F. Pawellek, and K. Wunderlich, *Angew. Chem.* **70**, 211 (1958); b) R. Werner and C. Rüchardt, *Tetrahedron Letters* p. 2407 (1969).
27. a) S. Kabuss, *Angew. Chem.* **80**, 81 (1968); *Angew. Chem. Intern. Ed. English* **7**, 64 (1968); b) S. Kabuss, Lecture to the German Chemical Society, Marburg, Dec. 12, 1968; c) W. Kirmse and S. Schneider, *Chem. Ber.* **102**, 2440 (1969); d) W. Kraus and W. Rothenwöhrer, *Tetrahedron Letters* p. 1013 (1968); W. Kraus, C. Chassin, and R. Chassin, *Tetrahedron Letters* p. 1277 (1970).
28. a) H. Hart and D. A. Tomalia, *Tetrahedron Letters* p. 3383 (1966); b) *Tetrahedron Letters* p. 1347 (1967); c) D. A. Tomalia and H. Hart, *Tetrahedron Letters* p. 3389 (1966); d) F. M. Beringer and S. A. Galton, *J. Org. Chem.* **32**, 2630 (1967).
29. J. W. Blunt, M. P. Hartshorn, and D. N. Kirk, *Chem. Ind. (London)* p. 1955 (1963).
30. S. Kabuss, *Angew. Chem.* **78**, 940 (1966); *Angew. Chem. Intern. Ed. English* **5**, 896 (1966).
31. G. A. Olah and J. M. Bollinger, *J. Am. Chem. Soc.* **89**, 4744 (1967).
32. H. Paulsen, W. P. Trautwein, F. Garrido Espinosa, and K. Heyns, *Chem. Ber.* **100**,

2822 (1967); *Chem. Ber.* **101,** 179 (1968); *Tetrahedron Letters* p. 4131 (1966); *Tetrahedron Letters* p. 4137 (1966).
33. D. J. Pettitt and G. K. Helmkamp, *J. Org. Chem.* **28,** 2932 (1963); *Org. Syn.* **46,** 122 (1966).
34. J. B. Lambert and D. H. Johnson, *J. Am. Chem. Soc.* **90,** 1349 (1968).
35. a) F. Klages, H. Meuresch, and W. Steppich, *Ann. Chem. Liebigs* **592,** 116 (1955); b) F. Klages and H. Meuresch, *Chem. Ber.* **86,** 1322 (1953).
36. a) E. Müller and H. Huber-Emden, *Ann. Chem. Liebigs* **649,** 76 (1961); b) A. Meuwsen and H. Mögling, *Z. Anorg. Allgem. Chem.* **285,** 262 (1956).
37. a) G. Hilgetag and H. Teichmann, *Chem. Ber.* **96,** 1446 (1963); H. Teichmann and G. Hilgetag, *Angew. Chem.* **79,** 1077 (1967); *Angew. Chem. Intern. Ed. English* **6,** 1013 (1967); b) R. A. Goodrich and P. M. Treichel, *J. Am. Chem. Soc.* **88,** 3509 (1966).
38. H. Meerwein, D. Delfs, and H. Morschel, *Angew. Chem.* **72,** 927 (1960).
39. S. Kabuss, *Angew. Chem.* **78,** 714 (1966); *Angew. Chem. Intern. Ed. English* **5,** 675 (1966).
40. a) H. Meerwein, G. Hinz, P. Hofmann, E. Kroning, and E. Pfeil, *J. Prakt. Chem.* **147,** 257 (1937); b) H. Meerwein, *Org. Syn.* **46,** 113, 120 (1966).
41. a) G. Köbrich and D. Wunder, *Ann. Chem. Liebigs* **654,** 131 (1962); b) A. T. Balaban and C. D. Nenitzescu, *Ann. Chem. Liebigs* **625,** 74 (1959).
42. K. Dimroth and K. H. Wolf, *in* "Newer Methods of Preparative Organic Chemistry" (W. Foerst, ed.), Vol. 3, p. 357. Academic Press, New York, 1964.
43. W. Schroth and G. Fischer, *Z. Chem.* **4,** 281 (1964); for recent review see A. T. Balaban, W. Schroth, and G. Fischer, *Advan. Heterocyclic Chem.* **10,** 241 (1969).
44. a) I. Degani, R. Fochi, and C. Vincenzi, *Gazz. Chim. Ital.* **94,** 203 (1964); see also b) K. Dimroth, W. Kinzebach, and M. Soyka, *Chem. Ber.* **99,** 2351 (1966); W. Kinzebach, Ph.D. Thesis, University of Marburg, 1964.
45. a) K. Dimroth and K. H. Wolf, *Angew. Chem.* **72,** 777 (1960); b) K. Dimroth and K. H. Wolf, see Reference 42, p. 410.
46. A. Schönberg and G. Schütz, *Chem. Ber.* **93,** 1466 (1960).
47. A. Baeyer and J. Piccard, *Ann. Chem. Liebigs* **384,** 208 (1911); *Ann. Chem. Liebigs* **407,** 332 (1915).
48. a) H. Decker and T. von Fellenberg, *Ann. Chem. Liebigs* **356,** 281 (1907); b) R. Willstätter, L. Zechmeister, and W. Kindler, *Ber. Deut. Chem. Ges.* **57,** 1938 (1924).
49. R. L. Shriner, W. H. Johnston, and C. E. Kaslow, *J. Org. Chem.* **14,** 204 (1949); R. L. Shriner and W. R. Knox, *J. Org. Chem.* **16,** 1064 (1951).
50. R. Wizinger, A. Grüne, and E. Jacobi, *Helv. Chim. Acta* **39,** 1 (1956).
51. J. N. Collie and T. Tickle, *J. Chem. Soc.* **75,** 710 (1899); J. Kendall, *J. Am. Chem. Soc.* **36,** 1222 (1914); W. A. Plotnikow, *Ber. Deut. Chem. Ges.* **39,** 1794 (1906); **42,** 1154 (1909); R. Willstätter and R. Pummerer, *Ber. Deut. Chem. Ges.* **37,** 3740 (1904); A. Werner, *Ann. Chem. Liebigs* **322,** 296 (1902).
52. a) R. M. Anker and A. H. Cook, *J. Chem. Soc.* p. 117 (1946); b) K. Hafner and H. Kaiser, *Ann. Chem. Liebigs* **618,** 140 (1958).
53. A. I. Kiprianov and A. L. Tolmachev, *Zh. Obshch. Khim.* **29,** 2868 (1959); *Chem. Abstr.* **54,** 12126 (1960).
54. H. Meerwein and H. Maier-Hüser, *J. Prakt. Chem.* **134,** 69 (1932).
55. W. Krafft, Diploma work (Dipl.-Chem.), University of Marburg, 1961.
56. L. C. King, F. J. Ozog, and J. Moffat, *J. Am. Chem. Soc.* **73,** 300 (1951); F. J. Ozog, V. Comte, and L. C. King, *J. Am. Chem. Soc.* **74,** 6225 (1952); G. Traverso, *Ann. Chim. (Rome)* **46,** 821 (1956); *Chem. Abstr.* **51,** 6622 (1957).
57. G. Traverso, *Ann. Chim. (Rome)* **47,** 3 (1957); *Chem. Abstr.* **51,** 10543 (1957).
58. L. L. Woods, *J. Am. Chem. Soc.* **80,** 1440 (1958); J. Kelemen and R. Wizinger, *Helv. Chim. Acta* **45,** 1908 (1962); see also G. Seitz, *Angew. Chem.* **79,** 96 (1967); *Angew. Chem. Intern. Ed. English* **6,** 82 (1967); F. Eiden, *Arch. Pharm.* **293,** 404 (1960).

59. a) H. Stetter and A. Reischl, *Chem. Ber.* **93**, 1253 (1960); b) M. Siemiatycki, *Ann. Chim. (Paris)* **2**, 189 (1957).
60. F. Klages and H. Träger, *Chem. Ber.* **86**, 1327 (1953).
61. M. Siemiatycki and R. Fugnitto, *Bull. Soc. Chim. France* p. 538 (1961).
62. W. Dilthey and T. Böttler, *Ber. Deut. Chem. Ges.* **52**, 2040 (1919).
63. A. T. Balaban and C. D. Nenitzescu, *J. Chem. Soc.* p. 3566 (1961).
64. W. Schroth and G. Fischer, *Angew. Chem.* **75**, 574 (1963); *Angew. Chem. Intern. Ed. English* **2**, 394 (1963); G. Fischer and W. Schroth, *Z. Chem.* **3**, 191 (1963); G. W. Fischer and W. Schroth, *Chem. Ber.* **102**, 590 (1969).
65. A. Roedig, M. Schlosser, and H.-A. Renk, *Angew. Chem.* **78**, 448 (1966); *Angew. Chem. Intern. Ed. English* **5**, 418 (1966).
66. W. Dilthey, *Ber. Deut. Chem. Ges.* **50**, 1008 (1917).
67. K. H. Wolf, Ph.D. Thesis, University of Marburg, 1961.
68. a) W. Dilthey, *J. Prakt. Chem.* **101**, 177 (1921); b) **94**, 53 (1916); c) **95**, 107 (1917); d) W. Dilthey and E. Floret, *Ann. Chem. Liebigs* **440**, 89 (1924).
69. A. T. Balaban, *Compt. Rend.* **256**, 4239 (1963); *Chem. Abstr.* **59**, 6353 (1963); see also: A. T. Balaban and N. S. Barbulescu, *Rev. Roumaine Chim.* **11**, 109 (1966); *Chem. Abstr.* **65**, 678 (1966).
70. a) A. T. Balaban, A. R. Katritzky, and B. M. Semple, *Tetrahedron* **23**, 4001 (1967); b) G. N. Dorofeenko, E. V. Kuznetzov, and V. E. Ryabinina, *Tetrahedron Letters* p. 711 (1969).
71. A. Treibs and H. Bader, *Chem. Ber.* **90**, 789 (1957).
72. K. Dimroth and W. Mach, *Angew. Chem.* **80**, 489 (1968); *Angew. Chem. Intern. Ed. English* **7**, 460 (1968).
73. W. Dilthey and B. Burger, *Ber. Deut. Chem. Ges.* **54**, 825 (1921).
74. a) R. C. Elderfield and T. P. King, *J. Am. Chem. Soc.* **76**, 5437 (1954); b) P. F. G. Praill and A. L. Whitear, *J. Chem, Soc.* p. 3573 (1961).
75. P. F. G. Praill and B. Saville, *Chem. Ind. (London)* p. 495 (1960).
76. G. I. Zhungietu and E. M. Perepelitsa, *Zh. Obshch. Khim.* **36**, 1858 (1966); *Chem. Abstr.* **66**, 55362 (1967).
77. a) A. T. Balaban, C. D. Nenitzescu, M. Gavat, and G. Mateescu, *J. Chem. Soc.* p. 3564 (1961); b) A. T. Balaban, D. Farcasiu, and C. D. Nenitzescu, *Tetrahedron* **18**, 1075 (1962); c) A. T. Balaban, M. Gavat, and C. D. Nenitzescu, *Tetrahedron* **18**, 1079 (1962); d) A. T. Balaban, E. Barabas, and M. Farcasiu, *Chem. Ind. (London)* p. 781 (1962); e) A. T. Balaban, *Tetrahedron Letters* p. 4643 (1968).
78. G. N. Dorofeenko, Y. A. Zhdanov, L. N. Yarova, and V. A. Palchkov, *Zh. Organ. Khim.* **3**, 955 (1967); *Chem. Abstr.* **67**, 53978 (1967).
79. a) W. Dilthey and J. Fischer, *Ber. Deut. Chem. Ges.* **57**, 1653 (1924); b) W. Schneider and A. Ross, *Ber. Deut. Chem. Ges.* **55**, 2775 (1923); c) W. Schneider and H. Sack, *Ber. Deut. Chem. Ges.* **56**, 1786 (1924).
80. a) R. Schmidt, *Angew. Chem.* **76**, 437 (1964); *Angew. Chem. Intern. Ed. English* **3**, 387 (1964); b) *Chem. Ber.* **98**, 337 (1965).
81. H. J. T. Bos and J. F. Arens, *Rec. Trav. Chim.* **82**, 845 (1963).
82. a) G. Fischer and W. Schroth, *Z. Chem.* **3**, 266 (1963); b) W. Schroth, G. W. Fischer, and J. Rottmann, *Chem. Ber.* **102**, 1202 (1969).
83. A. N. Nesmeyanov, N. K. Kochetkov, and M. I. Rybinskaya, *Izv. Akad. Nauk SSSR, Otd. Khim. Nauk* p. 479 (1953); *Chem. Abstr.* **48**, 10015 (1954).
84. a) W. Schroth and G. Fischer, *Z. Chem.* **3**, 147 (1963); *Chem. Ber.* **102**, 1214 (1969); b) S. V. Krivun, Z. V. Shiyan, and G. N. Dorofeenko, *Zh. Obshch. Khim.* **34**, 107 (1964); G. N. Dorofeenko, J. A. Shdanow, G. I. Shungijetu, and S. W. Kriwun, *Tetrahedron* **22**, 1821 (1966).
85. C. Bülow and H. Wagner, *Ber. Deut. Chem. Ges.* **34**, 1189 (1901).
86. A. Robertson and R. Robinson, *J. Chem. Soc.* p. 1951 (1926); F. M. Irvine and R. Robinson, *J. Chem. Soc.* p. 2088 (1927).

87. W. C. Dovey and R. Robinson, *J. Chem. Soc.* p. 1389 (1935); R. Lombard and J. P. Stephan, *Bull. Soc. Chim. France* p. 1458 (1958).
88. R. Wizinger, S. Losinger, and P. Ulrich, *Helv. Chim. Acta* **39**, 5 (1956).
89. J. A. van Allan and G. A. Reynolds, *J. Org. Chem.* **33**, 1102 (1968).
90. W. Dilthey and F. Quint, *J. Prakt. Chem.* **131**, 1 (1931).
91. H. Strzelecka and M. Simalty, *Bull. Soc. Chim. France* p. 4122 (1968).

5. Reactions of Oxonium Salts

In onium complexes the ligands are loosened cationically and can therefore be transferred to nucleophiles. Oxonium ions with hydrogen or organic residues as ligands are therefore proton or alkyl (aryl, acyl) donors. "Acidic" oxonium ions are protonic acids, but under suitable reaction conditions they can also split out alkyl cations, as shown schematically below:

$$[R^{\oplus}] + \underset{R}{\overset{R}{\diagdown}}\!\!\underset{}{O}\text{-H} \longleftarrow \underset{R}{\overset{R}{\diagdown}}\!\!\underset{}{\overset{\oplus}{O}}\text{-H} \longrightarrow \underset{R}{\overset{R}{\diagdown}}\!\!\underset{}{O}| + H^{\oplus}$$

(7)

Correspondingly, tertiary oxonium ions are alkylating agents; schematically:

$$\underset{R'}{\overset{R}{\diagdown}}\!\!\underset{}{\overset{\oplus}{O}}\text{-R} \longrightarrow \underset{R'}{\overset{R}{\diagdown}}\!\!\underset{}{O}| + [R^{\oplus}]$$

(8)

In general, the behavior of oxonium salts can be described as the reverse of the formation reactions; *cf.* Section 4.1.

Carboxonium ions can therefore react mainly in two directions: either the ligand attached to the oxonium center is transferred to the nucleophile Y^- (reversal of formation route A):

$$\underset{R^1}{\overset{R^2}{\diagdown}}\!\!\underset{}{=}\!\overset{\oplus}{O}\text{-H} \xrightarrow{+Y^{\ominus}} \underset{R^1}{\overset{R^2}{\diagdown}}\!\!\underset{}{=}\!\bar{O} + H\text{-}Y$$

(13)

$$\underset{R^1}{\overset{RO}{\diagdown}}\!\!\underset{}{=}\!\overset{\oplus}{O}\text{-R} \xrightarrow{+Y^{\ominus}} \underset{R^1}{\overset{RO}{\diagdown}}\!\!\underset{}{=}\!\bar{O} + R\text{-}Y$$

(9)

or nucleophiles Y^- add to the carbonium center (reversal of the formation route B), e.g.:

$$\underset{R^1}{\overset{RO}{\diagdown}}\!\!\underset{}{\overset{\oplus}{\diagup}}\!\text{OR} \xrightarrow{+Y^{\ominus}} \underset{R^1}{\overset{RO}{\diagdown}}\!\!\underset{Y}{\diagup}\!\!\overset{OR}{\diagup}$$

(9)

The carboxonium ion is therefore called an ambident cation.*

---

* The term "ambident ions" was introduced by N. Kornblum *et al.* [103] for mesomeric anions (such as enolates and nitroalkane anions) which can add an electrophilic particle X^+ to different centers; e.g.:

The reaction routes of the carboxonium ions will be described below in the same way as the corresponding formation reactions; route A therefore denotes dealkylation (deprotonation) of the oxonium center and route B the addition of a nucleophile to the carbonium center. Cyclic carboxonium ions also exhibit this reactive behavior; e.g.:

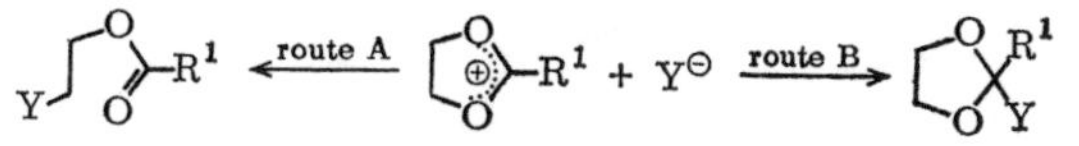

(66)

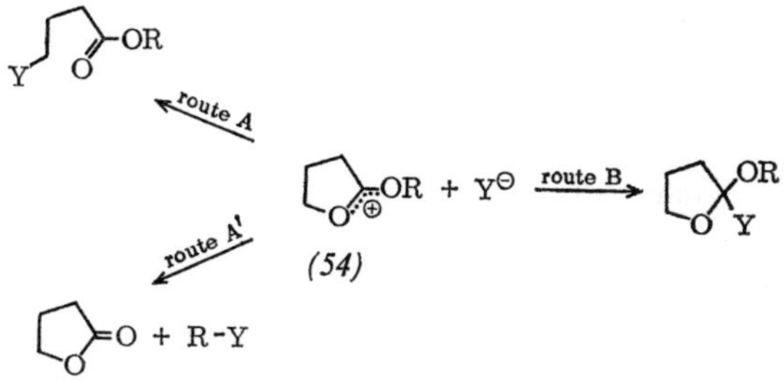

(54)

where $R = C_2H_5$.

In the case of lactonium ions such as *(54)*, dealkylation of the oxonium center can take place by two routes, with ring opening (route A) and by dealkylation at the exocyclic alkoxy group (route A'); cf. Section 4.1.1.3.

More rarely, in reversal of formation route C, detachment of a proton from the carbon atom in the β-position to the ethereal oxygen with the formation of vinyl ethers (or ketene acetals) is observed:

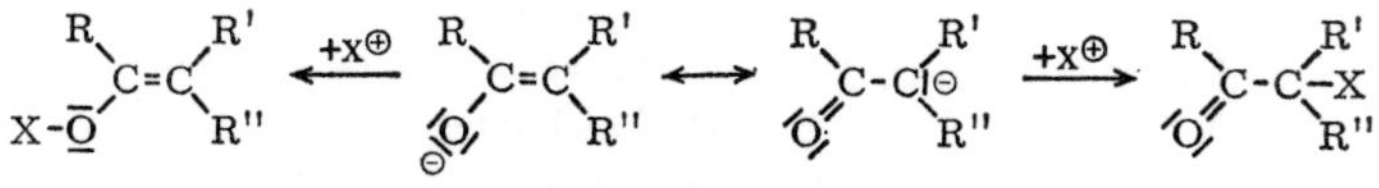

The exact cationic counterpart to enolates are mesomeric ox*e*nium–carbonium ions, which should add nucleophiles $Y^\ominus$ at the oxygen or the carbon atom:

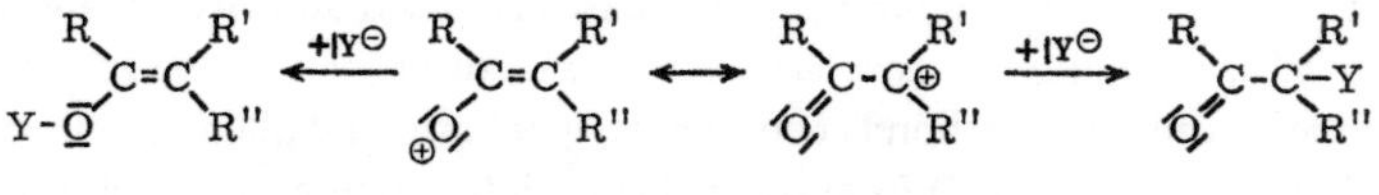

The analogy between ambident ions in this sense and carboxonium ions—e.g. *(9)*, *(54)*, *(66)*—is not strictly satisfied. As shown by the examples above, the attacks of nucleophiles $Y^\ominus$ by the two reaction routes A and B are not analogous mechanistically (route B is a nucleophilic addition whereas route A is an $S_N 2$ reaction). In spite of these mechanistic differences, the term "ambident cations" used by S. Hünig [71] for carboxonium ions will be retained here.

$$\underset{R^1}{\overset{H}{\underset{R^2}{C}}}\!\!\!=\!\!\overset{\oplus}{\underset{}{O}}\!\!-R\ +\ Y^{\ominus}\ \xrightarrow{\text{route C}}\ \underset{R^1}{\overset{R^2-C}{\underset{}{}}}\!\!\overset{H}{\underset{}{\diagdown}}\!\!-O-R\ +\ H-Y$$

The reaction has some importance in the case of the cyclic carboxonium ions; e.g.:

$$\text{(57)} + Y^{\ominus} \xrightarrow{\text{route C}} \text{(56)} + H\text{-}Y$$

It is desirable to treat the reactive behavior of oxonium salts separately for the following three groups: (a) acidic, (b) tertiary saturated, and (c) ambident tertiary oxonium salts.

5.1. Acidic Oxonium Salts

5.1.1. Deprotonation

As conjugate acids of the alcohols [1,2b], ethers [2a,b], aldehydes [3,4], ketones [4,5a,b,c], carboxylic acids [6a,b], carboxylic esters [6a,7], or carbonic acid derivatives [6a,8a,b], all types of acidic oxonium salts can act as proton donors in a reversal of their formation (*cf.* Section 4.1.1). Since the oxygen bases mentioned above are weak, oxonium ions are strong acids. Numerous investigations have been carried out on the differences in the basicities of oxygen bases [10a]. Nevertheless, in measurements of the dissociation constants of the conjugate acids in solution, not only marked deviations in the pK_a values occur, but also relative sequence changes when different measuring processes are used [10a,b]. For example, in the series of saturated oxonium ions it is to be expected that water is less basic than aliphatic alcohols, and that these in turn are less basic than dialkyl ethers. In the following example we find this sequence only if H_2SO_4 in dichloroacetic acid is used [10a]:

Acid used	Sequence of basicities
H_2SO_4 in $CHCl_2CO_2H$	$(C_2H_5)_2O > C_2H_5OH > H_2O$
HBr	$ROH > R_2O$
HCl (between 1 and 50 °C)	$n\text{-}C_4H_9OH > (n\text{-}C_4H_9)_2O > H_2O$
HCl (above 50 °C)	$n\text{-}C_4H_9OH > H_2O > (n\text{-}C_4H_9)_2O$

The most usual methods of determination (colored indicators in slightly diluted sulfuric acid, distribution equilibria of the bases between an organic phase and an aqueous acid, solubility of gaseous HCl in solutions of the oxygen bases, IR measurements of hydrogen bonds between oxygen bases and phenols, titration with $HClO_4$ in glacial acetic acid) frequently give no information on which protonated species is determined by the measurement [10a, b]*. In the case of dialkyl ethers the following particles, among others, can be envisaged:

---

* *Note in added proof (June 1970):* A newer method uses NMR measurements [10c].

$$\underset{R}{\overset{R}{>}}\overset{\oplus}{O}-H \quad \underset{R}{\overset{R}{>}}\overset{\oplus}{O}|\cdots H-\overset{R}{\underset{R}{O}} \quad \underset{R}{\overset{R}{>}}\overset{\oplus}{O}|\cdots H-\overset{H}{\underset{H}{O}} \quad \underset{R}{\overset{R}{>}}O|\cdots H-X$$

In addition to this, the sometimes considerable deviations of the pK_a values depend also on different degrees of solvation. Finally, in concentrated acids necessary for basicity determinations the protonated molecules are frequently unstable because of decomposition reactions such as the saponification of carboxylic esters, formation of acyl cations from protonated carboxylic acids, and splitting out of alkyl cations from protonated alcohols and ethers [10a]. Thus, Fig. 2 can give only an approximate idea of the pK_a values of oxygen bases; for details, reference should be made to review articles [10a,b].

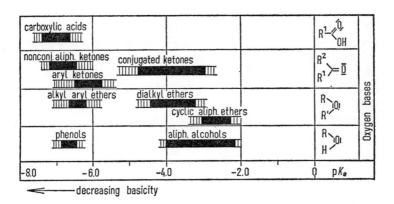

Fig. 2. pK_a values for the conjugate acids of oxygen bases, schematic. Numerical values from [10a].

Particularly in the series of protonated alcohols and ethers, a steric influence of the substituents on the pK_a values is observed in addition to the electronic influence. As an example we give the following sequence of increasing basicities for the case of ethers in aqueous sulfuric acid [10a]:

Ethers: C_6H_5-O-CH_3 < C_6H_5-O-C_2H_5 < (n-C_4H_9)$_2$O < (CH_3)$_2$O <
pK_a -6.54   -6.44   -4.40   -3.83

Ethers: < CH_3-O-C_2H_5 < (C_2H_5)$_2$O < i-C_3H_7-O-CH <
pK_a -3.82   -3.59   -3.47   -2.08

So far there has been no comparison of the pK_a values of acidic oxonium salts isolated as such. The salts with halogen complexes (of metals or nonmetals) as anions isolated at low temperatures are in equilibrium with the Lewis acid complexes of the conjugate oxygen bases and hydrogen halide. With rising temperature the equilibrium is displaced in favor of the dissociation products [2b, 4, 6a]:

$$\underset{R}{\overset{R}{>}}\overset{\oplus}{O}\text{-H SbCl}_6^{\ominus} \rightleftarrows \underset{R}{\overset{R}{>}}O \rightarrow \text{SbCl}_5 + \text{HCl}$$

$$\underset{R^1}{\overset{R^2}{>}}\overset{\oplus}{C}\text{-OH SbCl}_6^{\ominus} \rightleftarrows \underset{R^1}{\overset{R^2}{>}}C=O \rightarrow \text{SbCl}_5 + \text{HCl}$$

$$R^1\text{-}\overset{OH}{\underset{OH}{\overset{\oplus}{C}}} \text{SbCl}_6^{\ominus} \rightleftarrows R^1\text{-}\overset{O \rightarrow \text{SbCl}_5}{\underset{OH}{C}} + \text{HCl}$$

The hydrogen halide pressures at a given temperature are determined not only by the acidity of the "acidic" oxonium ions but also by the stability of the Lewis acid–base complexes [2b]. The decomposition pressures can therefore serve as a measure of acidity only to a very rough approximation, since the Lewis acid complexes derived from ethers, carbonyl compounds, and carboxylic acids probably have very different stabilities.

A comparison of acidities by base exchange reactions, e.g.,

$$\underset{R}{\overset{R}{>}}\overset{\oplus}{O}\text{-H SbCl}_6^{\ominus} + \underset{R'}{\overset{R'}{>}}O \rightarrow \underset{R}{\overset{R}{>}}O + \text{H-}\overset{\oplus}{\underset{R'}{\overset{R'}{O}}} \text{SbCl}_6^{\ominus}$$

is better. In ethylene chloride or methylene chloride solution, in this way methyloxonium hexachloroantimonate and diethyl ether give the diethyloxonium salt, phenyloxonium salt and dimethyl ether yield the dimethyloxonium salt, and dimethyloxonium salt and di-n-butyl ether give the di-n-butyloxonium salt [2b]. The following relative sequence of decreasing acidity was obtained by such experiments:

Phenyloxonium ion > alkyloxonium ion > dialkyloxonium ion

In the case of dialkyloxonium salts, the acidity increases with increasing chain length of the alkyl groups.

Acidic aldehydium and ketonium ions are easily deprotonated by dialkylethers; consequently aldehydes and ketones are less basic than dialkyl ethers [4].

Almost all acidic oxonium salts that have been isolated as such are converted by water into the oxygen bases from which they were derived [2b, 9a,b,c,d].

The alkylation of acidic oxonium salts with diazoalkanes should also be regarded with respect to base exchange reactions. In both cases, mono- and dialkyloxonium salts give trialkyloxonium salts with diazomethane [2b]; cf. Section 4.2.1. This reaction has been extended to acidic *carboxonium* salts only in the case of the O,O'-dialkyl carbonate acidium ion [6a]:

$$\underset{(24)}{\overset{RO}{\underset{RO}{>}}}\overset{\oplus}{-}OH\ SbCl_6^{\ominus} + CH_2N_2 \longrightarrow \underset{RO}{\overset{RO}{>}}\overset{\oplus}{-}OCH_3\ SbCl_6^{\ominus} + N_2$$

5.1.2. Splitting Out of Carbonium Ions

In addition to the cleavage of the O–H bond, heterolysis of O–C bonds is important in the case of acidic oxonium ions. Such reactions have recently been detected spectroscopically by low-temperature NMR measurements in *solution* in the strongly acidic medium FSO_3H–SbF_5 ("magic acid") or FSO_3H–SbF_5SO_2, especially by G. A. Olah *et al.* [1, 2a, 3, 5a,b,c, 6b–8a, 11a,b, 105].

5.1.2.1. Acidic Saturated Oxonium Ions

With rising temperature the *protonated alcohols (5)* observed in such acidic solutions at $-60\,°C$ give rise to alkyl cations and hydroxonium ions in a pseudomonomolecular reaction [1]:

$$\underset{(5)}{R-\overset{\oplus}{\underset{H}{O}}\diagup^{H}} \xrightarrow[(FSO_3H-SbF_5)]{+H^\oplus} H-\overset{\oplus}{\underset{H}{O}}\diagup^{H} + R^\oplus$$

This is in any case a long-established reaction for obtaining carbonium ions (see, for example, the sequence of formulas *(112)→(111)*, Section 4.2.3.2).

In accordance with expectations, the O-alkyl cleavage takes place more readily with increasing chain length and branching of the alkyl groups: Increasing rates of cleavage are found in *n*-alkyloxonium ions with an increasing number of C atoms (while the activation energies remain very stable) [1]. For the *n*-alkyloxonium ions *(5)*:

R	Stable in FSO_3H–SbF_5 up to (°C)	$k_1 \times 10^{-3}$ min^{-1} (at $+15\,°C$)
CH_3	+50	
C_2H_5	+30	
C_3H_7	0	20.5
C_4H_9	0	48.1
C_5H_{11}	0	68.4
C_6H_{13}	0	91.4

With an increasing degree of branching, cleavage proceeds at a lower temperature, as can be shown by using the protonated butanols as examples. Owing to the rapid isomerization of the butyl cations produced first, in all cases only the trimethylcarbonium ion can be detected [1]:

Butyloxonium ions	Unstable in FSO₃H–SbF₅ at (°C)	Carbonium ion expelled	Carbonium ion observed
(n-Bu, H₂O⁺)	0	(propyl cation)	
(iso-Bu, H₂O⁺)	−30	(isopropyl cation)	(t-Bu cation)
(sec-Bu, H₂O⁺)	−60	(propyl cation)	
(t-Bu, H₂O⁺)	cannot be observed at −60	(isopropyl cation)	

Glycols can be protonated in FSO₃H–SbF₅–SO₂ in the same way as monoalcohols. Again, with rising temperature the bisoxonium ions that exist at −60°C form carbonium ions that are frequently stabilized by subsequent rearrangement: thus, aldehydium ions are produced (by hydride shift) [11a] from the bisprotonated α- and β-glycols *(120a)* and *(120b)*:

$$H_2O^+\text{-CH}_2\text{-CH}_2\text{-}O^+H_2 \xrightarrow{-H_2O} [\text{intermediate}] \xrightarrow{-H^+} HO\text{-}\overset{+}{C}\text{-CH}_2\text{-CH}_3$$
(120a)

$$H_2O^+\text{-CH}_2\text{-CH}_2\text{-CH}_2\text{-}O^+H_2 \xrightarrow{-H_2O} [\text{intermediate}] \xrightarrow{-H^+} HO\text{-}\overset{+}{C}\text{-CH}_2\text{-CH}_3$$
(120b)

Bisprotonated di*primary* γ-, δ-, and ε-glycols *(121)*, $n = 2$–4, are stable in solution up to +40°C; the di*secondary* γ-bisoxonium ion *(122)* undergoes ring closure to yield a tetrahydrofuranium ion even at −30°C [11a]:

$$H_2\overset{\oplus}{\underset{OH_2}{C}}\text{-(CH}_2)_n\text{-}\overset{\oplus}{\underset{OH_2}{C}}H_2$$
(121)

$$\underset{\overset{\oplus}{O}H_2}{\overset{CH_3}{\underset{|}{H\text{-}C\text{-}}}}\text{(CH}_2)_2\text{-}\underset{\overset{\oplus}{O}H_2}{\overset{CH_3}{\underset{|}{C\text{-}H}}} \xrightarrow{-H_2O} \left[\underset{H}{\overset{CH_3}{H\text{-}C\text{-}}}\text{(CH}_2)_2\text{-}\overset{\oplus}{\underset{H}{C}}\overset{CH_3}{\underset{H}{}}\right] \xrightarrow{-H^+} H_3C\text{-tetrahydrofuranium-}CH_3$$
(122)

Protonated ethers (6) also give an O-alkyl cleavage in acid solution in a pseudo-monomolecular reaction [2a]:

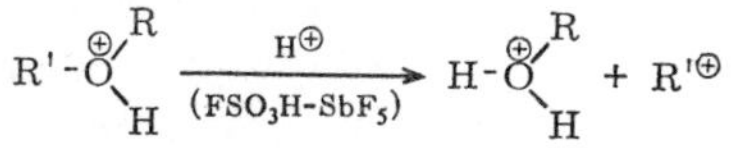

$$R'-\overset{\oplus}{\underset{H}{O}}{-}R \xrightarrow[(FSO_3H-SbF_5)]{H^\oplus} H-\overset{\oplus}{\underset{H}{O}}{-}R + R'^{\oplus}$$

(6)   *(5)*

Under these conditions the larger or more highly branched alkyl residue splits out cationically; simultaneously an alkyloxonium ion *(5)* is formed. In general, the cleavage begins at higher temperatures than in the case of the protonated alcohols. Dialkyloxonium ions with secondary and tertiary alkyl groups have, however, similar decomposition temperatures to protonated secondary and tertiary alcohols. On cleavage, the isomeric protonated butyl methyl ethers each yield the trimethylcarbonium and methyloxonium ions [2a]:

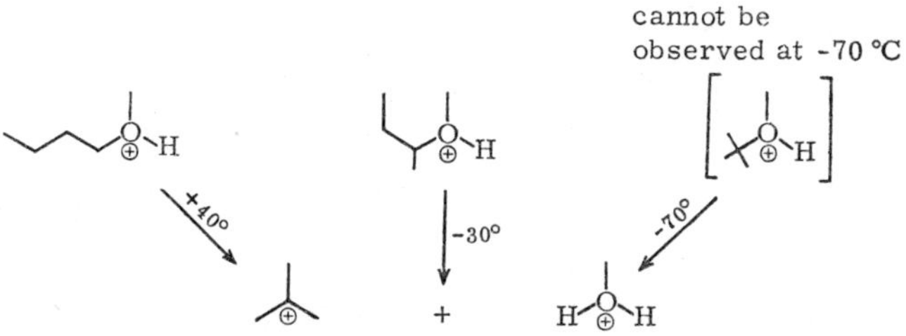

In these ether cleavages in the absence of nucleophiles the monomolecular decomposition of the protonated dialkyl ether proceeds with the formation of the most stable carbonium ion. In marked contrast to this are the preparatively important ether cleavages with hydrogen halides (HBr, HI), in which dialkyloxonium ions are again formed in situ. Here dialkyl ethers containing one methyl group give almost exclusively the methyl halide together with the alcohol of the larger alkyl residue; this also applies to methyl *sec*-alkyl ethers [20b]. Thus, on cleavage with HBr, 2-butyl methyl ether yields 2-butanol and methyl bromide; the optically active ether gives the pure optically active 2-butanol [20b]:

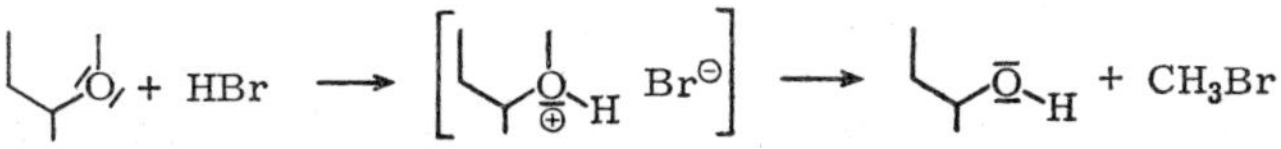

The only exceptions to this rule are ethers with a tertiary alkyl group, in which the tertiary alkyl halide is formed.

In the presence of the nucleophilic anions Br⁻ and I⁻, the cleavage of the dialkyloxonium ion in the case of ethers with primary and secondary alkyl groups therefore obviously takes place by a kind of S_N2 mechanism. Under these conditions, the attack of the nucleophile on the smaller alkyl group leads to a more favorable transition state and should be preferred. In the case of ethers with a tertiary alkyl group, however, the monomolecular decomposition of the protonated ether into a tertiary carbonium ion and alcohol takes place faster than the attack of a nucleophile on the smaller alkyl group by the S_N2 mechanism.

Bisprotonated alkoxyalcohols (123), such as those that arise in FSO_3H–SbF_5–SO_2 at $-60\,°C$, exhibit simultaneously the properties of dialkyloxonium ions and bisprotonated glycols. On the one hand, with rising temperature they split out water and thereby rearrange to alkoxycarbonium ions; on the other hand, they can yield alkyl cations, and this in two directions: either bisprotonated glycols or alkyloxonium and hydroxycarbonium ions are formed [11b]; schematically:

$$\underset{H}{\overset{R}{O^{\oplus}}}\!\!-\!\!\underset{H}{\overset{H}{C}}\!\!-\!(CH_2)_{n-1}\!-\!\underset{R'}{\overset{H}{C^{\oplus}}} \xrightarrow{-H^{\oplus}} \underset{H}{\overset{RO_{\oplus}}{C}}\!-(CH_2)_{n-1}\!-\!CH_2R'$$

(hydride displacement) (R primary, R' ≠ H)

$$\underset{H}{\overset{R}{O^{\oplus}}}\!-\!(CH_2)_n\!-\!\underset{R'}{\overset{H}{C}}\!-\!\underset{H}{\overset{H}{O^{\oplus}}} \xrightarrow{+H^{\oplus}} R^{\oplus} + \underset{H}{\overset{H}{O^{\oplus}}}\!-\!(CH_2)_n\!-\!\underset{R'}{\overset{H}{C}}\!-\!\underset{H}{\overset{H}{O}}$$ (R tertiary, secondary)

(123)

$$R\!-\!\underset{H}{\overset{H}{O^{\oplus}}} + \underset{H}{\overset{H}{{}^{\oplus}C}}\!-\!(CH_2)_{n-1}\!-\!\underset{R'}{\overset{H}{C}}\!-\!\underset{H}{\overset{H}{O}} \xrightarrow{-H^{\oplus}} H_3C\!-\!(CH_2)_{n-1}\!-\!\underset{R'}{\overset{\oplus\,OH}{C}}$$

(hydride displacement) (R primary)

5.1.2.2. Acidic Carboxonium Ions

While monohydroxycarbonium ions (acidic aldehydium and ketonium ions) are not capable of O–C cleavage, this reaction mode is important in dihydroxycarbonium [6b], alkoxyhydroxycarbonium [7], alkoxydihydroxycarbonium [8a], and dialkoxyhydroxycarbonium ions [8a].

With rising temperature, dihydroxycarbonium ions *(17)*—formed, for example, in FSO_3H–SbF_5–SO_2 at $-60\,°C$ by the protonation of carboxylic acids—cleave between $-30°$ and $0\,°C$, with dehydration, into oxocarbonium ions *(10)* in a first-order reaction [6b]:

$$R^1\!-\!\underset{OH}{\overset{OH}{C^{\oplus}}} \xrightarrow{H^{\oplus}} R^1\!-\!\overset{\oplus}{C}\!=\!O + H_3O^{\oplus}$$

(17) (10)

Such cleavages have also been studied in concentrated sulfuric acid [12a], H_2SO_4–SO_3 [13], and HF–BF_3 [14]. They are possible both with aliphatic and with aromatic [12a] and unsaturated [15] protonated carboxylic acids. Protonated dicarboxylic acids also undergo cleavage in the sense of the preceding equation. In concentrated sulfuric acid, monoprotonation together with a small proportion of diprotonation of dicarboxylic acids has been detected on the basis of cryoscopic investigations [12a,b]. In FSO_3H–SbF_5–SO_2 at a sufficiently low temperature, the bis(dihydroxycarbonium) ions *(124)* exist, as shown by NMR measurements. When temperature is raised, dehydration takes place in steps: first into dihydroxycarbonium–oxocarbonium ions *(125)* and then into dioxocarbonium ions *(126)* [16]:

$$HO_2C(CH_2)_nCO_2H \xrightarrow[(\leq -60°)]{+2\ H^\oplus} H_2\overset{\oplus}{O}_2C(CH_2)_n\overset{\oplus}{C}O_2H_2 \xrightarrow[-20°]{-H_2O} H_2\overset{\oplus}{O}_2C(CH_2)_n\overset{\oplus}{C}O$$
$$(124)(125)$$

$$\xrightarrow[-20°\ to\ +10°]{-H_2O} \overset{\oplus}{O}\overset{}{C}(CH_2)_n\overset{\oplus}{C}O$$
$$(126)$$

The course of this reaction sequence depends on the number n of methylene groups in *(124)*. For $n = 0$ and 1 the reaction ceases at *(124)*, and for $n = 2$ at *(125)*; dioxocarbonium ions *(126)* are first formed when $n = 3$ [17]. In connection with the appearance of *(125)*, the protonation of cyclic carboxylic anhydrides may be mentioned: Instead of bisprotonated anhydrides *(127)* only *(125)* can be detected so that an equilibrium between *(125)* and *(127)* is inferred [17]; e.g.:

$$H_2\overset{\oplus}{O}_2C(CH_2)_2\overset{\oplus}{C}O \rightleftarrows \underset{HO\overset{\oplus}{}O\overset{\oplus}{}OH}{\triangle} \rightleftarrows \overset{\oplus}{O}\overset{}{C}(CH_2)_2\overset{\oplus}{C}O_2H_2$$
$$(125)(127)(125)$$

In H_2SO_4 (instead of $FSO_3H-SbF_5-SO_2$), the monoprotonation [12a] of cyclic anhydrides at the most has been observed. Even at very low temperatures ($-80°C$) in $FSO_3H-SbF_5-SO_2$, acyclic carboxylic anhydrides form 1:1 mixtures of the oxocarbonium ion *(10)* and the dihydroxycarbonium ion *(17)* [16]:

$$(R^1CO)_2O \xrightarrow{+2\ H^\oplus} R^1C\overset{\oplus}{O}_2H_2 + R^1\overset{\oplus}{C}O$$

The same conclusion can be drawn from the results of cryoscopic measurements on acyclic anhydrides in concentrated sulfuric acid [12a].

In contrast to protonated carboxylic acids, protonated esters can undergo O–C cleavage in two different directions: the O-acyl cleavage analogous to that of the dihydroxycarbonium ions (corresponding to an $A_{AC}1$ mechanism)* or an O-alkyl cleavage (according to an $A_{AL}1$ mechanism)*:

$$R^1-C\underset{OR}{\overset{OH}{\overset{\oplus}{\diagup}}}$$
$$(18)$$

$$\nearrow^{H^\oplus} R^1-\overset{\oplus}{C}=O + R-\overset{\oplus}{O}\diagdown^H_H$$
$$(10)(5)$$

$$\searrow^{H^\oplus} R^1-C\underset{OH}{\overset{OH}{\overset{\oplus}{\diagup}}} + R^\oplus \longrightarrow \text{(sometimes with subsequent reactions).}$$
$$(17)$$
$$\downarrow H^\oplus$$
$$R^1-\overset{\oplus}{C}=O + H-\overset{\oplus}{O}\diagdown^H_H$$
$$(10)$$

---

* In the classification of ester hydrolysis (or esterification) mechanisms introduced by Ingold [104], A denotes that the initial compound takes part in the reaction in the form of its conjugate *acid*. In the present case, therefore, the protonated ester reacts. AC and AL denote cleavage at the *ac*yl oxygen and at the *al*kyl oxygen bond, respectively. The number 1 shows that the reaction takes place by a monomolecular mechanism in the rate-determining step.

Which of these routes is realized depends mainly on the nature of the alkyl residue R. Protonated esters of primary alcohols are stable in $FSO_3H–SbF_5–SO_2$ solution even at about 0 °C [7]; with a further rise in temperature, the methylesters decompose into protonated methanol and acyl cations.*

On the other hand, the protonated ethyl esters undergo cleavage at the O-alkyl bond (with the elimination of the ethyl cation, which is converted into the trimethylcarbonium ion in subsequent reactions), provided the molar ratio of FSO_3H to SbF_5 is maintained between 1:1 and 3:1. In a more weakly acidic system (with a molar ratio of 4:1), on the other hand, the slower O-acyl cleavage is found for ethyl esters. Protonated esters of secondary alcohols follow the $A_{AL}1$ mechanism exclusively (even between $-40°$ and $-20°C$). In the case of esters of tertiary alcohols, *(18)* cannot be detected even at $-80°C$, and only the tertiary carbonium ion and the protonated carboxylic acid are found [7].

The protonated esters of carbonic acids also fit into this scheme [8a]: protonated methyl hydrogen carbonate *(23)*, $R = CH_3$, decomposes in $FSO_3–SbF_5$ above $-50°C$ by the O-acyl cleavage route into CO_2 and methyloxonium ion:

$$RO\overset{\oplus}{-}\overset{OH}{\underset{OH}{C}} \xrightarrow{> -50°} CO_2 + ROH_2^{\oplus}$$

(23)   *(5)*

With rising temperature (to +10 and +20 °C, respectively), protonated dimethyl and diethyl carbonates are relatively stable in solution. Both cations can also be detected when the carbonates are dissolved in concentrated sulfuric acid [8b]. On the other hand, the corresponding di-*tert*-butyl-substituted cation *(129)* is very unstable and cannot be detected even at $-78°C$. The trimethylcarbonium ion and trihydroxycarbonium ion *(22)* arise through O-alkyl cleavage by the following mechanism:

$$\left[\begin{array}{c}+O\\+O\end{array}\!\!\!\!\!\!\overset{\oplus}{\diagup}\!\!\text{OH}\right] \xrightarrow[(-78°)]{H^{\oplus}} \text{\Large)}^{\oplus} + \left[\begin{array}{c}+O\\HO\end{array}\!\!\!\!\!\!\overset{\oplus}{\diagup}\!\!\text{OH}\right] \xrightarrow{H^{\oplus}} \text{\Large)}^{\oplus} + \begin{array}{c}HO\\HO\end{array}\!\!\!\!\!\!\overset{\oplus}{\diagup}\!\!\text{OH}$$

(129)   *(22)*

However, the diisopropyl cation can be detected below $-70°C$, but above this temperature it is decomposed into a mixture of hexyl cations (which arise from the dimethylcarbonium ion that is the primary cleavage product).

5.1.3. Reactions at the Carbonium Center of Acidic Carboxonium Ions

Numerous reactions of carbonyl compounds with nucleophiles occur under acid

---

* The intermediate stage *(128)* formulated in $A_{AC}1$ mechanisms cannot be detected by NMR spectroscopy; nevertheless it could exist in a very low equilibrium concentration together with *(18)* [7]:

$$R^1-C\overset{\displaystyle O}{\underset{\underset{H}{\overset{\oplus}{O}-R}}{\diagdown}} \quad (128)$$

catalysis, since the nucleophilic attack takes place much more readily at the hydroxycarbonium ions *(13)* that arise as intermediates:

$$\underset{R^1}{\overset{R^2}{>}}\overset{\oplus}{\text{OH}} + :Y^n \longrightarrow \left[\underset{R^1}{\overset{R^2}{\times}}\underset{\text{OH}}{Y}\right]^{(n+1)}$$

(13)

Since the addition of acid also leads to the establishment of a protolysis equilibrium between the nucleophile and its conjugate acid, there are optimum acid concentrations for such reactions. When these are exceeded, the nucleophile becomes inactive. Hydroxycarbonium salts isolated *as such* are very strong acids; so far, no nucleophilic additions to them have been described. Only in the case of the protonated carbonic acid *(22)*, which has been detected in strongly acid solution and is stable in FSO_3–$HSbF_5$ up to about 0 °C, did G. A. Olah *et al.* succeed in observing the carboxylation of amines [8a]:

$$>\text{NH} + \text{C(OH)}_3^{\oplus} \rightleftharpoons \left[\overset{H}{\underset{\oplus}{>\text{N}}}-\overset{OH}{\underset{OH}{C}}-OH\right] \xrightarrow{-H_2O} >\text{N}\overset{\oplus}{=}\overset{OH}{\underset{OH}{C}} \xrightarrow{-H^{\oplus}} >\text{N}-\overset{O}{\underset{OH}{C}}$$

(22)

Perhaps this reaction is important in biochemical processes. Hydride transfers to protonated chalcones may also be mentioned in this connection; *cf.* Section 4.2.3.2.

Intramolecular transfers of anions can also be regarded as reactions at the carbonium center of acidic carboxonium ions. The rearrangement of the pivalaldehydium ion *(130)* to the isopropyl methyl ketonium ion *(131)* belongs in this category. The reaction, which is formulated in the manner of a retropinacolone rearrangement, is possible in FSO_3H–SbF_5 at −70 °C [3] and has been known for a relatively long time in 70% sulfuric acid under severe conditions [18]:

(130) → *(131)*

A vinylogous analog of the first step of this rearrangement is probably to be found in the conversion of protonated chalcone derivatives *(132)* into cyclic carboxonium ions *(133)* [19]:

(132) → *(133)*

In connection with rearrangements of acidic carboxonium ions, reference may also be made to the equilibrium that possibly exists between the cyclopropanonium ion *(134)* and the 2-hydroxyallyl cation *(135)* [5b]:

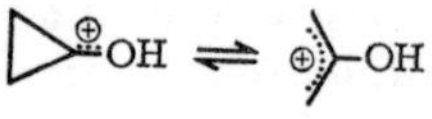

△⊕-OH ⇌ ⊕◇-OH
(134)   (135)

5.2. Saturated Tertiary Oxonium Salts

5.2.1. Trialkyloxonium Salts

Because of the ready elimination of alkyl cations, trialkyloxonium salts are strong alkylating agents. Thus, numerous inorganic and organic nucleophiles $:B^n$ can in general be alkylated smoothly even at room temperature. This is the basis of the great preparative importance of trialkyloxonium salts [20a, 21, 22]:

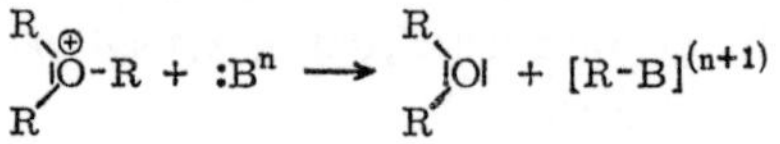

$$\underset{R}{\overset{R}{>}}\overset{\oplus}{O}-R + :B^n \longrightarrow \underset{R}{\overset{R}{>}}O| + [R-B]^{(n+1)}$$

(7)

Anionic nucleophiles react according to this scheme; e.g., hydroxyl and halide ions [21], anionic CH-active compounds (enolates [21, 23], malonic ester anions [21], and nitroalkane anions [24]), the sodium derivative of α-pyridone [25], and alkylphosphorothionate anions [26]. The alkylation of iodide may serve as an example of these reactions [21]:

$$R_3O^+ + I^- \rightarrow R_2O: + R-I$$

When the nucleophile is an *uncharged molecule*, the cations *(136)* arising as primary products can generally be isolated, often in the form of well-crystallized BF_4^- or $SbCl_6^-$ salts [20a]; *cf.* Table 8*:

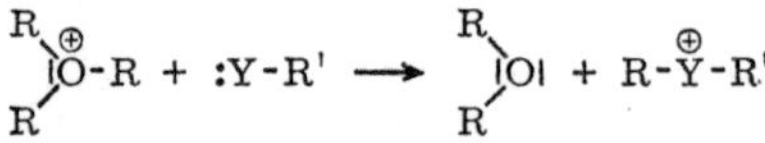

$$\underset{R}{\overset{R}{>}}\overset{\oplus}{O}-R + :Y-R' \longrightarrow \underset{R}{\overset{R}{>}}O| + R-\overset{\oplus}{Y}-R'$$

(7)

For example [27a,b,c]: $R_3O^\oplus + :N\equiv C-R^1 \longrightarrow R_2O: + R-\overset{\oplus}{N}\equiv C-R^1$

The nucleophilic centers Y include, in particular, functional compounds with O, S, N, and P. For many classes of substances listed in Table 8, particularly the weak nucleophiles, alkylation has been possible up to this time only by means of oxonium salts. However, the very weakly nucleophilic carbonyl oxygens of aryl aldehydes and aryl ketones or of carboxylic esters cannot be alkylated even with trialkyloxonium salts.

---

* *Note added in proof (1970):* For a recent review on the alkylation of weak bases by trialkyloxonium salts see [53b].

TABLE 8. Reactions of Trialkyloxonium Salts with Neutral Molecules :Y–R′

$$RO_3^{\oplus} + :Y-R' \rightarrow R_2O + R-\overset{\oplus}{Y}-R'$$

:Y–R′ Class of compound	Type of formula	$R-\overset{\oplus}{Y}-R'$ Isolable cationic alkylation product	References
Ethers	$R^1-\bar{O}-R^1$	$R^1-\overset{R}{\underset{}{\overset{\oplus}{O}}}-R^1$	[22]
Carbonyl compounds	$\overset{R^2}{\underset{R^1}{>}}=\bar{O}$	$\overset{R^2}{\underset{R^1}{>}}\overset{\oplus}{=}OR$	[21, 35a,b, 36]
Dialkyl carbonates	$\overset{R^1O}{\underset{R^1O}{>}}=\bar{O}$	$\overset{R^1O}{\underset{R^1O}{>}}\overset{\oplus}{=}OR$	[28]
Sulfides	$R^1-\bar{S}-R^1$	$R^1-\overset{R}{\underset{}{\overset{\oplus}{S}}}-R^1$	[21, 31b]
Disulfides	$R^1-\bar{S}-\bar{S}-R^1$	$R^1-\bar{S}-\overset{\oplus}{\underset{R^1}{S}}\overset{R}{}$	[31a, 32b]
Mercaptals	$R^1-\bar{S}-CR^1_2-\bar{S}-R^1$	$\overset{R}{\underset{R^1}{\overset{\oplus}{S}}}-CR^1_2-\overset{\oplus}{\underset{R^1}{S}}\overset{R}{}$	[32a,b]
Sulfoxides	$\overset{R^1}{\underset{R^1}{>}}S=\bar{O}$	$\overset{R^1}{\underset{R^1}{>}}S\overset{\oplus}{=}OR$	[21]
Amines	$\overset{R^1}{\underset{R^2}{>}}\bar{N}-R^3$	$\overset{R^1}{\underset{R^2}{>}}\overset{\oplus}{\underset{R^3}{N}}\overset{R}{}$	[21, 45]
Schiff bases	$>=\bar{N}-R^1$	$>\overset{\oplus}{\underset{R^1}{N}}\overset{R}{}$	[33]
Nitriles	$R^1-C\equiv N|$	$R^1-C\overset{\oplus}{\equiv}N-R$	[27a,b,c]
Amine oxides	$\overset{R^1}{\underset{R^2\,R^3}{>}}\overset{\oplus}{N}\bar{O}^{\ominus}$	$\overset{R^1}{\underset{R^2\,R^3}{>}}\overset{\oplus}{N}OR$	[22, 34]
Nitrosamines	$\overset{R^2}{\underset{R^1}{>}}N-\bar{N}=\bar{O}$	$\overset{R^2}{\underset{R^1}{>}}N\cdots\overset{\oplus}{N}\cdots OR$	[37a,b]
Carboxamides	$R^1-C\overset{\nearrow O}{\underset{N<}{}}$	$R^1-\overset{\oplus}{C}\overset{\nearrow OR}{\underset{N<}{}}$	[22, 30, 39, 40a,b,c,d]
Enamino carbonyl compounds	$\overset{|O|}{\underset{}{\overset{}{\diagup}}}\overset{\diagdown N'}{}$	$\overset{R\bar{O}}{\underset{}{\overset{}{\diagup}}}\overset{\diagdown N'}{\overset{\oplus}{}}$	[39, 41]
Urethanes	$\overset{>\bar{N}}{\underset{R^1\bar{O}}{>}}=\bar{O}$	$\overset{>N}{\underset{R^1\bar{O}}{>}}\overset{\oplus}{=}OR$	[39]
Ureas	$\overset{>\bar{N}}{\underset{>\bar{N}}{>}}=\bar{O}$	$\overset{>N}{\underset{>N}{>}}\overset{\oplus}{=}OR$	[22, 39]
Diazoacetic esters (derivatives)	$|N\equiv \overset{\oplus}{N}\underset{R^2}{\overset{}{-C-C}}\overset{\nearrow O}{\underset{\bar{O}R^1}{}}$	$|N\equiv N\underset{R^2}{\overset{}{-C}}\overset{\oplus}{=}\overset{\nearrow OR}{\underset{OR^1}{}}$	[42]

TABLE 8. Continued

$$R_3O^\oplus + :Y-R' \rightarrow R_2O + R-\overset{\oplus}{Y}-R'$$

:Y-R' Class of compound	Type of formula	$R-\overset{\oplus}{Y}-R'$ Isolable cationic alkylation product	References
Phosphines			[43]
Phosphine oxides			[44a,b]
Dialkylphosphinic acid esters			[44a,b]
Phosphorous acid triesters			[46]
Alkyl phosphonic acid diesters			[44a,b]
Phosphoric acid triesters			[44b]
Lactones			[21, 30]
Thiolactones			[30]
3-Oxo-1,2-dithioles			[38]
3-Thioxo-1,2-dithioles			[38]
Lactams			[30, 39, 47a,b,c,d,e,f]
Thiolactams			[47a, 48]
N-Heterocycles			[21, 49a,b, 50, 51a,b]
Thiophene			[52]
Azulene derivatives			[53a]

Reactions with other classes of compounds not mentioned in Table 8 are also known; with such compounds the isolation of the salts formed as primary products has not been possible. Specifically this applies to the action of trialkyloxonium salts on alcohols and phenols [20a, 21], carboxylic acids [21], acetals or ketals [30], orthocarbonic esters [30], azides [54], and cyanic acid esters [55].

In many cases the reaction with uncharged molecules takes place with the transfer of the anionically eliminated group Y to the alkyl cation expelled from the trialkyloxonium ion. Here we are generally dealing with the subsequent reaction of a primary alkylation product *(136)*, which leads to the formation of a more stable cation $R'^{\oplus}$ (carbonium ions, protons):

$$R_3O^{\oplus} + :Y-R' \xrightarrow{-R_2O} [R-\overset{\oplus}{Y}-R'] \longrightarrow R-Y: + R'^{\oplus}$$
(136)

[29]

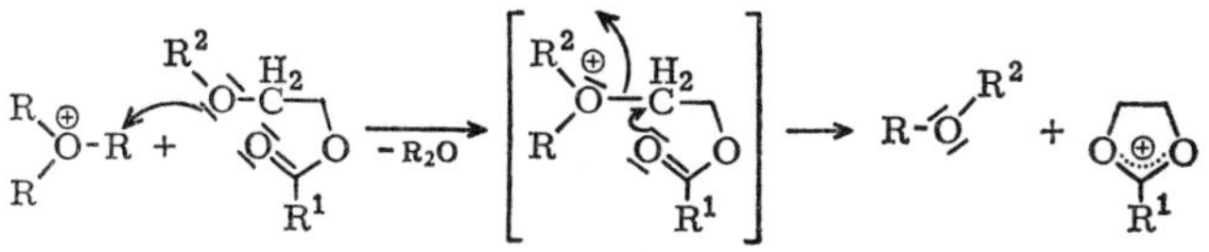

(66)

Smooth alkylation with trialkyloxonium salts takes place only when the complex anion is sufficiently stable with respect to the nucleophile to be alkylated. Thus, trialkyloxonium tetrafluoroborates react with water to give the corresponding alcohol and tetrafluoroboric acid [21, 22]:

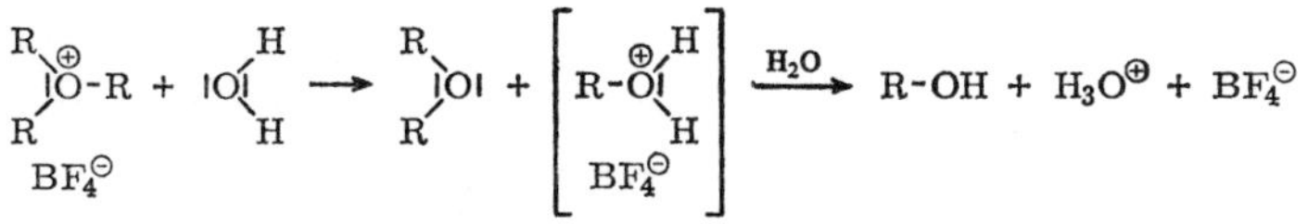

The tetrachloroferrates behave similarly. In contrast, tetrachloro*aluminates* yield the alkyl chloride instead of the alcohol. The rapid decomposition of the $AlCl_4^-$ complex with water gives rise to chloride ions, which can easily be alkylated [22]. Because of the stability of the anion and its high solubility, the tetrafluoroborates are particularly suitable for alkylation reactions [20a].

Until now, only a few investigations have dealt with the mechanistic course of the alkylation reaction, but an S_N2 mechanism must be assumed. Thus, in the thermal decomposition of solid ethyldimethyloxonium tetrafluoroborate *(137)*, the preferred splitting out of the methyl group (to the extent of 75%) is observed [21]. In the alkylation of 3,5-dinitrobenzoic acid with *(137)*, 70% of the methyl ester and 30% of the ethyl ester are produced [21]:

Reactions of Oxonium Salts

$$CH_3F + \begin{matrix} H_5C_2 \\ H_3C \end{matrix} O \to BF_3 \xrightarrow[25\%]{75\% \Delta} \begin{matrix} H_3C \\ H_3C \end{matrix} \overset{\oplus}{O}-C_2H_5 \, BF_4^{\ominus} \xrightarrow[-HBF_4]{+RCO_2H} \begin{matrix} 70\% \nearrow \\ \\ 30\% \searrow \end{matrix} \begin{matrix} RCO_2CH_3 + H_3C-O-C_2H_5 \\ \\ RCO_2C_2H_5 + (CH_3)_2O \end{matrix}$$

$$C_2H_5F + \begin{matrix} H_3C \\ H_3C \end{matrix} O \to BF_3$$

(137)   R = 3,5-$C_6H_3(NO_2)_2$

The attack of a nucleophile by an S_N2 mechanism should be more difficult at the ethyl group than at a methyl group because of the sterically more unfavorable transition state. This is shown very clearly by the Newman projection for the ethyl group, given below; the directions of attack of the nucleophile are marked by arrows:

(more difficult) Y: :Y (easier)

However, the result of the alkylation of 3,5-dinitrobenzoic acid comes very close to the ratio expected from a statistical attack of the nucleophile ($CH_3:C_2H_5 = 2:1$).

The idea of an S_N1 mechanism of the alkylation is opposed by the fact that the rate of the decomposition of trialkyloxonium salts with water *decreases* in the sequence [22]:

$$\begin{matrix} H_3C \\ H_3C \end{matrix} \overset{\oplus}{O}-CH_3 \gg \begin{matrix} H_5C_2 \\ H_5C_2 \end{matrix} \overset{\oplus}{O}-C_2H_5 > \begin{matrix} H_7C_3 \\ H_7C_3 \end{matrix} \overset{\oplus}{O}-C_3H_7 > \bigcirc \overset{\oplus}{O}-C_2H_5$$

Recently, Allred and Winstein studied the reactivity of cyclic O-methyloxonium ions formed in situ [56, 57]. According to these investigations, five-membered O-methyloxonium ions react with nucleophiles almost exclusively with ring cleavage, whereas six-membered compounds react with the transfer of methyl to the nucleophile together with a small degree of ring opening:

Even earlier, Meerwein *et al.* found preferential ring opening in the reaction of O-ethyltetrahydrofuranium tetrafluoroborate *(138)* with water [21]. Under these conditions the dimeric ether is formed in addition to the γ-glycol monoethyl ether:

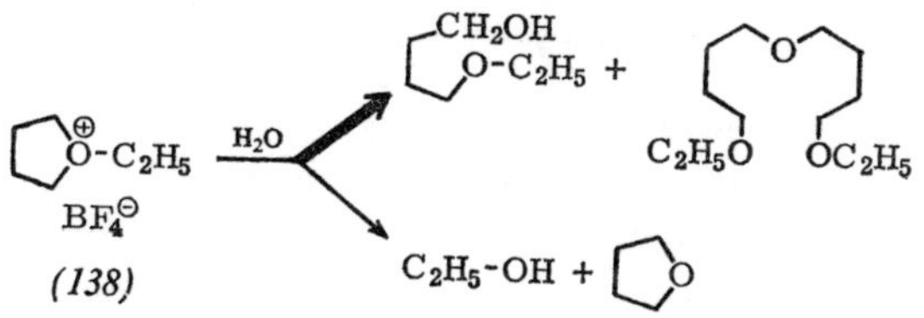

(138)

These results can also be harmonized with an S_N2 alkylation mechanism, as shown by a consideration of Newman projections: In a five-membered ring the O-methylene ring carbons are less deterred from nucleophilic attack (direction of the arrow) than in a six-membered ring, where the equatorial H-atoms of the neighboring methylene groups lie precisely in the plane of attack. Consequently, a more favorable transition state for ring cleavage can be formed in the five-membered ring than in the six-membered ring [56]:

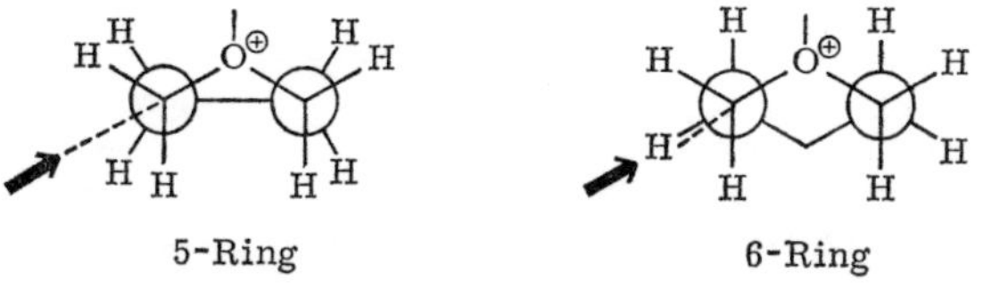

5-Ring   6-Ring

In the six-membered ring the steric situation is similar to that in the ethyldimethyloxonium ion mentioned above; consequently, demethylation takes place preferentially [56].

5.2.2. Triaryloxonium Salts

The triphenyloxonium salt that can be obtained by the action of benzendiazonium tetrafluoroborate on diphenyl ether in very low yield (2%) is extremely inert toward nucleophiles in comparison with trialkyloxonium salts. This is also exhibited by the remarkable stability of triphenyloxonium halides (Cl⁻, Br⁻, I⁻) [58a,b].

Although there have been no kinetic studies on reactions with nucleophiles, it can hardly be assumed that arylation (like the heterolysis of diazonium ions) takes place in a monomolecular reaction:

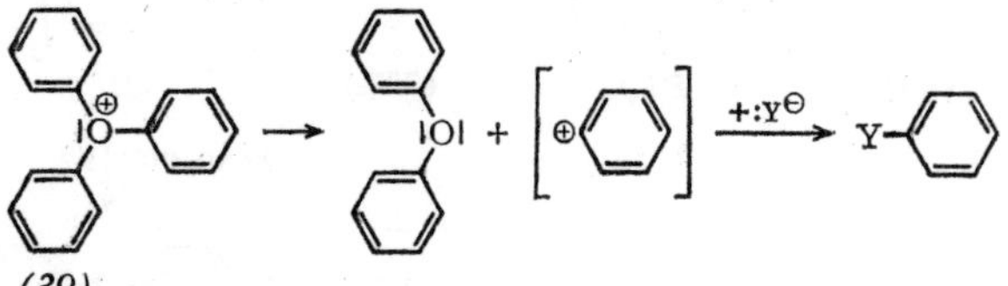

(39)

A bimolecular mechanism is more likely:

In any case the assumption of a monomolecular mechanism is opposed by the fact that the tris(p-nitrophenyloxonium) ion is more readily attacked by nucleophiles than is *(139)* [60]. The lack of reactivity of *(139)* will be due above all to the poor approach of the nucleophile to the site of substitution because of the steric situation. Special stabilization of the cation may also be responsible.*

Long times and severe conditions are generally necessary for reactions with *nucleophiles*. The use of these compounds as arylating agents is limited. Thus, even after boiling with water for 25 h, only 50% of the tetrafluoroborate of *(139)* is converted into the phenol [58a, 61]. The following reactions with nucleophiles have also been described [58a, 61]:

With *electrophiles*, substitution in the phenyl nuclei takes place. Nitration with HNO_3 or NO_2BF_4 has been studied, leading after very long reaction times to good yields of the tris(p-nitrophenyloxonium) salt already mentioned (together with a small amount of the ortho isomers). Like the halogens, the oxonium oxygen atom has an inhibiting, but an *o,p*-orientating, effect on electrophilic substitution [60, 62].

The nitro group can be reduced without cleavage of the aryl nuclei to give the tris(p-aminophenyloxonium) ion, and this can be converted after diazotisation under suitable conditions into the tris(p-iodophenyloxonium) ion. However, this reaction can also be directed to give the di(p-iodophenyl) ether with the elimination of an aryl nucleus [63].

---

* The O-atom is assumed to have not a pyramidal but a planar structure [59].

The o,o'-biphenylylenephenyloxonium ion *(140)* is a cyclic representative of the triaryloxonium ions, whose preparation takes place similarly to that of *(139)*. Here, again, the iodide can be isolated, but when this is heated the ring opens to form 2-iodo-2'-phenoxybiphenyl [64]:

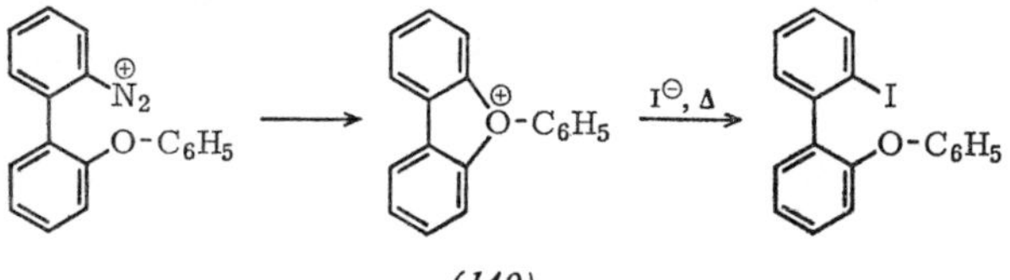

$$\text{Ar-N}_2^{\oplus}\text{-O-C}_6\text{H}_5 \longrightarrow \text{[cyclic oxonium]}^{\oplus}\text{-O-C}_6\text{H}_5 \xrightarrow{I^{\ominus}, \Delta} \text{2-I-2'-O-C}_6\text{H}_5\text{-biphenyl}$$

(140)

5.2.3. Acyldialkyloxonium Salts

As cationic ether *and* carboxylic acid derivatives, acyldialkyloxonium ions *(141)* can act as alkyl or acyl donors [65]. So far there have been no examples of rearrangements to dialkoxycarbonium ions *(9)* in the case of the acyclic cations*:

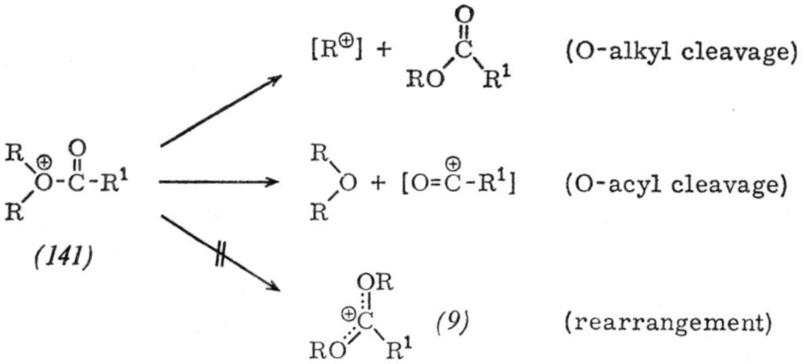

$$R_2\overset{\oplus}{O}\text{-}\overset{O}{\underset{\|}{C}}\text{-}R^1 \quad (141)$$

- $[R^{\oplus}] + RO\text{-}\overset{O}{\underset{\|}{C}}\text{-}R^1$ (O-alkyl cleavage)
- $R_2O + [O=\overset{\oplus}{C}\text{-}R^1]$ (O-acyl cleavage)
- $RO\text{-}\overset{\overset{\oplus}{O}R}{\underset{\|}{C}}\text{-}R^1$ *(9)*   (rearrangement)

The hexachloroantimonates of *(141)*, which can be obtained by the action of an acyl chloride ($R^1 = CH_3$, $n\text{-}C_3H_7$) and $SbCl_5$ on a dialkyl ether ($R = CH_3, C_2H_5$), prove to be stable only at low temperatures ($-20\,^\circ$C) [65].

An O-*alkyl cleavage* is found to take place in the thermal decomposition of the hexachloroantimonates in accordance with the following scheme [65]:

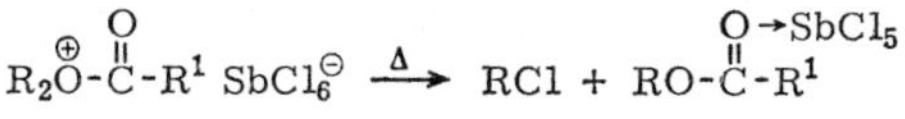

$$R_2\overset{\oplus}{O}\text{-}\overset{O}{\underset{\|}{C}}\text{-}R^1 \; SbCl_6^{\ominus} \xrightarrow{\Delta} RCl + RO\text{-}\overset{O \to SbCl_5}{\underset{\|}{C}}\text{-}R^1$$

(141)

The long-known ether cleavages with acyl halides catalyzed by metal or nonmetal halides (e.g., $ZnCl_2$, $AlCl_3$, $FeCl_3$, BF_3 [66, 67a,b]), which lead to carboxylic esters (or their adducts with Lewis acids) and alkyl halides, are formulated in analogy with this reaction. Ether cleavage with the aid of carboxylic acid anhydrides and Lewis acids [67a,c] or strong protonic acids [67a, 68] probably includes intermediate acyldialkyloxonium ions *(141)*.

---

* However, such rearrangements are occasionally formulated for cyclic cations formed in situ; see Section 6.3.1.

The O-acetyltetrahydrofuranium tetrafluoroborate that can be obtained from the tetrahydrofuran–BF$_3$ complex and acetyl fluoride is more stable than the acyclic hexachloroantimonates [69]. The O-alkyl cleavage of *(142)* takes place with ring opening by the reaction of suitable nucleophiles, as in the attempt to prepare the HgI$_3$ salt in which 4-iodobutyl acetate is formed [70], or by the action of tetrahydrofuran by which the polymerization of the latter can be initiated [69]:

$$H_3C-\overset{O}{\underset{\|}{C}}-O-(CH_2)_4-I + HgI_2 + BF_4^{\ominus}$$

$$H_3C-\overset{O}{\underset{\|}{C}}-O^{\oplus}\underset{BF_4^{\ominus}}{\bigcirc}$$
(142)

$$H_3C-\overset{O}{\underset{\|}{C}}-O-(CH_2)_4-O^{\oplus}\underset{BF_4^{\ominus}}{\bigcirc} \longrightarrow H_3C-\overset{O}{\underset{\|}{C}}-O-(CH_2)_4-O-(CH_2)_4-O^{\oplus}\underset{BF_4^{\ominus}}{\bigcirc}$$

O-acyl cleavage is the preferred mode of reaction with nucleophiles in the case of acyclic acyldialkyloxonium hexachloroantimonates. It takes place even at $-20\,^\circ$C. Thus, *(141)*, R$_1$ = CH$_3$, gives acetyl chloride with chloride ions, *p*-methoxyacetophenone (50% yield) with anisole, and acetophenone (19% yield) with benzene [65]; even the acetylation of chlorobenzene takes place to a smaller extent.

Thermal decomposition of the hexachlorantimonate *(143)* also takes place at room temperature with cleavage of the O-acyl bond, in addition to the O-alkyl cleavage mentioned above. The dialkyl ether that then remains can be detected as the dialkyloxonium salt (or its etherate), whose appearance may be connected with simultaneous ketene formation [65]:

$$H_3C-\overset{O}{\underset{\|}{C}}-\overset{\oplus}{O}\underset{SbCl_6^{\ominus}}{\overset{R}{\diagdown_R}} \longrightarrow [H_2C=C=O] + H-\overset{\oplus}{O}\underset{R}{\overset{R}{\diagdown}} \; SbCl_6^{\ominus}$$
(143)   *(6)*

5.3. Ambident Tertiary Oxonium Salts

5.3.1. General Considerations on Modes of Reaction

Carboxonium ions [e.g., *(9)*] can react with nucleophiles :Y$^{\ominus}$ mainly in two different directions, for which we shall retain the same symbols (route A and route B) we used for the syntheses of carboxonium ions that represent their reversals. In the case of acidic carboxonium ions (*cf.* Section 5.1), route A is preferred because of the greater ease of proton transfer; nevertheless, in the case of the tertiary carboxonium ions we may expect both routes, since in each case carbonium ions are treated with the nucleophile:

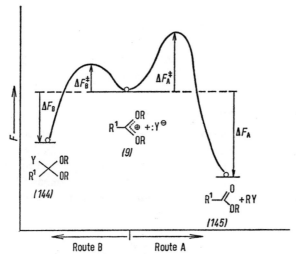

The behavior of *(9)* as an *oxonium* ion is, however, limited to route A.

In a review of ambident cations, Hünig [71] studied how the results of the reaction are affected especially by the type of the nucleophile and the stability of the cation. The reactive behavior can best be explained on the basis of an energy profile (Fig. 3). The primary reaction step is probably the addition of the nucleo-

Fig. 3. General energy profile for the reaction sequence *(144)* ⇌ *(9)* → *(145)* (schematic) according to [71].

phile :$Y^\ominus$ to the carbonium center of *(9)*, the site of lowest electron density. This addition product by route B can be isolated when the gain in energy on the formation of the bond is large enough; the result of the reaction is then controlled *kinetically*.

If, on the other hand, an equilibrium between the addition product and the carboxonium ion can be established, *(144)* ⇌ *(9)*, then *(9)* reacts with :$Y^\ominus$, by route A, with dealkylation to give the *thermodynamically* controlled final products *(145)*.

80   Reactions of Oxonium Salts

The course of the reaction depends not only on the nucleophile and the energy of the ambident cation but also on the temperature, the solvent, and the time of the reaction. If for the moment we neglect steric factors, the following facts may generally be assumed:

The activation energy $\Delta F_B^\ddagger$ for route B, will be smaller than $\Delta F_A^\ddagger$, since the combination of *(9)* and $:Y^\ominus$ in route B consists simply of an ionic association. Here the position of the equilibrium in the ambident cation *(9)* and the addition product *(144)* will be determined by the energy difference ΔF_B. In the reaction of *(9)* by route A, in addition to charge neutralization, additional resonance energy is gained for the carboxylic ester *(145)*, so that this reaction route leads to thermodynamically stable products. Here we must put $\Delta F_A > \Delta F_B$. Since in general we may assume ΔF_A and $\Delta F_A^\ddagger$ to be fairly large, this reaction is practically irreversible.

The kinetically controlled product *(144)* will therefore be obtained if
(*a*) The nucleophile is a strong one, and
(*b*) The ambident cation is high in energy.
Both factors contribute to a large value of ΔF_B.

Correspondingly, the reaction will take place in a thermodynamically controlled manner if
(*a*) The nucleophile is a weak one, and
(*b*) The ambident cation is low in energy,
under which conditions ΔF_B assumes small values.

For the moment we shall disregard changes in the ratio $\Delta F_B^\ddagger / \Delta F_A^\ddagger$, which is responsible for the distribution of the primary products.

5.3.2. Reactions with Various Nucleophiles

The influence of the nucleophile on the result of a reaction will be illustrated with the 2-methyl-1,3-dioxolenium ion *(66)*, $R^1 = CH_3$, prepared by H. Meerwein *et al.* The highly nucleophilic anions CN^- and $OC_2H_5^-$ react by addition to the carbonium center (by route B), while less nucleophilic anions Cl^-, Br^-, I^- or uncharged molecules react with *(66)*, with opening of the ring to give the stable β-substituted ethyl acetates (by route A) [29, 72]:

(146)   *(66)*   *(147)*

Y = $CN^\ominus$, $C_2H_5O^\ominus$ Y = $Cl^\ominus$, $Br^\ominus$, $I^\ominus$,
                                              C_2H_5OH, (H_2O),
                                              $N(CH_3)_3$

The reaction products *(146)* arising in a kinetically controlled process are linked by the sequence of reactions given below with the thermodynamically stable chlorine derivative *(151)* [29, 71]:

Ambident Tertiary Oxonium Salts

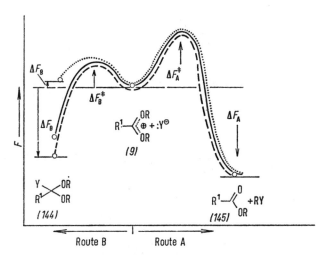

The alkoxy group can be displaced from the cyclic ortho ester *(148)* by hydrocyanic acid with the formation of the cation *(149)*, which immediately reacts further with the cyanide ions to give the 2-cyano-1,3-dioxolane derivative *(150)*. The cyano group in turn can be split out with hydrogen chloride by the reaction of the protons, and the cation *(149)*, again liberated, reacts with the chloride ions to give the stable β-chloroethylacetate *(151)*. The transformation of the ortho ester *(148)* into the stable β-alkoxyethylacetate can be brought about, for example, by the treatment of *(148)* with *catalytic* amounts of BF_3 [29]. (The cation *(149)* formed as an intermediate in this process can be isolated as 1,3-dioxolenium tetrafluoroborate with an excess of BF_3; *cf.* Section 4.2.2.2):

Fig. 4. Influence of the nucleophile on the energy of the addition product *(144)* according to [71] (schematic).
.... $Y^\ominus$ weak ⎫ nucleophile
------ $Y^\ominus$ strong ⎭
──────── general curve (Fig. 3).

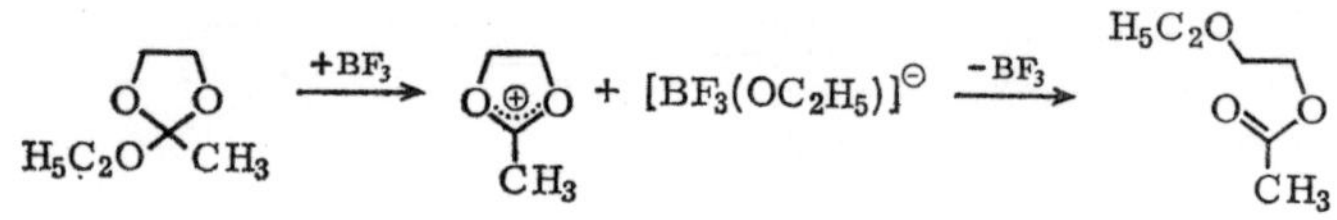

$$\text{(scheme 149)}$$

(149)

Figure 4 shows schematically the influence of the nucleophile on the energy of the addition product *(144)*. The result of the reaction is therefore in favor of the addition product *(144)* in the case of strong nucleophiles (CN^-, $OC_2H_5^-$) because of a high gain in energy ΔF_B. If, on the other hand, in the case of weak nucleophiles (Cl^-), ΔF_B is small or if the energy of the addition product exceeds the energy of *(9)*, Y can be easily eliminated from such adducts.

Halogen derivatives of type *(144)* and *(146)* can be isolated only in exceptional cases, since halide ions are weak nucleophiles and the cations *(9)* present in the equilibrium are dealkylated by the halide [71, 76]. Normally, in reactions in which 2-halo-1,3-dioxolanes (or 1,3-dioxolenium halides) should occur, reaction by route A is to be expected, as in the bromination of 2-aryl-1,3-dioxolane [77a],

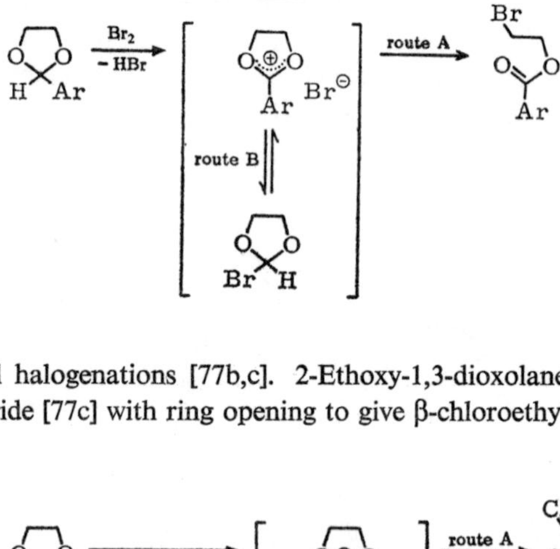

and other acetal halogenations [77b,c]. 2-Ethoxy-1,3-dioxolane reacts similarly with acetyl chloride [77c] with ring opening to give β-chloroethyl formate:

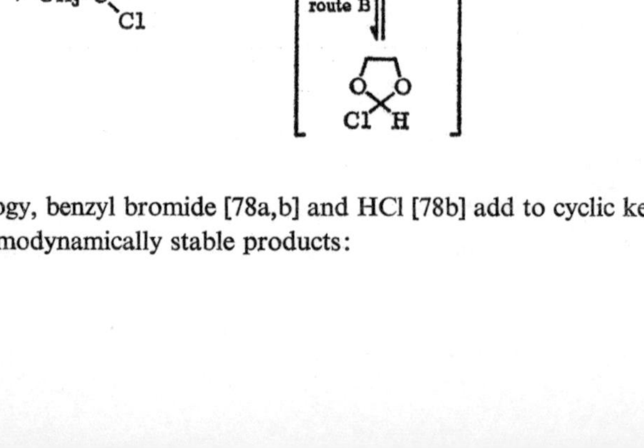

In full analogy, benzyl bromide [78a,b] and HCl [78b] add to cyclic ketene acetals to give thermodynamically stable products:

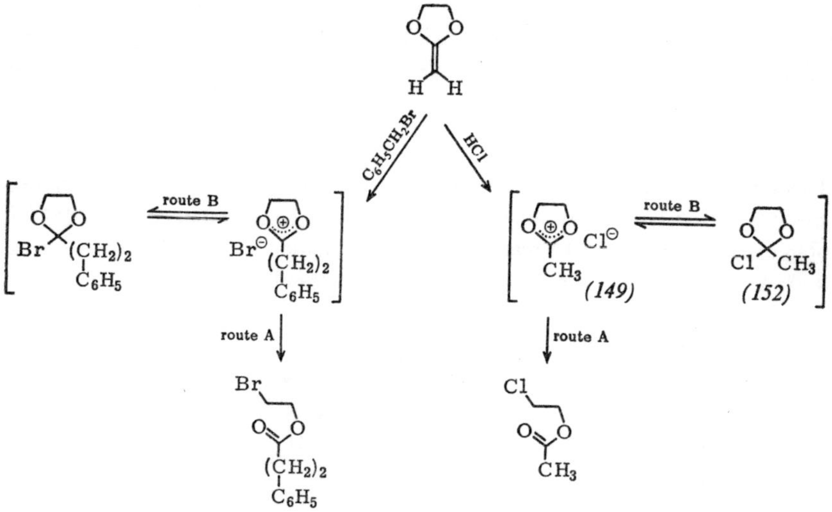

[Reaction scheme showing 1,3-dioxolane derivatives reacting via route A and route B with C₆H₅CH₂Br and HCl, forming brominated and chlorinated products (149) and (152)]

Only recently has it been possible to detect in solution at $-60°C$ an HCl adduct with a ketene acetal analogous to *(152)*, i.e., 2-chloro-2-dichloromethyl-1,3-dioxolane. At this temperature the activation barrier $\Delta F_A^{\ddagger}$ for the thermodynamically controlled reaction of the cation is obviously not exceeded; above $-30°C$, however, a rearrangement by route A is observed by NMR spectroscopy [79a]:

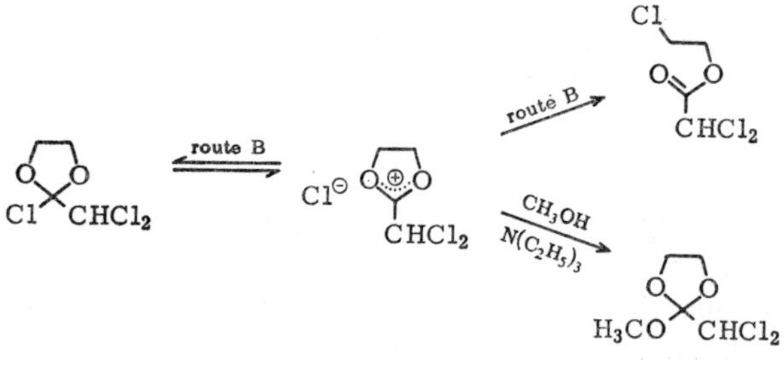

[Reaction scheme showing 2-chloro-2-dichloromethyl-1,3-dioxolane rearranging via route B to an oxonium cation with Cl⁻, which reacts via route B to give an open-chain chloro ester with CHCl₂ group, and with CH₃OH/N(C₂H₅)₃ to give the 2-methoxy-2-dichloromethyl-1,3-dioxolane]

The existence of the 2-chloro-1,3-dioxolane derivative shown by the NMR spectrum is confirmed by its conversion into the corresponding 2-methoxydioxolane with methanol–triethylamine (at $-60°C$) [79a]. It is true that other cyclic ortho-ester halides have also been described by various authors, but their structures—e.g., in in the case of acetohalo sugar derivatives [73, 74]—have had to be revised [75].

Besides the *anionic* nucleophiles treated so far, *uncharged molecules* (H_2O, C_2H_5OH, $N(CH_3)_3$) can react with *(66)* [29]. The cationic primary products arising in this way are, in the case of water or ethanol, deprotonated in a subsequent step. Since reaction products are obtained by ring opening (e.g., *(154)*, [29]) one might assume reaction by route A. Nevertheless, with high probability, the addition of water takes place primarily to give an ortho-ester derivative *(153)*, which undergoes transformation into *(154)* only by transprotonation and ring opening:

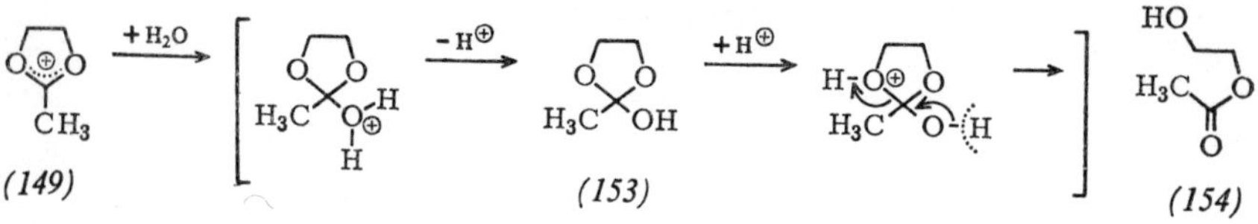

This conclusion can be drawn because, by the action of water on the analogous *cis*-cyclohexene-1,2-acetoxonium ion *(155)*, only the *cis*-monoacetate *(157)* is obtained. This course of the reaction can be understood if the intermediate formation of an ortho ester *(156)* and its transprotonation [80] is assumed:

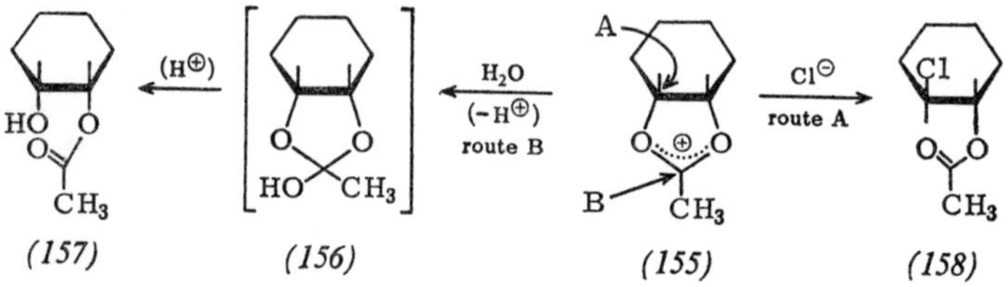

The reaction of *(155)* by route A (as with Cl^-), on the other hand, leads to *trans* products, e.g., *(158)* [80]. As a further example of the stereochemistry of route A we may also mention the bromination of the bicyclic phenyl-1,3-dioxolane derivative *(159)* [81]:

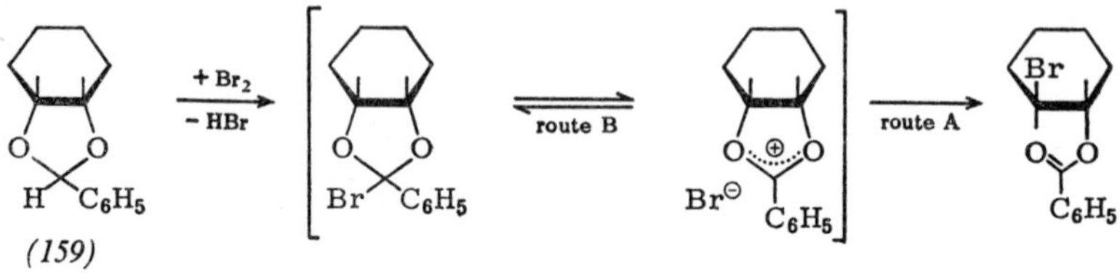

The action of trimethylamine on the dioxolenium cation *(66)* ($R^1 = CH_3$ or C_6H_5) enables only the thermodynamically stable end product *(147)* to be isolated [29]. On the other hand, the amide acetal *(160)* obtained in another manner can be quaternized with methyl iodide at the amino group [39]. The alkylation product proves to be thermally unstable, and when heated will change with ring cleavage into compounds of the type *(147)*, formylcholine iodide, and β-iodoethylformate [39]:

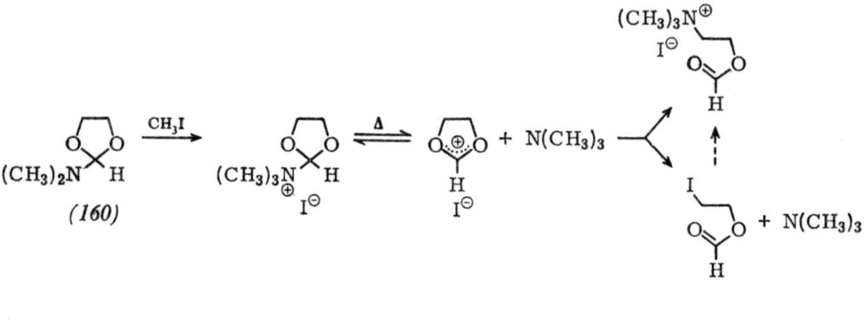

Other dialkoxycarbonium ions also behave like 1,3-dioxolenium ions. In the case of acyclic cations *(9)*, alkoxides give ortho esters in a kinetically controlled reaction [28, 30]. With halide ions, on the other hand, here again *no* ortho-ester halides *(144)* can be normally obtained. Reactions in which their appearance is expected take place generally with dealkylation by route A.

The following may serve as examples:

(*a*) The addition of HCl (or alkyl halides) to ketene acetals [82a,b,c] or the addition of alcohols to α-chlorovinyl ethers *(161)* [83]:

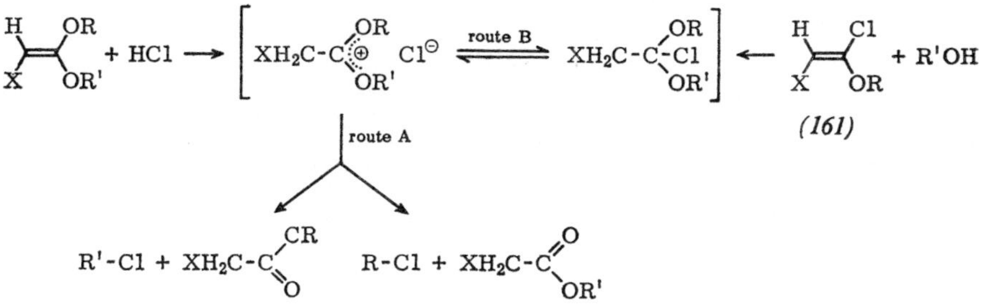

$$\underset{X}{\overset{H}{\diagdown}}C=C\underset{OR'}{\overset{OR}{\diagup}} + HCl \longrightarrow \left[XH_2C-C\underset{OR'}{\overset{OR}{\diagdown}}\overset{\oplus}{} \quad Cl^{\ominus} \underset{\longleftarrow}{\overset{\text{route B}}{\longrightarrow}} XH_2C-\underset{OR'}{\overset{OR}{\overset{|}{C}}}-Cl \right] \longleftarrow \underset{X}{\overset{H}{\diagdown}}C=C\underset{OR}{\overset{Cl}{\diagup}} + R'OH$$

$$\downarrow \text{route A}$$

$$R'-Cl + XH_2C-C\underset{O}{\overset{CR}{\diagup}} \qquad R-Cl + XH_2C-C\underset{OR'}{\overset{O}{\diagdown}}$$

(161)

(*b*) The oxidation of tetramethoxyethylene *(162)* with iodine, which can be regarded as the addition of iodine to a bis(ketene acetal) [84]:

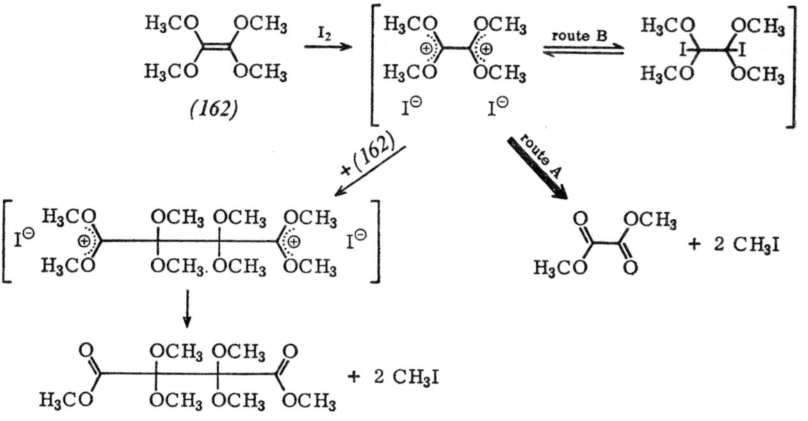

[chemical scheme showing reaction of (162) with I_2 via route B and route A, producing $H_3CO-C(=O)-C(=O)-OCH_3 + 2\ CH_3I$, and via +(162) forming an extended intermediate that gives $H_3CO-C(=O)-C(OCH_3)_2-C(OCH_3)_2-C(=O)-OCH_3 + 2\ CH_3I$]

(*c*) Acetal halogenation, with the bromination of dialkoxyacetic acid *(163)* as a relatively new example [85]. Here the main reaction leads to the oxalic acid diester:

Reactions of Oxonium Salts

$$\underset{(163)}{\overset{H}{\underset{HO_2C}{\diagdown}}\overset{OR}{\underset{OR}{\diagup}}} \xrightarrow{Br_2} \left[\overset{HO}{\underset{O}{\diagdown}}\overset{OR}{\underset{OR}{\diagup}} \overset{\oplus}{C} Br^{\ominus} \right] \underset{\text{route B}}{\rightleftarrows} \left[\overset{Br}{\underset{HO_2C}{\diagdown}}\overset{OR}{\underset{OR}{\diagup}} \right]$$

with branches $-CO_2$ / route A leading to:

$$\left[\overset{H}{\underset{Br}{\diagdown}}\overset{OR}{\underset{OR}{\diagup}} \right] \underset{\text{route B}}{\rightleftarrows} \left[H-\overset{OR}{\underset{OR}{C^{\oplus}}} Br^{\ominus} \right]$$

and

$$\underset{O}{\overset{RO}{\diagdown}}\overset{O}{\underset{OR}{\diagup}}$$

route A ↓

$$H-\overset{OR}{\underset{O}{C}} + RBr$$

Recently it has been shown by NMR spectroscopy that at −80°C in SO$_2$, dialkoxycarbonium ions are formed by the reaction of bromine with ortho esters [79b]:

$$R^1-\overset{OCH_3}{\underset{OCH_3}{C-OCH_3}} + Br_2 \xrightarrow{\text{liq SO}_2} R^1-\overset{OCH_3}{\underset{OCH_3}{C^{\oplus}}} Br^{\ominus} + CH_3OBr$$

where R^1 = H, CH$_3$. The cations are stable up to about −30°C in the presence of the halide; above this temperature they are dealkylated in the known manner.

The preparation of a dialkoxymethyl chloride has also been achieved by the action of acetyl chloride on tri(neopentyloxy)methane. In this case the successful synthesis must be ascribed to the inhibited approach of the chloride ion to the methylene carbon in the neopentyl group [79c]:

$$H-\overset{OR}{\underset{OR}{C-OR}} + CH_3COCl \longrightarrow H-\overset{OR}{\underset{OR}{C-Cl}} + CH_3CO_2R$$

where R = H$_2$C–C(CH$_3$)$_3$.

Normally, ortho esters react with acetyl chloride [86] under thermodynamic control, while with acetyl cyanide [87] (with the strong nucleophile CN$^-$) the kinetically controlled product is obtained:

$$R^1-\overset{OR}{\underset{OR}{C-CN}} + CH_3COOR \xleftarrow[\text{route B}]{CH_3COCN} R^1-\overset{OR}{\underset{OR}{C-OR}} \xrightarrow[\text{route A}]{CH_3COCl} R^1-\overset{O}{\underset{OR}{C}} + R-Cl + CH_3COOR$$

In addition, such weak nucleophiles as compounds with ether, carbonyl, and nitrile functions are capable of dealkylating dialkoxycarbonium salts *(9)* by route A.

The alkylation products arising in this way—trialkyloxonium [88]* alkoxycarbonium [88], or nitrilium salts [88]—can be isolated under suitable conditions; e.g.:

$$R^1-C{\overset{OR}{\underset{OR}{\oplus}}}\ SbCl_6^{\ominus} + O=C{\overset{C_6H_5}{\underset{C_6H_5}{}}} \xrightarrow{\text{route A}} R^1-C{\overset{O}{\underset{OR}{}}} + R-O\overset{\oplus}{=}C{\overset{C_6H_5}{\underset{C_6H_5}{}}}\ SbCl_6^{\ominus}$$

The N,N-disubstituted amides, diphenyl sulfoxide, and even higher carboxylic esters are alkylated in the same way [88]. This way of dealkylating dialkoxycarbonium ions was used even earlier in the reaction with phosphorus triesters [48]:

$$R^1-C{\overset{OR'}{\underset{OR'}{\oplus}}}\ BF_4^{\ominus} + \overset{OR}{\underset{OR}{|P-OR}} \longrightarrow R^1-C{\overset{O}{\underset{OR'}{}}} + R'-\overset{OR}{\underset{OR}{\overset{\oplus}{P}-OR}}\ BF_4^{\ominus}$$

The behavior of lactonium ions with respect to nucleophiles is similar to that of the dialkoxycarbonium ions treated above [30, 72]:

(164) with OC₂H₅, OC₂H₅ ← C₂H₅O⁻ — A → (54) with OC₂H₅ — C₂H₅OH → (166) H₅C₂O, OC₂H₅

(165) with OC₂H₅, CN ← CN⁻ — (54) — Cl⁻ → (167) Cl, OC₂H₅

The conversion of the ortho ester *(164)* with HCN into the nitrile *(165)* and its reaction with HCl to give ethyl-γ-chlorobutyrate *(167)*, in each case with the intermediate formation of the cation *(54)*, corresponds completely to the reaction sequence *(148)–(150)–(151)* [72]. The reaction of the lactonium ion *(54)* with ethoxide gives the kinetically controlled product *(164)*, but reaction with ethanol produces only the thermodynamically stable ethyl-γ-ethoxybutyrate *(166)*. This is also formed from *(164)* by the addition of catalytic amounts of a strong acid; cf. the rearrangement of *(148)*, [72]. However, the reaction of *(54)* with isopropanol leads in a kinetically controlled reaction to the replacement of the isopropoxy group by the ethoxy group to give the cation *(168)*, which precipitates because of

---

* The alkylation of ethers to oxonium salts with *(9)* takes place under certain conditions partially by route B, in the following way [89]:

$$H-C{\overset{OR}{\underset{OR}{\oplus}}} + |\overset{R}{\underset{R}{O}}| \xrightleftharpoons{\text{route B}} H-C{\overset{OR}{\underset{OR}{\overset{\oplus}{-O}{\overset{R}{\underset{R}{}}}}}} \xrightarrow{+R_2O} H-C(OR)_3 + R-\overset{\oplus}{O}{\overset{R}{\underset{R}{}}}$$

the low solubility of its tetrafluoroborate. The preferred elimination of the ethoxy group from the mixed ortho ester *(169)* is also observed in its reaction with catalytic amounts of acid or BF_3 to give the thermodynamically stable butyric ester *(170)* [30]. The action of isopropyl alcohol on *(168)* gives, in addition to the analogous ring cleavage to *(171)* by route A, diisopropyl ether and butyrolactone (route A′) [30]:

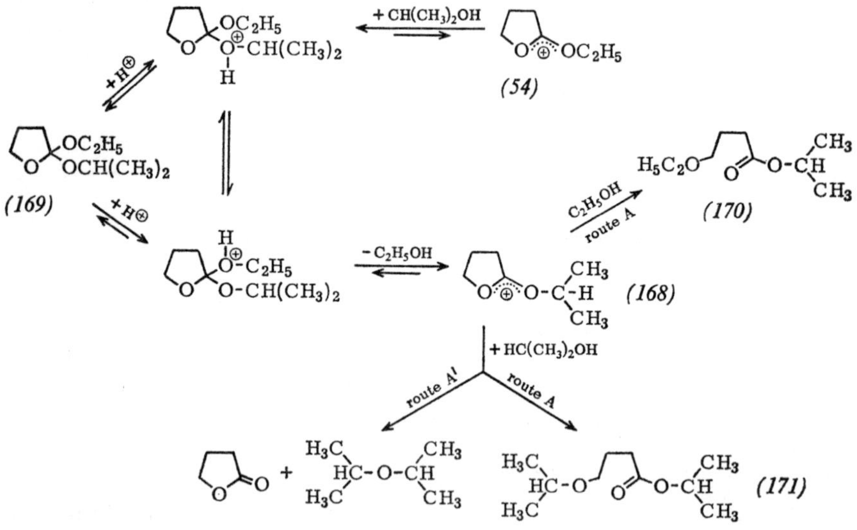

An exchange reaction, as in the conversion of *(54)* into *(168)*, also takes place in the reaction of *(54)* with primary arylamines. The imidic ester cation *(172)* can be isolated in some cases, but reacts further with an excess of amine with ring opening by route A [30]:

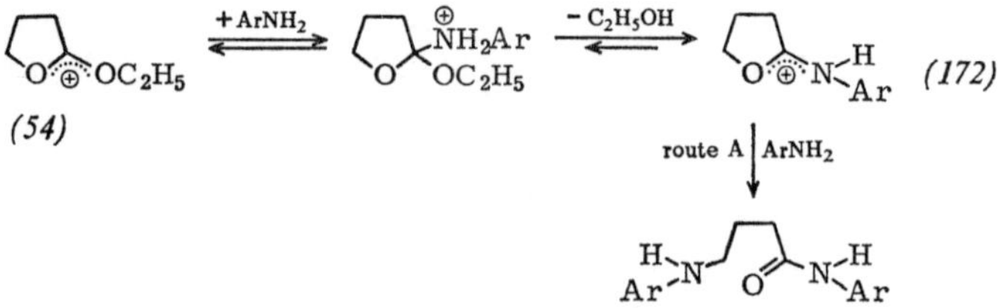

The dealkylation of lactonium ions can basically take place not only by ring cleavage (route A) but also at the exocyclic alkoxy grouping with the formation of the lactone (route A′). In the case of the butyrolactonium ion *(54)*, this route is observed only rarely (e.g., in the reaction of *(168)* with isopropanol, which has been mentioned). On the other hand, route A′ is the usual reaction mode of the O-ethylphthalidium cation *(173)* [30,90]:

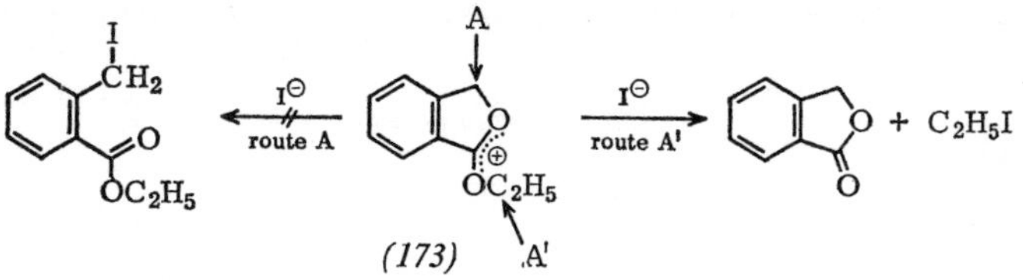

An additional example of reaction route A' besides A is found in the addition of alcohol to the cyclic α-chlorovinyl ether *(174)* proceeding via the intermediate lactonium ion *(175)* [91]:

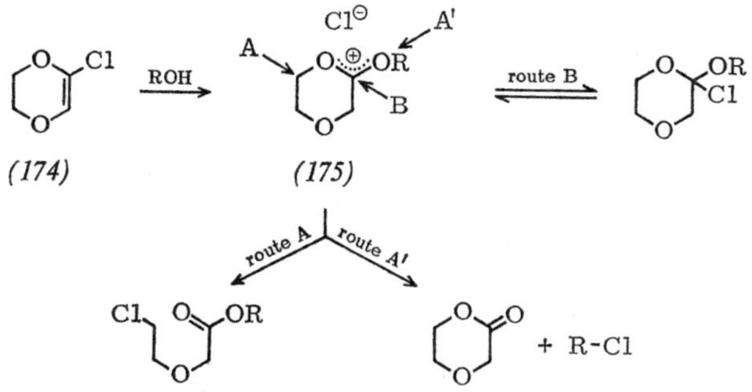

Table 9 gives a review of the reactions of carboxonium ions with various nucleophiles. It includes monoalkoxycarbonium ions *(176)*, *(177)*, and also trialkoxycarbonium ions *(51)*, *(178)*.

Sometimes the steric requirements of the nucleophiles do not permit the addition of carbonium ions but only that of protons to the nucleophilic center. In these cases the reactions do not proceed by routes A or B, but via proton abstraction from the carbon atom in β-position of the oxonium center, as mentioned in the beginning of Chapter 5. The schematic for this route C is

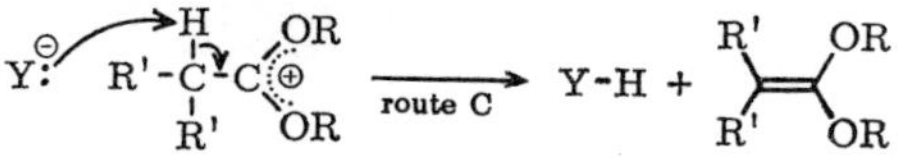

In this way a cyclic ketene acetal is formed by reaction of the lactonium ion *(54)* with *N*-methylpiperidine [90, 92a]:

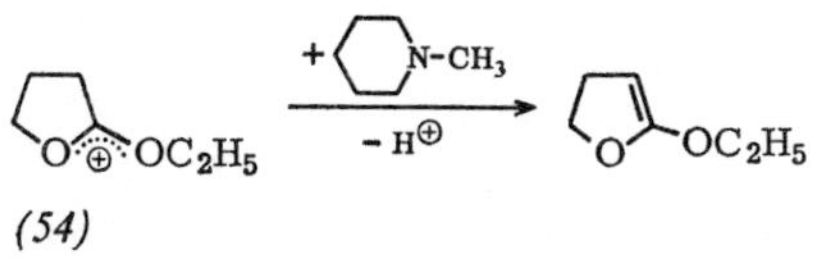

If the carbonium center of a dialkoxycarbonium ion is substituted by a hydrogen atom, carbene intermediates should be available in an analogous manner:

TABLE 9. REACTIONS OF TERTIARY CARBOXONIUM IONS WITH VARIOUS NUCLEOPHILES

Route A (and route A′)		Carboxonium ion	Route B	
Reaction products	Y		Y	Reaction products
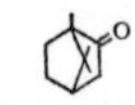 + C_2H_5I	$I^{\ominus}$ [93]	(176) [20a, 28, 30] 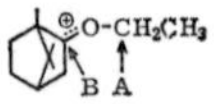	$C_2H_5O^{\ominus}$ [20a, 30] $P(C_6H_5)_3$ [43]	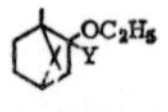
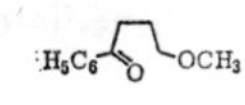	$CH_3OH/H^{\oplus}$ [94b]	(177) [94a, b] 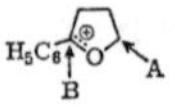	$CH_3O^{\ominus}$ [94b] (CH_3OH)	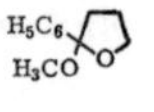
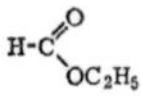 + C_2H_5-Y	R_2O $(C_6H_5)_2SO$ RR′C=O R-CO_2R R-CN [88] $P(OR)_3$ [46]	(67) [29, 46, 88] 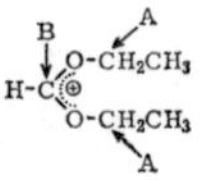	$C_2H_5O^{\ominus}$ [29]	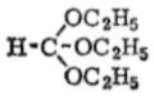
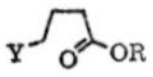 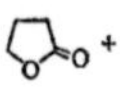 + $[(CH_3)_2CH]_2O$	$Cl^{\ominus}$ $Br^{\ominus}$ $I^{\ominus}$ $CH_3CO_2^{\ominus}$ $C_6H_5O^{\ominus}$ C_2H_5OH $ArNH_2$ [30] [72] $(CH_3)_2CHOH$ [30]	[30, 72] 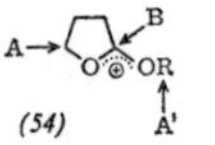 R = C_2H_5 (54) R = $(CH_3)_2CH$ (168)	$C_2H_5O^{\ominus}$ $CN^{\ominus}$ ($ArNH_2$) [30] [72] $P(C_2H_5)_3$ [43] $P(C_6H_5)_3$ [43]	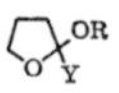
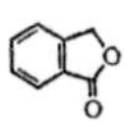 + C_2H_5Y	$I^{\ominus}$ [90] C_2H_5OH [30]	(173) [30, 90] 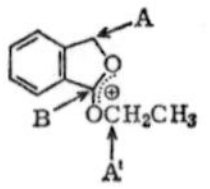	$C_2H_5O^{\ominus}$ [30] $P(C_2H_5)_3$ [43]	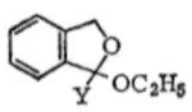
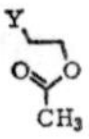	$Cl^{\ominus}$ $Br^{\ominus}$ $I^{\ominus}$ C_2H_5OH $(CH_3)_3N$ (H_2O) [29,71]	(149) [29, 71] 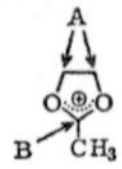	$C_2H_5O^{\ominus}$ $CN^{\ominus}$ [29, 71]	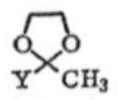
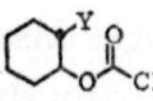 (trans)	$Cl^{\ominus}$ $Br^{\ominus}$ $CH_3CO_2^{\ominus}$ $OTs^{\ominus}$ [80]	(155) [80] 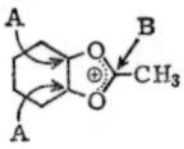	H_2O CH_3CO_2H [80]	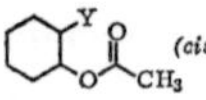 (cis)
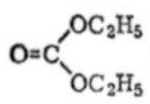 + C_2H_5-Y	$Br^{\ominus}$ [95] $P(C_2H_5)_3$ [43]	(51) [20a, 30] 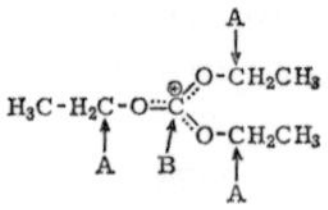	$C_2H_5O^{\ominus}$ [30]	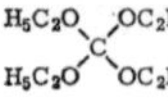
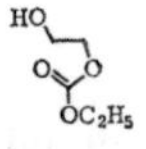 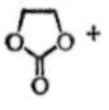 + C_2H_5OH	H_2O [96]	(178) [96] 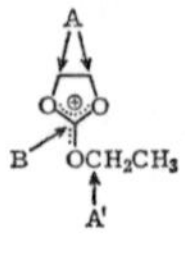	$C_2H_5O^{\ominus}$ [96]	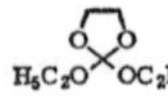

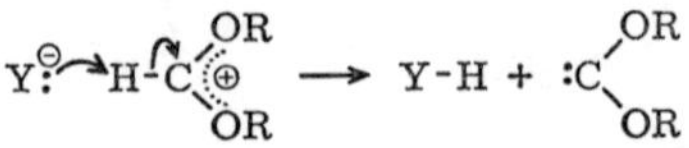

This reaction route was in fact confirmed by recent investigations using diisopropylethylamine and dialkoxycarbonium salts [92b]. The final reaction products are to be explained via dialkoxycarbene intermediates, e.g.:

Tetramethoxyethylene *(162)* [84] was not isolated in the present case; a corresponding bis(ketene acetal), however, is available, *inter alia*, with 3,3,4,4-tetramethyl-1,3-dioxolenium salt by the analogous reaction [92b]

5.3.3. Influence of the Ambident Cation on the Result of the Reaction

5.3.3.1. Cations of Different Energies

Whether a kinetically controlled product (by route B) can be isolated depends on the relative position of the cation and of the addition product in the energy profile, i.e., on ΔF_B (*cf.* Fig. 3 in Chapter 5). Up to this point we have considered only the effect of the type of nucleophile on ΔF_B. Figure 5 shows how ΔF_B is affected by the energy of the carboxonium ion, too [71].

If the cation *(9)* has very high energy, the gain in energy ΔF_B may be large even on reaction with weak nucleophiles and the kinetically controlled product *(144)* can be isolated. On the other hand, in the case of cations with low energy, ΔF_B will be small (or change its sign) even when strong nucleophiles react with *(9)*; thermodynamically stable products *(145)* should then arise by route A [71].

Carboxonium Ions with High Energy

The isolation or detection of dialkoxyhalomethane derivatives is possible in only a few cases under special conditions [79a,c]. In divinyloxycarbonium and diaryloxycarbonium ions the free-electron pairs of the oxygen atoms are available for charge delocalization to a smaller extent than in dialkoxycarbonium ions, since

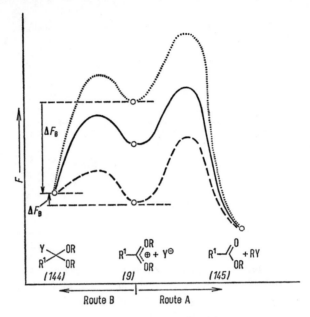

Fig. 5. Energy profiles for cations of different energies, after [71] (schematic).
·········· cation of high ⎫
------ cation of low ⎬ energy
———— general curve (Fig. 3).

they are also involved in O-vinyl or O-phenyl mesomerism [71]. The cations *(48)* and *(181)* therefore have relatively high energy and their halide adducts can therefore be isolated.

The following reactions may serve as examples:

(*a*) The addition of HCl to ketene divinyl acetal *(179)* [82c], where the addition product can react further with ethanol/pyridine to give the ortho ester:

$$H_2C=C\begin{matrix}O-CH=CH_2\\O-CH=CH_2\end{matrix} \xrightarrow{HCl} H_3C-C^{\oplus}\begin{matrix}O-CH=CH_2\\O-CH=CH_2\end{matrix} Cl^{\ominus} \underset{route\ B}{\rightleftarrows} H_3C-C\begin{matrix}O-CH=CH_2\\-Cl\\O-CH=CH_2\end{matrix}$$

(179)   (48)

$$\downarrow C_2H_5OH/\text{pyridine}$$

$$H_3C-C\begin{matrix}O-CH=CH_2\\-OC_2H_5\\O-CH=CH_2\end{matrix}$$

(*b*) The action of acyl chlorides on triphenylorthoformate *(181)* [87]:

$$H-C\begin{matrix}OC_6H_5\\-OC_6H_5\\OC_6H_5\end{matrix} + R-C\begin{matrix}O\\Cl\end{matrix} \xrightarrow[-R-C\begin{smallmatrix}O\\OC_6H_5\end{smallmatrix}]{} H-C^{\oplus}\begin{matrix}OC_6H_5\\OC_6H_5\end{matrix} Cl^{\ominus} \underset{route\ B}{\rightleftarrows} H-C\begin{matrix}OC_6H_5\\-Cl\\OC_6H_5\end{matrix}$$

(180)   (181)

In addition to the energy considerations given above, the thermodynamically controlled nucleophilic attack of the halide ion on the vinyl or aryl group should in any case take place with greater difficulty than on alkyl groups.

Monoalkoxycarbonium ions in which the positive charge is distributed over only two centers have higher energy than dialkoxycarbonium ions. Consequently, in general monoalkoxycarbonium ions react with the weakly nucleophilic chloride and bromide ions by addition to form α-haloethers. For this reason, the reactions in which alkoxycarbonium halides (e.g., *(182)*) occur as intermediates are suitable for their preparation, such as the addition of hydrogen halide or halogen to vinyl ethers, the action of hydrogen halide or acyl halides on acetals, and halogenations of ethers [97]; e.g.:

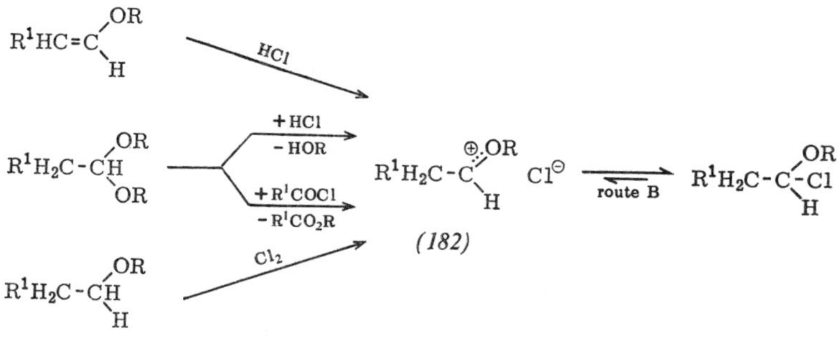

$$R^1HC=C\begin{matrix}OR\\H\end{matrix} \xrightarrow{HCl}$$

$$R^1H_2C-CH\begin{matrix}OR\\OR\end{matrix} \xrightarrow[-HOR]{+HCl}$$

$$R^1H_2C-CH\begin{matrix}OR\\H\end{matrix} \xrightarrow[-R^1CO_2R]{+R^1COCl}$$

$$R^1H_2C-\overset{\oplus}{C}\begin{matrix}OR\\H\end{matrix} Cl^{\ominus} \underset{\text{route B}}{\rightleftarrows} R^1H_2C-C\begin{matrix}OR\\H\end{matrix}Cl$$

$$(182)$$

$$R^1H_2C-CH\begin{matrix}OR\\H\end{matrix} \xrightarrow{Cl_2}$$

But in the reaction of iodide with the carboxonium ion *(176)* derived from camphor dealkylation takes place rather than formation of the α-iodoether (Table 9). Many α-haloethers react by route A, too, when the temperature is raised. This is made possible by cleavage to the carboxonium halide [76]; e.g.:

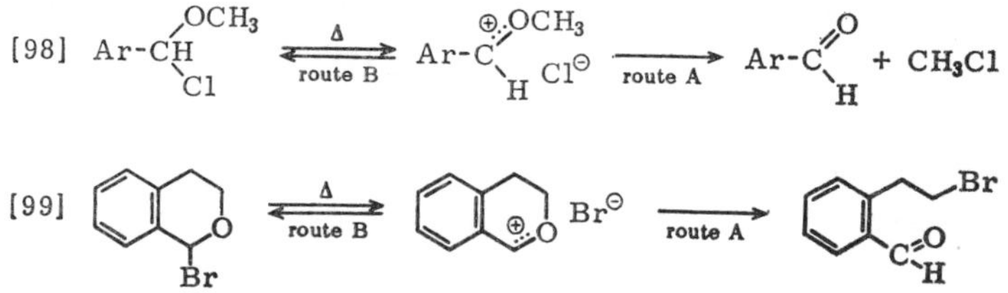

[98] $Ar-CH\begin{matrix}OCH_3\\Cl\end{matrix} \underset{\text{route B}}{\overset{\Delta}{\rightleftarrows}} Ar-\overset{\oplus}{C}\begin{matrix}OCH_3\\H\end{matrix}Cl^{\ominus} \xrightarrow{\text{route A}} Ar-C\begin{matrix}O\\H\end{matrix} + CH_3Cl$

[99] (isochromane-Br) $\underset{\text{route B}}{\overset{\Delta}{\rightleftarrows}}$ (isochromanyl cation) $Br^{\ominus} \xrightarrow{\text{route A}}$ (ortho-CH₂CH₂Br benzaldehyde)

Low-Energy Carboxonium Ions

Although it may be expected that *tri*alkoxycarbonium ions (e.g. *(51)*, Table 9) have lower energy than *di*alkoxycarbonium ions, this is not detectable in their reactive behavior with respect to such a strong nucleophile as ethoxide; with *(51)*, ethoxide yields orthocarbonic ester in a kinetically controlled reaction [30]. On the other hand, the action of triethylphosphine on *(51)* gives thermodynamically stable products with dealkylation [43]:

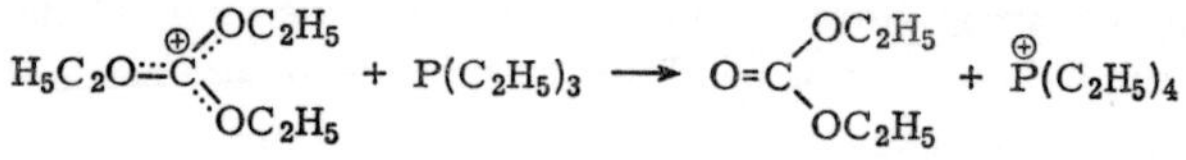

Reactions of Oxonium Salts

$$H_5C_2O\overset{\oplus}{\cdots}C\begin{smallmatrix}OC_2H_5\\OC_2H_5\end{smallmatrix} + P(C_2H_5)_3 \longrightarrow O=C\begin{smallmatrix}OC_2H_5\\OC_2H_5\end{smallmatrix} + \overset{\oplus}{P}(C_2H_5)_4$$

(51)

The reaction of triethylphosphine with *mono-* and *di*alkoxycarbonium ions, however, proceeds with kinetical control to yield addition products (*cf.* Table 9) [43].

No dialkoxycarbonium ions have yet been prepared in which the energy is lowered to such an extent that the usual kinetically controlled addition reactions with the strong nucleophiles (alkoxide, cyanide) do not take place. Even the 2-(2′,6′-dimethoxyphenyl)-1,3-dioxolenium ion *(184)*, for which a substantially lower energy was expected than for the simple phenyl derivative *(183)*, reacts with alkoxide to give the ortho ester [100]

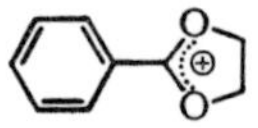

(183) 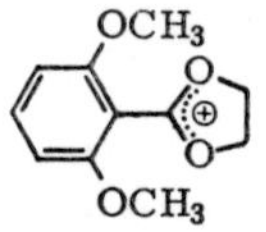   *(184)*

However, changes in the substitution of the phenyl nucleus of 2-aryl-1,3-dioxolenium ions *(183)* have only a slight effect on delocalization of the positive charge; *cf.* Section 3.1.2. In the case of *(184)*, moreover, the steric requirements due to the *o,o′*-substituents probably prevent more pronounced delocalization.

An example of a cation with a very low energy and with a structure similar to *(183)* is given by the tricyclic imidic ester cation *(187)* studied by Hünig [71]. Normally, such imidic ester cations—*(185)*, *(186)*—react with alkoxides to form amide acetals [30, 39]:

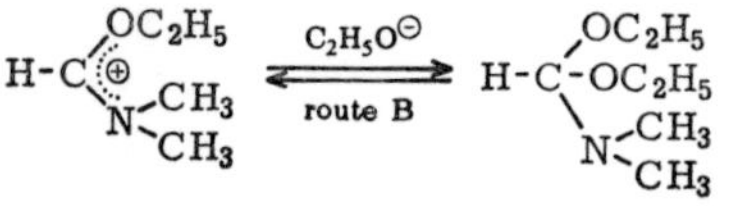

(185) 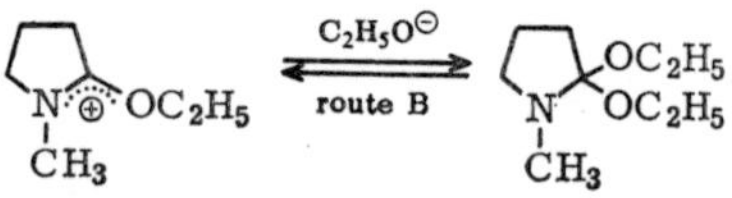   *(186)*

In the case of *(187)*, however, the reaction with ethoxide takes place in the sense of the thermodynamically stable end products with ring opening [71]:

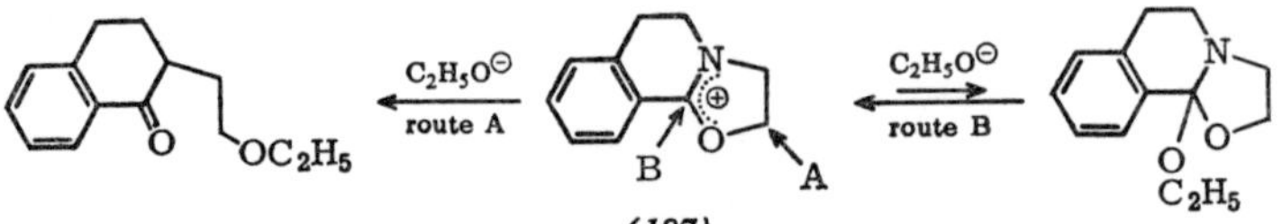

(187)

5.3.3.2. Steric Effects

In the reaction of carboxonium ions with nucleophiles, we have ignored for the present the possible influence of steric factors. If, however, the substituent attached to the oxonium center of a carboxonium ion is somewhat inaccessible to nucleophilic attack because of the steric situation at the cation, the reaction does not take place by route A but by route B, even with weak nucleophiles. The successful synthesis of the dineopentyloxymethyl chloride already mentioned is based on a steric hindrance of this type [79c].

The situation is exactly the reverse when a carbonium center is somewhat inaccessible to nucleophilic attack. Then, even with relatively strong nucleophiles, only thermodynamically stable products can be isolated.

The dimethoxy(phenyl)carbonium ion *(188)* reacts with catechol to form 2-phenylbenzodioxolium ion *(189)* [102], which is converted by water into catechol monobenzoate *(190)*. The reaction is initiated by the addition of the catechol to the carbonium center of *(188)* by route B [96, 101]:

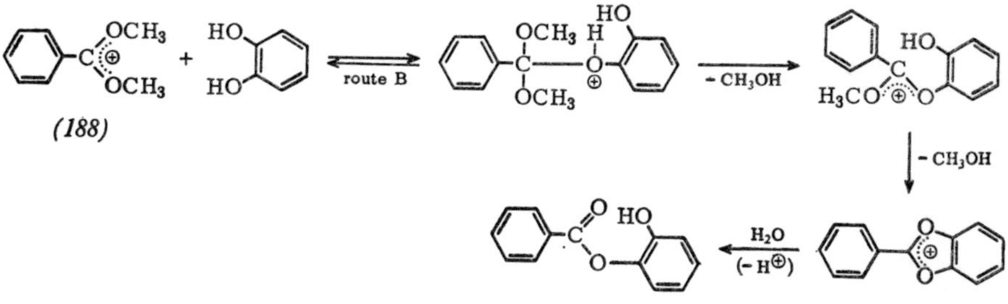

With the analogous dimethoxy(2,4,6-trimethylphenyl)carbonium ion, in the same reaction, the catechol monoester corresponding to *(190)* is not obtained, but instead the alkylation product of catechol, guaiacol [96, 101], is produced:

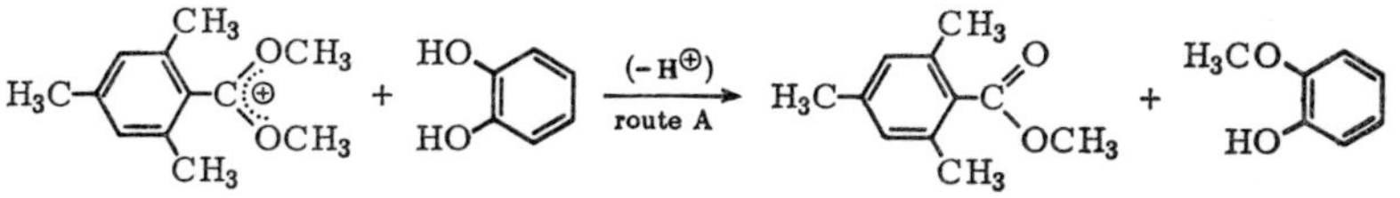

However, the sterically inhibited dialkoxycarbonium ion reacts with the smaller methoxide ion to give the ortho ester in a kinetically controlled reaction. In the cation *(57)* even the attack of the methoxide ion on the carbonium center is hindered so that only elimination products *(56)*, *(191)*, and *(192)*, are obtained [19]. On the other hand, the analogous cation *(177)* yields the expected cyclic acetal by route B [94a,b].

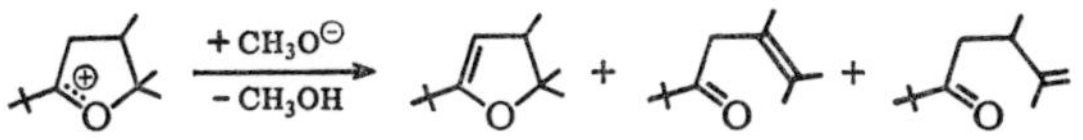

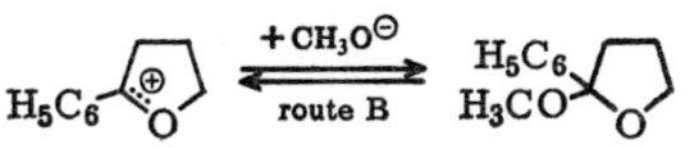

References

1. G. A. Olah, J. Sommer, and E. Namanworth, *J. Am. Chem. Soc.* **89**, 3576 (1967).
2. a) G. A. Olah and D. H. O'Brien, *J. Am. Chem. Soc.* **89**, 1725 (1967); b) F. Klages, H. Meuresch, and W. Steppich, *Ann. Chem. Liebigs* **592**, 81 (1955).
3. G. A. Olah, D. H. O'Brien, and M. Calin, *J. Am. Chem. Soc.* **89**, 3582 (1967).
4. F. Klages, H. Träger, and E. Mühlbauer, *Chem. Ber.* **92**, 1819 (1959).
5. a) G. A. Olah, M. Calin, and D. H. O'Brien, *J. Am. Chem. Soc.* **89**, 3586 (1967); b) G. A. Olah and M. Calin, *J. Am. Chem. Soc.* **90**, 938 (1968); c) M. Brookhart, G. C. Levy, and S. Winstein, *J. Am. Chem. Soc.* **89**, 1735 (1967).
6. a) F. Klages and E. Zange, *Chem. Ber.* **92**, 1828 (1959); b) G. A. Olah and A. M. White, *J. Am. Chem. Soc.* **89**, 3591 (1967).
7. G. A. Olah, D. H. O'Brien, and A. M. White, *J. Am. Chem. Soc.* **89**, 5694 (1967).
8. a) G. A. Olah and A. M. White, *J. Am. Chem. Soc.* **90**, 1884 (1968); b) B. G. Ramsey and R. W. Taft, *J. Am. Chem. Soc.* **88**, 3058 (1966).
9. a) O. Wallach and W. Brass, *Ann. Chem. Liebigs* **225**, 298 (1884); b) A. Baeyer and V. Villiger, *Ber. Deut. Chem. Ges.* **34**, 2694 (1901); c) D. Vorländer and E. Mumme, *Ber. Deut. Chem. Ges.* **36**, 1481 (1903); d) P. Pfeiffer, *Ann. Chem. Liebigs* **412**, 327 (1917).
10. a) E. M. Arnett, *Progr. Phys. Org. Chem.* **1**, 285, 351 (1963); b) S. Searles, Jr., and M. Tamres, in "The Chemistry of the Ether Linkage" (S. Patai, ed.), p. 260. Wiley (Interscience), New York, 1967; c) G. C. Levy, *Chem. Commun.* p. 1257 (1969); G. C. Levy and J. D. Cargioli, *Tetrahedron Letters* p. 919 (1970).
11. a) G. A. Olah and J. Sommer, *J. Am. Chem. Soc.* **90**, 927 (1968); b) *J. Am. Chem. Soc.* **90**, 4323 (1968).
12. a) R. J. Gillespie and J. A. Leisten, *Quart. Rev. (London)* **8**, 40 (1954); b) A. Wiles, *J. Chem. Soc.* p. 996 (1953).
13. N. C. Deno, C. U. Pittman, Jr., and M. J. Wisotsky, *J. Am. Chem. Soc.* **86**, 4370 (1964).
14. H. Hogeveen, *Rec. Trav. Chim.* **86**, 289 (1967).
15. G. A. Olah and M. Calin, *J. Am. Chem. Soc.* **90**, 405 (1968).
16. G. A. Olah and M. B. Comisarow, *J. Am. Chem. Soc.* **88**, 3313 (1966).
17. G. A. Olah and A. M. White, *J. Am. Chem. Soc.* **89**, 4752 (1967).
18. S. Daniloff and E. Venus-Danilova, *Ber. Deut. Chem. Ges.* **59**, 377 (1926).
19. K. Dimroth and W. Mach, *Angew. Chem.* **80**, 490 (1968); *Angew. Chem. Intern. Ed. English* **7**, 461 (1968); see also W. Rundel and K. Besserer, *Tetrahedron Letters* p. 4333 (1968).
20. a) H. Meerwein, in "Methoden der organischen Chemie (Houben-Weyl)" (E. Müller, ed.), 4. Auflage, Bd. VI/3: Sauerstoffverbindungen I, p. 359. Thieme, Stuttgart, 1965; b) H. Meerwein, see Reference 20a, p. 147.
21. H. Meerwein, G. Hinz, P. Hofmann, E. Kroning, and E. Pfeil, *J. Prakt. Chem.* **147**, 257 (1937).
22. H. Meerwein, E. Battenberg, H. Gold, E. Pfeil, and G. Willfang, *J. Prakt. Chem.* **154**, 83 (1939).
23. T. A. Mastryukova, A. E. Shipov, V. V. Abalyaeva, E. E. Kugucheva, and M. I. Kabachnik, *Dokl. Akad. Nauk SSSR* **164**, 340 (1965); *Chem. Abstr.* **63**, 17890 (1965); see also H. O. House and B. M. Trost, *J. Org. Chem.* **30**, 2502 (1965); G. J. Heiszwolf and H. Kloosterziel, *Chem. Commun.* p. 51 (1966).
24. N. Kornblum and R. A. Brown, *J. Am. Chem. Soc.* **85**, 1359 (1963); see also L. G. Donamura, *J. Org. Chem.* **22**, 1026 (1957).
25. N. Kornblum and G. P. Coffey, *J. Org. Chem.* **31**, 3447 (1966).
26. T. A. Mastryukova, A. E. Shipov, V. V. Abalyaeva, E. M. Popov, and M. I. Kabachnik, *Dokl. Akad. Nauk SSSR* **158**, 1372 (1964); *Chem. Abstr.* **62**, 2700 (1965).
27. a) H. Meerwein, P. Laasch, R. Mersch, and J. Spille, *Chem. Ber.* **89**, 209 (1956); b)

H. Meerwein, P. Laasch, R. Mersch, and J. Nentwig, *Chem. Ber.* **89**, 224 (1956); c) R. F. Borch, *Chem. Commun.* p. 442 (1968).
28. H. Meerwein, *Angew. Chem.* **67**, 374 (1955).
29. H. Meerwein, K. Bodenbenner, P. Borner, F. Kunert, and K. Wunderlich, *Ann. Chem. Liebigs* **632**, 38 (1960).
30. H. Meerwein, P. Borner, O. Fuchs, H.-J. Sasse, H. Schrodt, and J. Spille, *Chem. Ber.* **89**, 2060 (1956).
31. a) G. K. Helmkamp, H. N. Cassey, B. A. Olsen, and D. J. Pettitt, *J. Org. Chem.* **30**, 933 (1965); G. K. Helmkamp, B. A. Olsen, and D. J. Pettitt, *J. Org. Chem.* **30**, 676 (1965); b) D. J. Pettitt and G. K. Helmkamp, *J. Org. Chem.* **28**, 2932 (1963); **29**, 2702 (1964).
32. a) C. P. Lillya and P. Miller, *J. Am. Chem. Soc.* **88**, 1559 (1966); b) H. Meerwein, K.-F. Zenner, and R. Gipp, *Ann. Chem. Liebigs* **688**, 67 (1965).
33. S. Hünig and J. Utermann, *Chem. Ber.* **88**, 1485 (1955).
34. C. Reichardt, *Chem. Ber.* **99**, 1769 (1966).
35. a) K. Hafner, H. Riedel, and M. Danielisz, *Angew. Chem.* **75**, 344 (1963); b) C. E. Forbes and R. H. Holm, *J. Am. Chem. Soc.* **90**, 6884 (1968).
36. R. Breslow, T. Eicher, A. Krebs, R. A. Peterson, and J. Posner, *J. Am. Chem. Soc.* **87**, 1320 (1965).
37. a) S. Hünig, L. Geldern, and E. Lücke, *Angew. Chem.* **75**, 476 (1963); b) A. Schmidpeter, *Angew. Chem.* **75**, 1111 (1963).
38. C. Bouillon and J. Vialle, *Bull. Soc. Chim. France* p. 4560 (1968).
39. H. Meerwein, W. Florian, N. Schön, and G. Stopp, *Ann. Chem. Liebigs* **641**, 1 (1961).
40. a) L. Weintraub, S. R. Oles, and N. Kalish, *J. Org. Chem.* **33**, 1679 (1968); b) R. F. Borch, *Tetrahedron Letters* p. 61 (1968); c) U. Schöllkopf and F. Gerhart, *Angew. Chem.* **79**, 990 (1967); *Angew. Chem. Intern. Ed. English* **6**, 970 (1967); d) French Patent 1,381,790 (Cl. C 08g D 06mp), Dec. 11, 1964, CIBA Ltd.; *Chem. Abstr.* **63**, 728 (1965).
41. S. G. McGeachin, *Can. J. Chem.* **46**, 1903 (1968).
42. K. Bott, *Angew. Chem.* **76**, 992 (1964); *Angew. Chem. Intern. Ed. English* **3**, 804 (1964); *Tetrahedron* **22**, 1251 (1966).
43. L. Horner and B. Nippe, *Chem. Ber.* **91**, 67 (1958).
44. a) A. Rhomberg and P. Tavs, *Monatsh. Chem.* **98**, 105 (1967); b) L. V. Nesterov and R. I. Mutalapova, *Tetrahedron Letters* p. 51 (1968).
45. D. R. Brown, J. McKenna, and J. M. McKenna, *Chem. Commun.* p. 186 (1969).
46. K. Dimroth and A. Nürrenbach, *Chem. Ber.* **93**, 1649 (1960).
47. a) S. Petersen and E. Tietze, *Ann. Chem. Liebigs* **623**, 133 (1959); b) L. A. Paquette, *J. Am. Chem. Soc.* **86**, 4096 (1964); c) L. A. Paquette and T. J. Barton, *J. Am. Chem. Soc.* **89**, 5480 (1967); P. Wegener, *Tetrahedron Letters* p. 4985 (1967); d) T. Oishi, M. Ochiai, M. Nagai, and Y. Ban, *Tetrahedron Letters* p. 497 (1968); e) L. A. Paquette and T. Kakihana, *J. Am. Chem. Soc.* **90**, 3897 (1968); f) see also E. Bertele, H. Boos, J. D. Dunitz, F. Elsinger, A. Eschenmoser, I. Felner, H. P. Gribi, H. Gschwend, E. F. Meyer, M. Pesaro, and R. Scheffold, *Angew. Chem.* **76**, 393 (1964); *Angew. Chem. Intern. Ed. English* **3**, 490 (1964).
48. Japanese Patent 9,420 ('65), Dec. 27, 1965; Konishiroku Photo Ind. Co., Ltd.; *Chem. Abstr.* **63**, 18321 (1965).
49. a) H. Balli and F. Kersting, *Ann. Chem. Liebigs* **647**, 1 (1961); G. F. Duffin, *Advan. Heterocyclic Chem.* **3**, 9 (1964); H. Balli, *Textilveredlung* **4**, 37 (1969); b) see also N. Greif, Ph.D. Thesis, University of Marburg, 1967.
50. T. J. Curphey, *J. Am. Chem. Soc.* **87**, 2063 (1965).
51. a) D. S. Kemp and R. B. Woodward, *Tetrahedron* **21**, 3019 (1965); R. B. Woodward and R. A. Olofson, *Tetrahedron Suppl.* **7**, 415 (1966); b) B. D. Wilson and D. M. Burness, *J. Org. Chem.* **31**, 1565 (1966).
52. G. C. Brumlick, H. J. Kosak, and R. Pitcher, *J. Am. Chem. Soc.* **86**, 5360 (1964).

53. a) K. Hafner, A. Stephan, and C. Bernhard, *Ann. Chem. Liebigs* **650**, 42 (1961); b) T. Oishi, *Yuki Gosei Kagaku Kyokai Shi* **27**, 889 (1969); *Chem. Abstr.* **72**, 11708z (1970).
54. W. Pritzkow and G. Pohl, *J. Prakt. Chem.* **20**, 132 (1963); see also N. Wiberg and K. H. Schmid, *Angew. Chem.* **76**, 381 (1964); *Angew. Chem. Intern. Ed. English* **3**, 444 (1964).
55. D. Martin and A. Weise, *Chem. Ber.* **100**, 3736 (1967).
56. E. Allred and S. Winstein, *J. Am. Chem. Soc.* **89**, 4012 (1967).
57. E. Allred and S. Winstein, *J. Am. Chem. Soc.* **89**, 3991 (1967).
58. a) A. N. Nesmeyanov and T. P. Tolstaya, *Dokl. Akad. Nauk SSSR* **117**, 626 (1957); *Chem. Abstr.* **52**, 9005 (1958); b) A. N. Nesmeyanov, L. G. Makarova, and T. P. Tolstaya, *Tetrahedron* **1**, 145 (1957).
59. Y. K. Syrkin, *Zh. Strukt. Khim.* **1**, 389 (1960); *Chem. Abstr.* **55**, 26808 (1961).
60. A. N. Nesmeyanov, *Bull. Soc. Chim. France* p. 897 (1965).
61. A. N. Nesmeyanov, "Selected Works in Organic Chemistry", p. 772. Macmillan (Pergamon), New York, 1963.
62. A. N. Nesmeyanov, T. P. Tolstaya, L. S. Isaeva, and A. V. Grib, *Dokl. Akad. Nauk SSSR* **133**, 602 (1960); *Chem. Abstr.* **54**, 24529 (1960).
63. A. N. Nesmeyanov, T. P. Tolstaya, and A. V. Grib, *Dokl. Akad. Nauk SSSR* **139**, 114 (1961); *Chem. Abstr.* **56**, 1374 (1962).
64. A. N. Nesmeyanov and T. P. Tolstaya, *Izv. Akad. Nauk SSSR, Otd. Khim. Nauk* p. 647 (1959); *Chem. Abstr.* **53**, 21796 (1959).
65. F. Klages, E. Mühlbauer, and G. Lukaszyk, *Chem. Ber.* **94**, 1469 (1961).
66. H. Meerwein and H. Maier-Hüser, *J. Prakt Chem.* **134**, 51 (1932).
67. a) R. L. Burwell, Jr., *Chem. Rev.* **54**, 637 (1954); b) S. Searles, Jr. and M. Tamres, see Reference 10b, p. 269; c) see also M. H. Karger and Y. Mazur, *J. Am. Chem. Soc.* **90**, 3878 (1968).
68. S. Coffi-Nketsia, A. Kergomard, and H. Tautou, *Bull. Soc. Chim. France* p. 2788 (1967).
69. H. Meerwein, D. Delfs, and H. Morschel, *Angew. Chem.* **72**, 927 (1960).
70. H. Meerwein, see Reference 20a, p. 344.
71. S. Hünig, *Angew. Chem.* **76**, 400 (1964); *Angew. Chem. Intern. Ed. English* **3**, 548 (1964).
72. K. Wunderlich, Ph.D. Thesis, University of Marburg, 1957.
73. K. Freudenberg and H. Scholz, *Ber. Deut. Chem. Ges.* **63**, 1969 (1930).
74. B. Helferich and H. Jochinke, *Ber. Deut. Chem. Ges.* **74**, 719 (1941).
75. K. Heyns, W.-P. Trautwein, F. Garrido Espinosa, and H. Paulsen, *Chem. Ber.* **99**, 1183 (1966).
76. H. Gross and E. Höft, *Z. Chem.* **4**, 417 (1964).
77. a) A. Rieche, E. Schmitz, and E. Beyer, *Chem. Ber.* **91**, 1935 (1958); b) L. A. Cort and R. G. Pearson, *J. Chem. Soc.* p. 1682 (1960); c) H. Baganz and L. Domaschke, *Chem. Ber.* **91**, 653 (1958).
78. a) S. M. McElvain, *Chem. Rev.* **45**, 453 (1949); b) S. M. McElvain and M. J. Curry, *J. Am. Chem. Soc.* **70**, 3781 (1948).
79. a) H. Gross, J. Freiberg, and B. Costisella, *Chem. Ber.* **101**, 1250 (1968); b) C. H. V. Dusseau, S. E. Schaafsma, H. Steinberg, and T. J. de Boer, *Tetrahedron Letters* p. 467 (1969); c) J. W. Scheeren, *Tetrahedron Letters* p. 5613 (1968).
80. C. B. Anderson, E. C. Friedrich, and S. Winstein, *Tetrahedron Letters* p. 2037 (1963).
81. A. Rieche, E. Schmitz, W. Schade, and E. Beyer, *Chem. Ber.* **94**, 2926 (1961).
82. a) S. M. McElvain, R. E. Kent, and C. L. Stevens, *J. Am. Chem. Soc.* **68**, 1922 (1946); b) S. M. McElvain and D. Kundiger, *J. Am. Chem. Soc.* **64**, 258 (1942); c) S. M. McElvain and A. N. Bolstad, *J. Am. Chem. Soc.* **73**, 1988 (1951).
83. H. Crompton and P. L. Vanderstichele, *J. Chem. Soc.* **117**, 691 (1920).
84. R. W. Hoffmann and J. Schneider, *Chem. Ber.* **100**, 3689 (1967).
85. H. Gross and J. Freiberg, *Chem. Ber.* **100**, 3777 (1967).

86. H. W. Post, *J. Org. Chem.* **1,** 231 (1937).
87. H. Böhme and R. Neidlein, *Chem. Ber.* **95,** 1859 (1962).
88. S. Kabuss, *Angew. Chem.* **78,** 714 (1966); *Angew. Chem. Intern. Ed. English* **5,** 675 (1966).
89. H. Meerwein, see Reference 20a, p. 341.
90. H.-J. Sasse, Ph.D. Thesis, University of Marburg, 1955.
91. M. J. Astle and J. D. Welks, *J. Org. Chem.* **26,** 4325 (1961).
92. a) H. Meerwein, see Reference 20a, p. 105; b) R. A. Olofson, S. W. Walinsky, J. P. Marino, and J. L. Jernow, *J. Am. Chem. Soc.* **90,** 6554 (1968).
93. H. Schrodt, Ph.D. Thesis, University of Marburg, 1953.
94. a) H. R. Ward and P. D. Sherman, Jr., *J. Am. Chem. Soc.* **89,** 4222 (1967); b) *J. Am. Chem. Soc.* **90,** 3812 (1968).
95. B. G. Ramsey and R. W. Taft, *J. Am. Chem. Soc.* **88,** 3058 (1966).
96. C. P. Heinrich, Ph.D. Thesis, University of Marburg, 1966.
97. H. Meerwein, see Reference 20a, pp. 123–131.
98. F. Straus and H. J. Weber, *Ann. Chem. Liebigs* **498,** 101 (1932).
99. A. Rieche and E. Schmitz, Chem. Ber. **89,** 1254 (1956); E. Schmitz, *Chem. Ber.* **91,** 1133 (1958).
100. F. M. Beringer and S. A. Galton, *J. Org. Chem.* **32,** 2630 (1967).
101. K. Dimroth and P. Heinrich, *Angew. Chem.* **78,** 714 (1966); *Angew. Chem. Intern. Ed. English* **5,** 676 (1966).
102. K. Dimroth, P. Heinrich, and K. Schromm, *Angew. Chem.* **77,** 863 (1965); *Angew. Chem. Intern. Ed. English* **4,** 873 (1965).
103. N. Kornblum, R. Smiley, R. K. Blackwood, and D. C. Iffland, *J. Am. Chem. Soc.* **77,** 6269 (1955).
104. C. K. Ingold, "Structure and Mechanism in Organic Chemistry", p. 754. Cornell University Press, Ithaca, 1953.
105. For a recent review see G. A. Olah, A. M. White, and D. H. O'Brien, *Chem. Rev.* **70,** 561 (1970).

6. Tertiary Oxonium Ions as Intermediates in Chemical Reactions

Tertiary oxonium ions play an important role as reactive intermediates in many organic reactions. In Chapter 5 we came across numerous examples of this, as in reactions of acetals, ketals, or orthocarboxylic esters with protonic or Lewis acids. Here we shall deal particularly with the occurrence of tertiary oxonium ions in the neighboring-group participation of oxygen functions.

In the solvolysis of ω-methoxy-*n*-alkyl-*p*-bromobenzenesulfonates *(193)*,* a considerable increase in the rate of solvolysis is observed for the butyl and pentyl derivatives as compared with the lower and higher homologs [1a], as shown in the accompanying tabulation. A similar statement applies to ω-alkoxyalkyl halides [2].

Compounds *(193)** $H_3CO-(CH_2)_n-OBs$	Relative rate of solvolysis [referred to $H_3C-(CH_2)_3-OBs = 1.00$]	
n	C_2H_5OH (25 °C)	CH_3CO_2H (75 °C)
2	0.25	0.28
3	0.67	0.63
4	20.4	657.0
5	2.8	123.0
6	1.19	1.16

In the solvolysis of the homologous ω-chloroketones *(194)* with silver perchlorate in 80% aqueous ethanol, a maximum in the reaction rate is again found at 4-chlorobutyrophenone ($n = 3$) and another clear increase in the rate for 5-chlorovalerophenone ($n = 4$) [3], as shown in the following table. The corresponding bromo ketones behave similarly [2].

Compound *(194)* $H_5C_6-CO-(CH_2)_n-Cl$	Relative rate of solvolysis [referred to $H_3C-(CH_2)_3-Cl = 1.00$] (56.2 °C)
n	
1	1.3
2	7.9
3	759.0
4	21.3
5	2.7

Winstein and Buckles [4] have given an explanation of these results on the basis of "neighboring-group participation" in nucleophilic substitutions. According to

---

* In what follows, the usual abbreviation "brosylate" for the *p*-bromobenzenesulfonate group will be used; in formulas this residue will be denoted by OBs.

this, a nucleophilic neighboring group, provided it is in a sterically favorable position in the same molecule, can attack at the reaction center from the rear and thus assist the elimination of a leaving group. The cyclic reactive intermediates formed are then accessible to the attack of nucleophiles (e.g., the solvent) [5a,b].

Oxygen functions are particularly important as neighboring groups of this type. The high rate of solvolysis of 4-methoxybutyl brosylate *(193)*, $n=4$, should be ascribed, according to this explanation, to the fact that the brosylate group is not displaced by a normal S_N2 mechanism. Rather, the reactive cyclic trialkyloxonium ion *(195)* is formed, and this undergoes ring cleavage under the action of nucleophiles [1a,b,c,d]:

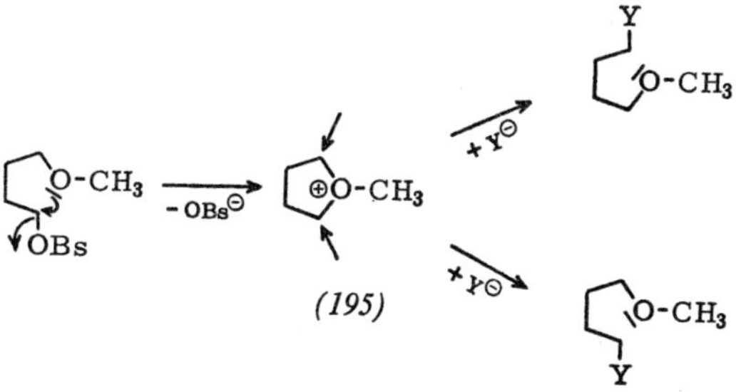

In the other cases mentioned with an increased rate of solvolysis, the six-membered cyclic trialkyloxonium ion *(196)* is formulated correspondingly, and so are the cyclic carboxonium ions *(177)* and *(197)* with the neighboring-group participation of the benzoyl group [3, 6a,b]:

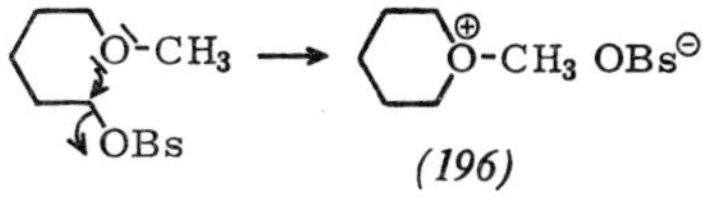

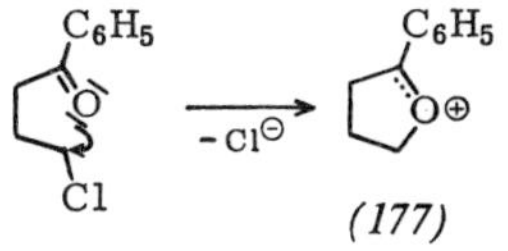

It is striking here that the neighboring-group participation in the case of primary alkoxy or benzoyl groups is obviously effective only if five- or six-membered rings arise, but not if three-, four-, or seven-membered heterocycles have to be formed; i.e., the same ring systems are favored as are preparatively available by intramolecular alkylation (see Section 4.1.1.3).

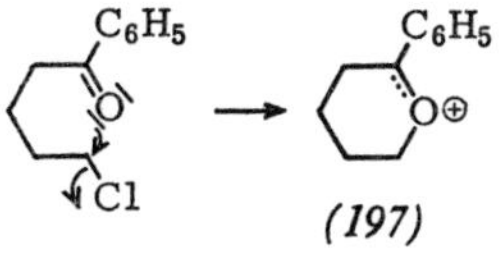

Apart from the sometimes surprisingly high increase in reaction rates with nucleophilic substitutions, neighboring-group participation can also be recognized on the basis of the stereochemistry of the reaction products or from the analysis of the reaction mixture, as will be shown in more detail below.

6.1. Neighboring-Group Participation of Ether Groups

In the reaction of the optically active brosylate *(198)* with lithium chloride in pyridine and in the action of brosyl chloride on 2-hydroxy-5-methoxypentane the stereochemistry of the reactions has been determined [1d]. The formation of a tetrahydrofuranium ion *(199)* from the brosylate *(198)* via methoxy neighboring-group participation must take place with inversion at carbon atom 2. With the intermediate occurrence of *(199)*, therefore, two compounds are to be expected in which C_2 exhibits inversion in relation to the initial product *(198)*: (*a*) by demethylation: 2-methyltetrahydrofuran; (*b*) by ring opening at C_5: 2-methoxy-5-pentyl chloride (4-methoxypentyl chloride). The 5-methoxy-2-pentyl chloride to be expected with ring opening at C_2, (*c*), should arise from *(199)* by a second inversion at C_2 and should therefore, with respect to *(198)*, exhibit retention at C_2, as shown in the following scheme

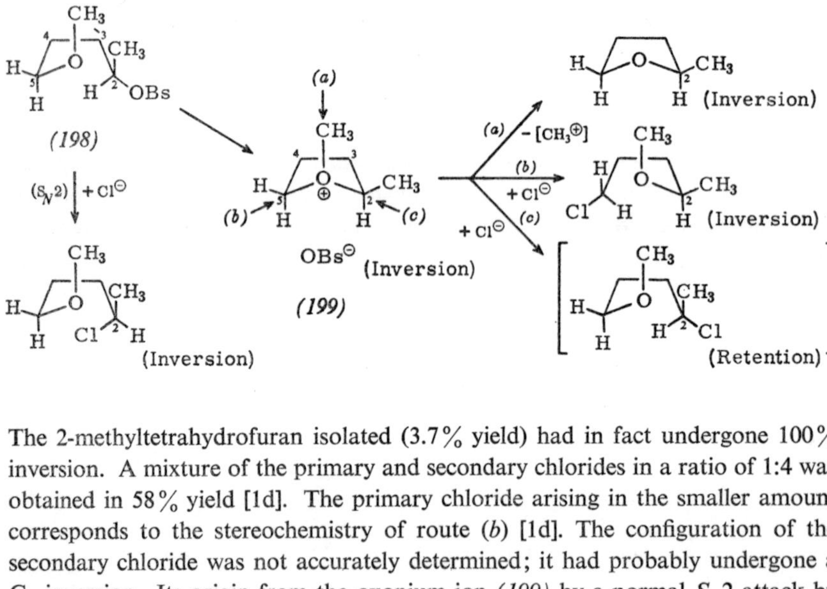

The 2-methyltetrahydrofuran isolated (3.7% yield) had in fact undergone 100% inversion. A mixture of the primary and secondary chlorides in a ratio of 1:4 was obtained in 58% yield [1d]. The primary chloride arising in the smaller amount corresponds to the stereochemistry of route (*b*) [1d]. The configuration of the secondary chloride was not accurately determined; it had probably undergone a C_2 inversion. Its origin from the oxonium ion *(199)* by a normal S_N2 attack by route (*c*) would then be impossible. It is true that nucleophilic exchange at *(199)* by chloride attack from the front side is fundamentally conceivable, but direct formation by S_N2 attack on C_2 of the initial compound *(198)* should rather be assumed [1d]. 2-Methyltetrahydrofuran and 4-methoxypentyl chloride can, on the other hand, be derived unambiguously from *(199)*.

The intermediate formation of *(199)* is also assumed in the acetolysis of the isomeric brosylates *(198)* and *(200)* in order to explain why the same isomeric mixture of primary and secondary acetates (ratio 2:3) is obtained in both cases [1a,b]:

Neighboring-Group Participation of Ether Groups

[Reaction scheme showing tetrahydrofuran derivative losing CH_3^+ to give intermediate, with pathways to (200), (199), (198), and with $+CH_3CO_2H$, $-H^+$ leading to products]

In this case the demethylation of the tetrahydrofuranium ion *(199)* to 2-methyltetrahydrofuran can hardly be detected experimentally. However, such a demethylation is important in the acetolysis of 5-methoxypentyl brosylate *(201)*, in which the tetrahydropyranium ion *(196)* is formed as an intermediate [1c]. Here, in contrast to *(199)*, demethylation takes place preferentially:

[Reaction scheme: (201) → (196) with $OBs^⊖$; with CH_3CO_2H, branching (28%), (54%), (16%) to three products]

Ion pairs from *(196)* and the brosylate anion, which are also important as well as separate ions, are responsible for the formation of methyl brosylate [1c].

The surprising difference between tetrahydrofuranium ions (e.g. *(199)*) and tetrahydropyranium ions *(196)* is explained by the different stereochemistries of five-membered and six-membered cyclic oxonium ions already mentioned [1c]. In the 5-ring the O-methylene carbon atoms are more readily accessible for nucleophilic attack than in the 6-ring; cf. Section 5.2.1.

In the action of iron(III) halides on ω-methoxybutyl or ω-methoxypentyl halides, cyclic ethers are obtained on heating [7a,b,c]. Cyclic oxonium ions are passed through as intermediates. In the case of the five-membered cyclic oxonium ion, demethylation is the main reaction; e.g.:

Tertiary Oxonium Ions as Intermediates in Chemical Reactions

$$Cl\text{-}(CH_2)_n\text{-}OCH_3 + FeCl_3 \rightleftharpoons (H_2C)_n \overset{\oplus}{O}\text{-}CH_3 \longrightarrow (H_2C)_n O + CH_3Cl + FeCl_3$$
$$n = 4, 5 \qquad\qquad\qquad FeCl_4^{\ominus}$$

where $n = 4, 5$. 2-Alkyl-substituted derivatives of tetrahydrofuran and tetrahydropyran [7b], including the 2-methyltetrahydrofuran [7c] mentioned above, are also accessible in this way. Small amounts of ω-unsaturated alkyl methyl ethers are often formed as by-products [7b].

The oxonium salt *(195)*, which has been isolated as such [8a], also undergoes thermal decomposition into tetrahydrofuran and methyl halide [8a]:

$$\underset{Br}{\overset{\displaystyle\frown}{O}}\text{-}CH_3 + SbCl_5 \longrightarrow [\overset{\oplus}{O}\text{-}CH_3\ (SbCl_5Br)^{\ominus}] \xrightarrow[150\text{-}180\,°C]{\Delta} \overset{\displaystyle\frown}{O} + CH_3Br$$
$$\qquad\qquad\qquad\qquad (195) \qquad\qquad\qquad\qquad (CH_3Cl)$$
$$\qquad\qquad\qquad\qquad\qquad\qquad\qquad\qquad\qquad\qquad + SbCl_5$$
$$\qquad\qquad\qquad\qquad\qquad\qquad\qquad\qquad\qquad\qquad (SbCl_4Br)$$

In the presence of the halide acceptor $SbCl_5$ (or $FeCl_3$), the product of ring opening—the δ-halobutyl methyl ether—that is formed from *(195)* in competition with demethylation is obviously reconverted to the cyclic oxonium ion. On the other hand, cleavage to the cyclic ether is favored at higher temperatures, since the methyl halide formed simultaneously is volatile. Furthermore, Meerwein observed that the *intra*molecular alkylation of ω-halo ethers with Lewis acids (which proceeds as the reverse of the cleavage reactions) takes place rapidly, whereas the *inter*molecular alkylation of ethers with alkyl halides and Lewis acids is slow [8b].

In the case of alkoxy groups on primary C atoms, no neighboring-group participation with the formation of three-membered cyclic oxonium ions can be detected; it does, however, take place when the alkoxy group is attached to a secondary or tertiary C atom [9a,b]. Thus, in the hydrolysis of the tertiary methyl ether *(202)*, the secondary methyl ether *(202a)* is produced, and in the case of *(203)*, isobutyraldehyde *(203a)* is formed:

$$\underset{(202)}{\overset{H_3CO\diagdown\ H}{\underset{CH_3(Br)}{H_3C\text{-}C\text{---}C\text{-}CH_3}}} \xrightarrow{-Br^{\ominus}} \left[\overset{\overset{CH_3}{\overset{\oplus}{O}}}{\underset{H_3C}{H_3C}\diagup C\diagdown_{C\diagdown CH_3}^{\diagup H}}\right] \xrightarrow[-H^{\oplus}]{+H_2O} \underset{(202a)}{\overset{H_3C\quad OCH_3}{\underset{HO\quad H}{H_3C\text{-}C\text{---}C\text{-}CH_3}}}$$

$$\underset{(203)}{\overset{H_3CO\diagdown\ H}{\underset{CH_3(OBs)}{H_3C\text{-}C\text{---}C\text{-}H}}} \xrightarrow{-OBs^{\ominus}} \left[\overset{\overset{CH_3}{\overset{\oplus}{O}}}{\underset{H_3C}{H_3C}\diagup C\diagdown_{C\diagdown H}^{\diagup H}}\right] \longrightarrow \underset{H_3C}{\overset{H_3C}{\diagdown}}\overset{\oplus}{C}\text{-}\underset{H}{\overset{OCH_3}{C}}\text{-}H \longrightarrow H\text{-}\underset{CH_3}{\overset{CH_3}{C}}\text{-}\overset{\oplus\ OCH_3}{\underset{H}{C}}$$
$$\qquad\qquad\qquad\qquad\qquad\qquad\qquad\qquad\qquad\qquad\qquad\qquad\quad \xleftarrow[-CH_3OH,\,-H^{\oplus}]{+H_2O}$$
$$\qquad\qquad\qquad\qquad\qquad (203a)\ H\text{-}\underset{CH_3}{\overset{CH_3}{C}}\text{---}\overset{O}{\underset{H}{C}}$$

The preparative production of three-membered cyclic oxonium ions by intramolecular alkylation has not in fact so far been achieved, but an oxonium ion of this type has been obtained from ethylene oxide with alkyl halide–AgBF$_4$ (although only in solution) [10].

Neighboring-group participation through a four-membered oxonium ring has recently been formulated to explain the bicycloheptanone derivative *(205)* arising among other products in the acetolysis of the *endo*tosylate *(204)* [11]:

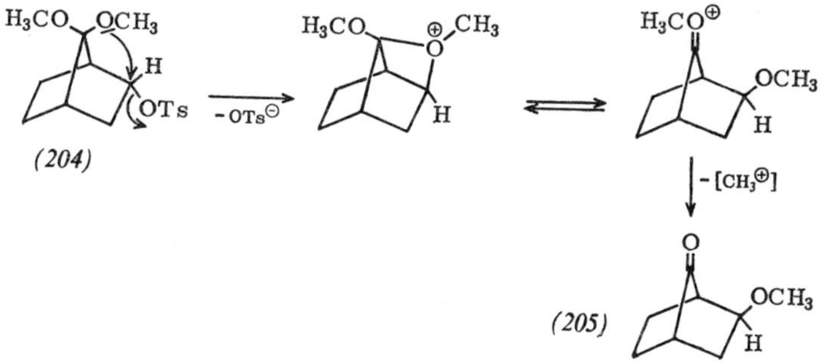

This assumption is supported by the fact that the corresponding *exo*tosylate *(206)* gives only solvolysis products in which the ketal group is preserved.

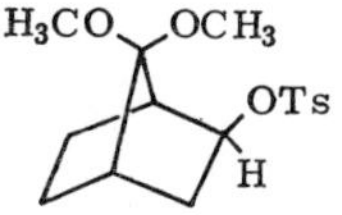

6.2. Participation of Keto Carbonyl Groups

The high rate of solvolysis of 4-halobutyrophenones and 5-halovalerophenones (see table on p. 100, compound *(194)*, $n = 3,4$), as already mentioned, gives rise to the assumption of an intermediate formation of the carboxonium ions *(177)* and *(197)* [2,3]. These two cations—which also arise in the solvolysis of the brosylates of 4-hydroxybutyrophenone and 5-hydroxyvalerophenone in trifluoroacetic acid— can be detected by NMR spectroscopy [6a,b]. The five-membered cyclic cation *(177)* has also been obtained by protonation of the corresponding vinyl ethers [6a,b]:

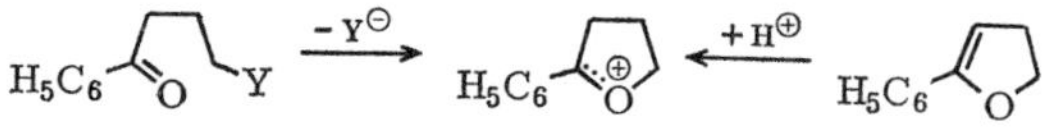

where Y = OBs, Cl.   *(177)*

Finally, the stable hexachloroantimonate of *(177)* has actually been obtained by the action of SbCl$_5$ on 4-chlorobutyrophenone [6b]. The methyl-substituted cation

(207) has also been prepared in solution, by analogy with *(177)*, from the corresponding brosylate and trifluoroacetic acid [6b]. Compound *(207)* may play a part as an intermediate in the action of copper(I) cyanide on 5-halo-2-pentanones [6c]:

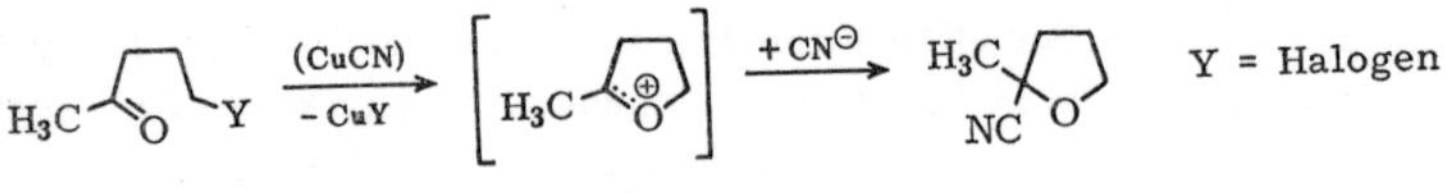

$$H_3C-CO-(CH_2)_3-Y \xrightarrow[-CuY]{(CuCN)} \left[\text{(207)} \right] \xrightarrow{+CN^\ominus} \underset{NC}{\overset{H_3C}{\diagup}}\!\!\diagdown\!\!\diagup\!\!O \qquad Y = \text{Halogen}$$

(207)

This reaction generally leads, with γ- or δ-haloketones, to good yields of 2-cyanotetrahydrofuran or tetrahydropyran derivatives [6c].

Analysis of the products of the anodic oxidation of acetylacetonate in the presence of alkenes also leads to the conclusion of the intermediate appearance of a cyclic ketonium ion *(208)* [6d]:

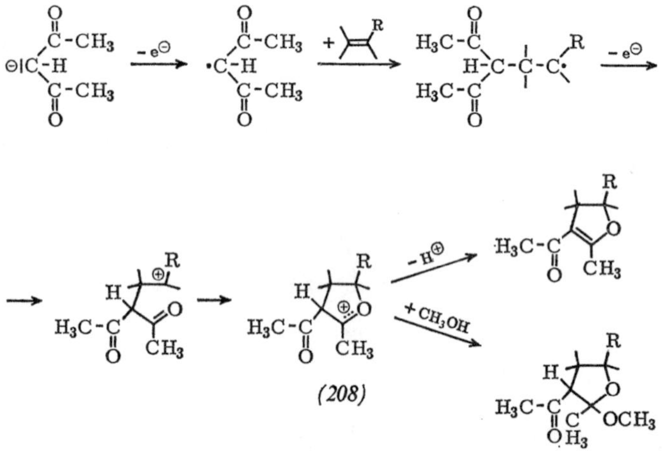

(208)

As a further example of keto neighboring-group participation we may mention the solvolysis of 10-acetyl-1α-bromo-*trans*-decalin [12], which probably takes place via the five-membered ring carboxonium ion *(209)*:

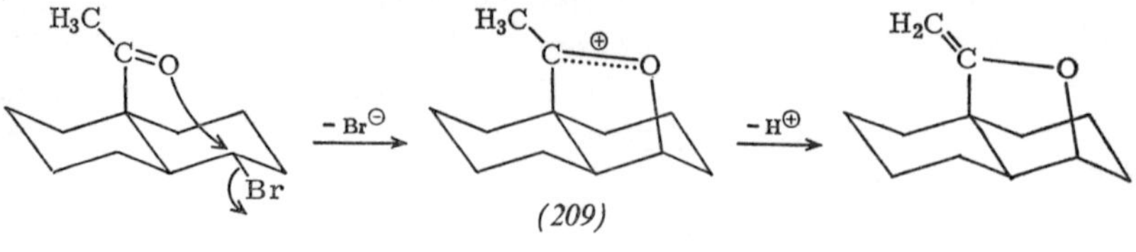

(209)

6.3. Participation of Ester Groups as Neighboring Groups

Three types of cyclic oxonium ions are to be expected by the neighboring-group participation of carboxylic ester groups:

(a) Nucleophilic attack of the carbonyl group at a carbon of the *acyl residue* leads to "lactonium ions" *(20)*:

$$\text{(20)}$$

(b) Correspondingly, nucleophilic attack of the alkoxy group at a carbon of the *acyl residue* leads to cyclic acyldialkyloxonium ions *(210)*:

$$\text{(210)}$$

(c) If the carbonyl group attacks a carbon of the *alkoxy group*, two oxygen atoms are included in the ring and 1,3-dioxacycloalkenium ions *(27)* are produced:

$$\text{(27)}$$

Here, again, five- and six-membered rings are formed preferentially.

6.3.1. Formation of Lactonium Ions

The solvolysis of homologous ethyl ω-bromocarboxylic esters *(211)* shows a striking rise in the reaction rate for ethyl-γ-bromobutyrate, $n=3$ [2].

In analogy with the results of the solvolysis of haloethers [2] or haloketones *(194)* [2,3], again a neighboring-group participation of an ester group may be assumed. It is particularly active in the formation of a heterocyclic five-membered ring. For the moment we leave open the question of whether the carbonyl or the alkoxy group assists solvolysis, i.e., whether the dialkoxycarbonium ion *(54)* or the cyclic acyldialkyloxonium ion *(212)* is formed as an intermediate.

$$\text{Br-(CH}_2)_n\text{-C}\begin{smallmatrix}\diagup\text{O}\\\diagdown\text{OC}_2\text{H}_5\end{smallmatrix}$$

(211)   *(54)*   *(212)*

Intermediates such as *(54)* or *(212)* should also occur in the *thermal* cleavage of γ-halocarboxylic esters to lactones and alkyl halides [13]. The mechanism of the reaction has been elucidated satisfactorily in one case. When the carbonyl group of the ester is labeled with ^{18}O, the labeling is found exclusively on the *ethereal* oxygen of the lactone [14]. This shows that the *carbonyl* group attacks nucleophilically, and therefore only the dialkoxycarbonium ion *(54)* and not *(212)* can be an intermediate:

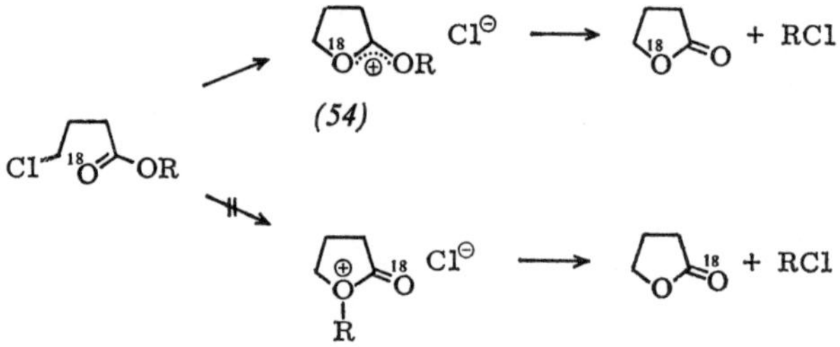

where R = cetyl. The following butenolide synthesis takes place analogously [15]:

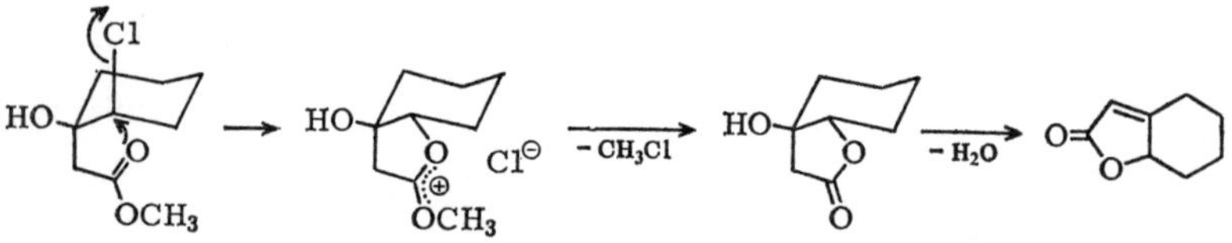

The preparative isolation of lactonium ions of type *(54)* in the form of their stable salts was achieved by Meerwein et al. [16a]; one of the reaction routes, the intramolecular alkylation of γ-halobutyric esters by means of $SbCl_5$ or $AgBF_4$ as halide acceptors, corresponds completely to the neighboring-group participations described here (see Table 5 in Chapter 3):

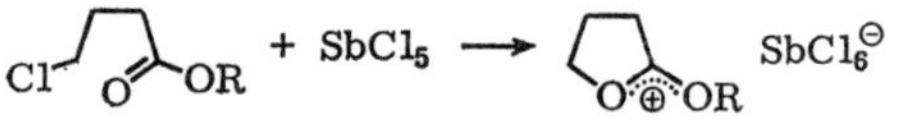

where R = C_2H_5. Structure *(54)* follows unambiguously from the fact that in the reaction with alkoxide, the cyclic ortho ester *(213)* is produced. With molar amounts of BF_3, this is reconverted (*cf.* Section 5.4.2) into the cation *(54)* [16a,b]:

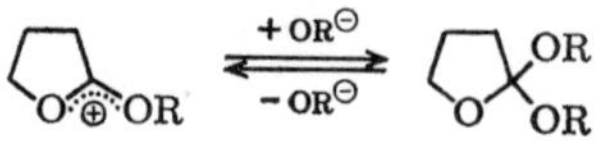

In the case of butyric esters, therefore, in all the cases known so far only the carbonyl group acts as the neighboring group. The cation *(54)*, more stable than *(212)*, arises in this way. In this connection, however, we shall also go into reactions in which the formation of acyldialkyloxonium ions *(212)* is to be expected, such as the action of $SbCl_5$ on γ-alkoxybutyryl chloride [16a,b]. However, here again it is *not (212)* that is obtained but only the dialkoxycarbonium ion *(54)* [16a,b]:

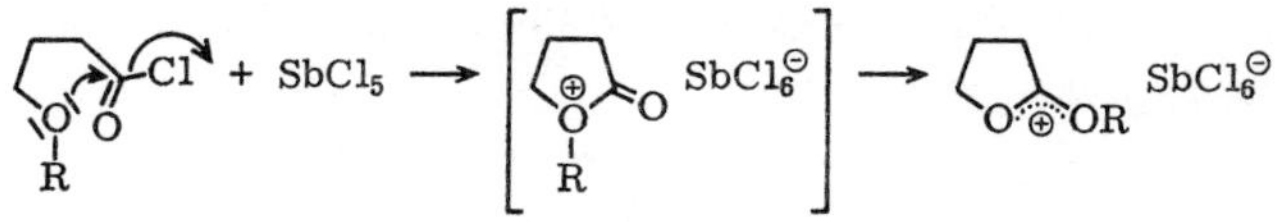

Participation of Ester Groups as Neighboring Groups 109

(212) (54)

No investigation has been made as to whether in this reaction rearrangement takes place with transfer of the alkyl residue R to the carbonyl oxygen or by ring opening and ring reclosure; for example:

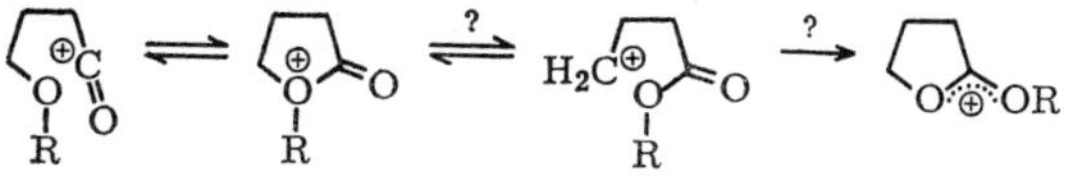

(212) (54)

The reaction to give *(54)* could, moreover, also precede the long-known rearrangement of γ-alkoxycarboxylic acid halides into γ-halocarboxylic esters [17a,b]. Here again the intermediate formation of an acyldialkyloxonium ion could be important. In the reaction of *cis*-3-methoxycyclohexanecarboxylic acid *(214)* with $SOCl_2$ to give, among other products, methyl-*trans*-3-chlorocyclohexanecarboxylate, for example, the formation of such an oxonium ion is necessary, particulary since the *trans*-carboxylic acid *(215)* yields only the corresponding acid chloride [18]:

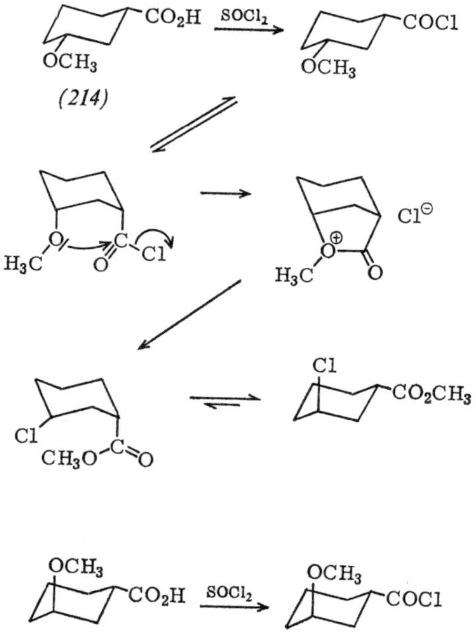

(214)

(215)

6.3.2. Formation of 1,3-Dioxacycloalkenium Ions

6.3.2.1. 1,3-Dioxolenium Ions [39]

Winstein and Buckles [4] found that the solvolysis of 2-acetoxy-3-bromobutanes in dry acetic acid with silver acetate takes place with the retention of the original configuration. The *threo* compound *(216)* gives the D, L-diacetate *(218)*, and the *erythro* compound *(219)* gives the *meso* diacetate *(221)*. In both cases the retention is almost quantitative. This result can be understood only by a neighboring-group participation of the ester carbonyl function with intermediate formation of 1,3-dioxolenium ion [4].

Under these conditions the *threo* compound *(216)* forms the symmetrical cyclic cation *(217)*. This is accessible to the nucleophilic attack of the solvent at both carbon atoms, C_2 and C_3:

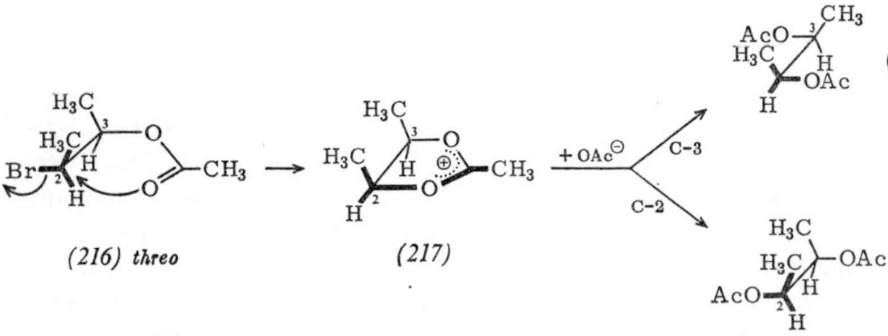

(216) threo *(217)* *(218b)* *(218a)*

D, L-Diacetate: *(a)* ≠ *(b)*

The reaction with acetate anions then leads, with attack at C_2, to the diacetate *(218a)*, which has the original configuration at both centers C_2 and C_3. On the other hand, the attack of the acetate at C_3 gives a diacetate *(218b)* that has undergone inversion at both centers C_2 and C_3 and is the enantiomer of *(218a)*. Like the initial compound *(216)*, compounds *(218a)* and *(218b)* have the configuration of *threo* compounds, but because of the occurrence of the symmetrical intermediate cation *(217)*, an optically pure bromoacetate *(216)* gives a racemic end product.

Quite analogously, in the case of the *erythro* compound *(219)*, acetolysis leads through the cyclic asymmetric cation *(220)* to a diacetate with the *erythro* configuration. Here *(221a)* and *(221b)* are identical. Consequently the *meso*-diacetate *(221)* is obtained.

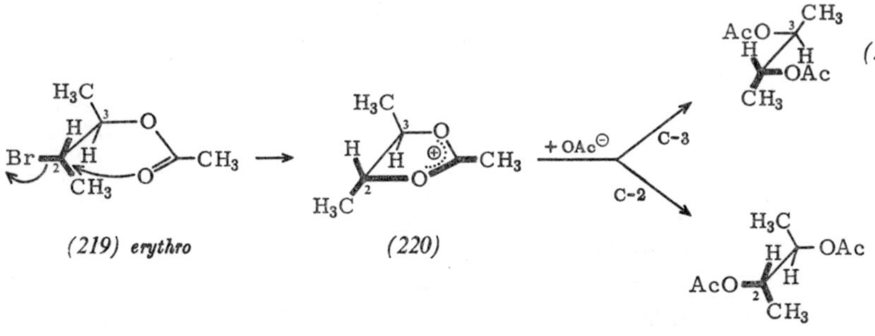

(219) erythro *(220)* *(221b)* *(221a)*

meso-Diacetate: *(a)* ≡ *(b)*

Participation of Ester Groups as Neighboring Groups   111

When 3-acetoxy-2-halo-2,3-dimethylbutanes *(222)* are dissolved in SbF_5–SO_2 or SbF_5–FSO_3H–SO_2 (at $-60°C$), the 1,3-dioxolenium ions can be detected directly by NMR spectroscopy [20]. The spectra show that under these conditions the halogen is active as a neighboring group in competition with the acetoxy group [4], giving halonium ions [20]. The ratio of the dioxolenium ion to the halonium ion depends on the nature of the halogen, decreasing with increasing atomic weight of the halogen [20]:

	Y = Cl	Br	I
dioxolenium	95%	30%	10%
halonium	5%	70%	90%

In this way Meerwein *et al.* achieved the preparative production of stable 1,3-dioxolenium salts [21]; cf. Table 7 in Chapter 3; e.g.:

(66)

Since the attack of the carbonyl oxygen takes place by an S_N2 reaction in the case of such neighboring-group participations, the acetate group and the leaving group must be oriented in the *anti* position with respect to one another; cf., for example, *(216)* and *(217)*. The necessary preorientation is shown particularly well with stereoisomers in the cyclohexane series: thus, *trans*-2-bromocyclohexyl acetate *(223a)* readily reacts with silver acetate in dry glacial acetic acid to give the *trans*-1,2-diacetate *(224)*. On the other hand, the *cis* compound does not react even under more vigorous conditions [4]. Compound *(224)* is also obtained by acetolysis of the *trans*-tosylate *(223b)* [19] or the *trans*-brosylate *(223c)* [22a] in glacial acetic acid and potassium acetate. Here the *trans*-brosylate *(223c)* is solvolyzed 10^3 times faster than its *cis* isomer [22a]. The acetolysis of *(223a–c)* takes place, as in the case of *(216)* and *(219)*, with retention of configuration. The optically active starting materials *(223a,b)* give the inactive diacetate *(224)* [4, 19]; cf. the reaction *(216)*–*(218)*:

(223a–c)   *(155)*   *(224)*

where in *(223a)*, Y = Br; in *(223b)*, Y = p-CH_3–C_6H_4–SO_2–O–; and in *(223c)*, Y = p-Br–C_6H_4–SO_2–O–.

Tertiary Oxonium Ions as Intermediates in Chemical Reactions

The course of the reaction through a 1,3-dioxolenium ion is further confirmed by the following labeling experiment: If in the acetolysis reaction a tosylate *(223b)* is used in which the carbonyl oxygen of the acetate group is labeled with ^{18}O, the *trans*-diacetate *(224)* obtained still contains all of the labeled oxygen. However, this is distributed equally between a carbonyl and an ethereal oxygen atom. The hydrolysis of *(224)* in fact gives a *trans*-diol with only about half (46%) of the original labeling [22b], which can be shown schematically in the following way:

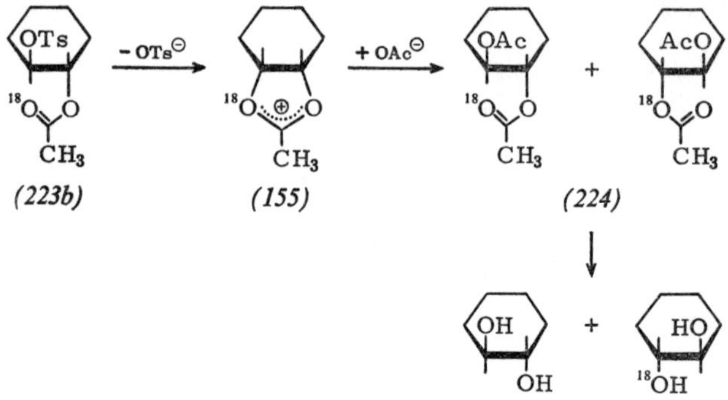

As an ambident cation, *(155)* can take up nucleophiles not only by ring opening to form *trans*-cyclohexane derivatives (route A) but also by additon to the carbonium center (route B); *cf.* Section 5.4.2. Thus the methanolysis of *trans*-2-benzoyloxy-cyclohexyl tosylate gives the corresponding bicyclic ortho ester (the possible directions of attack of the nucleophile are denoted by arrows) [23]:

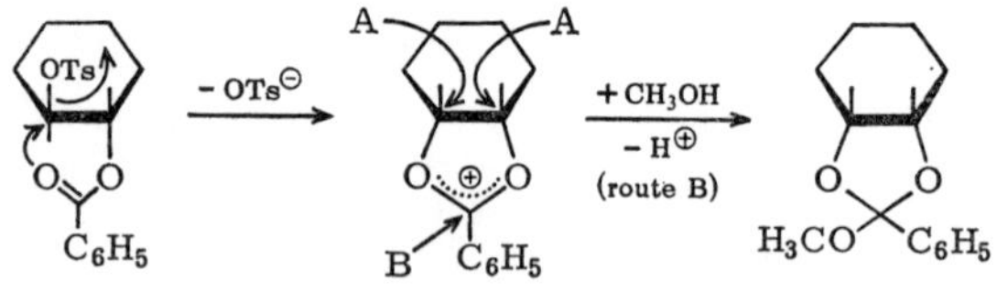

The acetolysis of *(223a)* or *(223b)* in *moist* glacial acetic acid, leading to *cis*-1-acetoxy-2-hydroxycyclohexane *(157)*, also takes place by route B. Here the 1,3-dioxolenium ion *(155)* to be assumed as an intermediate first takes up a molecule of water to form the ortho-ester derivative *(156)*. This is then converted with transprotonation and ring opening into *(157)*; *cf.* Section 5.4.2 [19, 22a]. Moreover, in glacial acetic acid the presence of an ortho diester *(225)* in equilibrium with *(155)* has been deduced [19, 24a,b]. The *cis*-diacetate *(226)* that is observed on acetolysis in the presence of a strong acid will be formed from this by rearrangement [19, 24a,b]. The cations *(225a)* and *(225b)* have been discussed as possible intermediate stages [24b]:

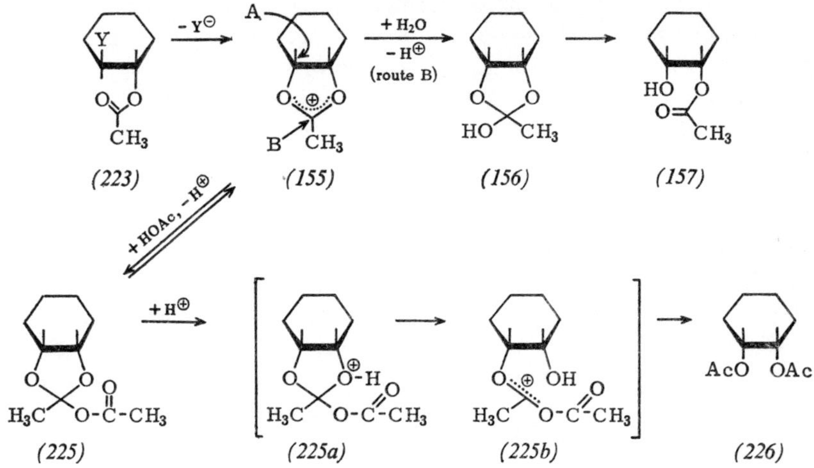

Participation of Ester Groups as Neighboring Groups   113

The validity of the assumed solvolysis mechanisms was finally demonstrated by Winstein et al., who succeeded in preparing the tetrafluoroborate of (155) by the action of $AgBF_4$ on the *trans*-bromide (223a). The salt isolated exhibited the expected properties [24a]; cf. Table 9.

Until now the use of stereo formulas for the cyclohexane molecule has been deliberately avoided. In the *cis* linkage of a 1,3-dioxolenium ring with the cyclohexane system, one oxygen atom is oriented axially (*a*) and the second equatorially (*e*); e.g. (155):

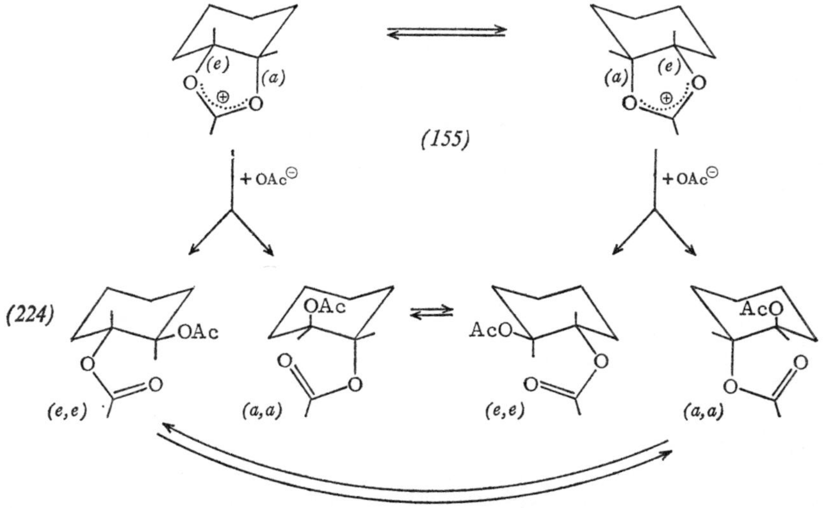

In the nucleophilic attack of acetate ions, the ring opening of (155) by route A can therefore yield two *trans*-diacetates of different conformations. The primary product is either a diaxial diacetate (*a,a*) or a diequatorial compound (*e,e*), depending on whether the attack of the nucleophile is on the C-atom with the equatorial or with the axial oxygen in *(155)*. In the case of flexible cyclohexane derivatives, no statement concerning the primary products of ring opening is possible. If, however, the conformation is fixed—as in *trans*-decalin or cholestane derivatives—the stereochemical course of the reaction can be determined [25a,b, 26]. For this purpose the corresponding 1,3-dioxolenium salts *(226)–(228)* were obtained by the reaction of the diaxial *trans*-bromhydrin esters (e.g., *(229)*, with AgSbF$_6$), i.e., by the preparative application of neighboring-group participation.

Ring opening by route A takes place by the action of tetramethylammonium bromide (in methylene chloride); under these conditions for the cation *(226)*, up to 95% of *diaxial* ring opening to the *trans*-bromhydrin ester *(229)* is found; only 5% of the diequatorial ester *(230)* is formed [25a,b]. On the basis of a very careful investigation of the origin of the preference for diaxial ring opening, it is concluded that the transition state for diequatonial opening is one of higher energy [25b].

where $R^1 = p\text{-}CH_3O\text{-}C_6H_4$ in *(226)*, *(227a)*, *(228)*, and $R^1 = C_6H_5$ in *(227b)*.

Compounds *(227a)* and *(227b)* behave in exactly the same way. On the other hand, the dioxolenium ion *(228)* exhibits predominantly *diequatorial* ring opening to *(231)* [25a,b]:

The difference is understandable, since for the formation of the diaxial ester *(232)*, attack of the bromide at C-2 in *(228)* must take place, but here this is inhibited by the neighboring methyl group (C-19), while the C-3 atom is "more open" to nucleophilic attack [25b].

Ring opening as a subsequent reaction to nucleophilic addition by route B takes place in the reaction of 1,3-dioxolenium ions with water. This reaction of *(226)*–*(228)* with *moist* glacial acetic acid takes place with remarkable stereospecificity [26]: ring-opening products in which the hydroxy group is in the *equatorial* position and the acetate grouping in the *axial* position—*(233)*, *(235)*, and *(236)*—are obtained almost exclusively. A C-19 methyl group in *(227a)* does not change the result of the reaction at all:

where $R^1 = p\text{-CH}_3\text{O–C}_6\text{H}_4$.

In an acid-catalyzed rearrangement, *(233)* is converted into the thermodynamically more stable equatorial ester *(234)*. If *(233)* or *(234)* is treated with camphor-10-sulfonic acid, in each case an equilibrium 1:2 mixture of *(233)* and *(234)* is obtained. The analogous situation is found for the cholestane derivatives *(235)* and *(236)*. The formation of the axial esters *(233)*, *(235)*, and *(236)* on ring opening is therefore probably kinetically controlled.

Partial hydrolysis of the cyclic ortho esters *(237a–c)* under mild conditions also gives almost exclusively *axial* esters of type *(233)*, with a very small amount of *(234)*. Here again the intermediate formation of 1,3-dioxolenium ions *(226)* may be assumed; *cf.* Tables 5 and 7.*

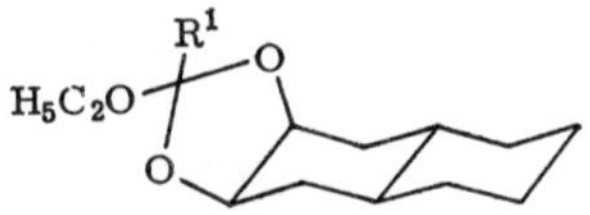

(a) $R^1 = C_6H_5$
(b) $R^1 = C_2H_5$
(c) $R^1 = CH_3$
(d) $R^1 = H$

(237a–d)

This stereospecificity of ring opening has been explained on the basis of different energies of the transition states [26]. Here the steric arrangement *(238)* will lead to the axial ester *(233)*, and the transition state *(239)* to the equatorial ester *(234)*:

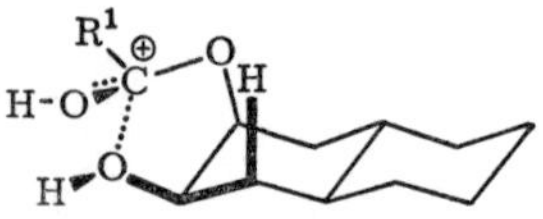

(238a) 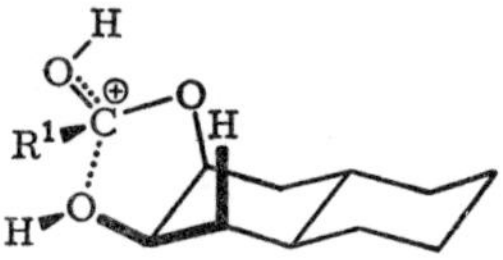   *(238b)* ⟶ *(233)*

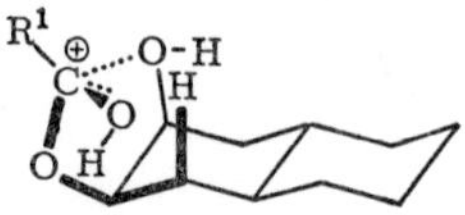

(239a) 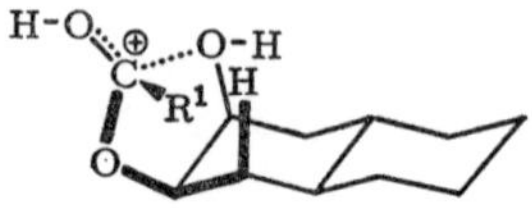   *(239b)* ⟶ *(234)*

The preferred ring opening to *(233)* is ascribed to the fact that the unfavorable, nonbonding interaction with the adjacent, axial *cis* hydrogen atom in *(238)* which is shown in the structural formulas given, is smaller than in *(239)*, regardless of whether the OH group and R^1 are oriented as in *(239a)* or in *(239b)*.** This assumption is well supported by the result of a mild hydrolysis of the orthoformate *(237d)*. Here, with a small substituent $R^1 = H$, the transition state *(239b)* should be as favorable as *(238)*; actually in this case the equatorial and axial esters are found in a ratio of 2:3.

In the chemistry of the carbohydrates neighboring-group participation of ester groups has been known for a long time, and the intermediate occurrence of 1,3-

---

* This method permits selective esterification of the *axial* hydroxy group of a *cis*-glycol if this is first treated with an acyclic ortho ester to give *(237)* [21] and the latter is then hydrolyzed under mild corditions [26].

** As a further possibility of the preferred formation of *(233)* and not *(234)*, it has been suggested that the equatorial oxygen may possess a somewhat higher basicity than the axial oxygen, and is therefore protonated preferentially [26].

Participation of Ester Groups as Neighboring Groups 117

dioxolenium ions has also been assumed, particularly in the formation of cyclic ortho-ester derivatives from acylated halogen-substituted sugars by the following scheme [27a,b]:

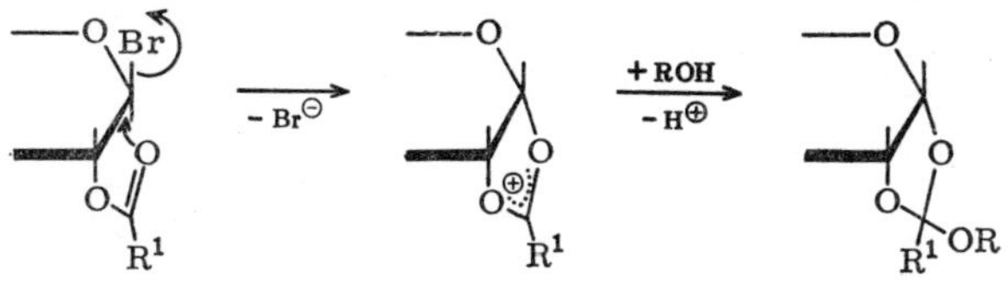

Such mechanisms have been finally proved by the isolation of crystalline 1,3-dioxolenium salts from acylated hexoses and pentoses [28a,b,c,d]. These salts can be prepared by the proven method of internal alkylation by the action of $SbCl_5$ on β-acetylglucosyl chloride *(240)* or β-pentaacetylglucose *(241)* [28c]. Surprisingly, however, the same acetoxonium salt can also be obtained from α-acetylglucosyl chloride *(242)*, but not from α-pentaacetylglucose *(243)*:

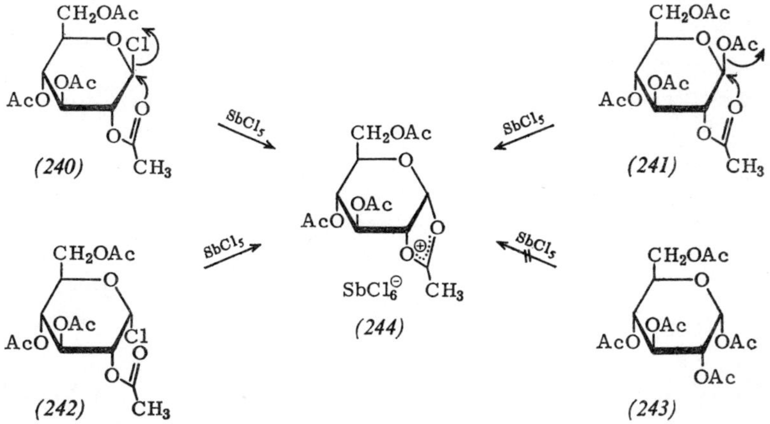

In fact, the hexachloroantimonate that precipitates in crystalline form is in *no* case the expected five-membered cyclic 1,2-acetoxonium salt *(244)* of α-D-glucopyranose but, as shown by NMR spectroscopy, is the 4,6-acetoxonium salt of α-D-*ido*pyranose *(245)*, a six-membered 1,3-dioxenium salt. The conversion of *(244)* into *(245)* obviously takes place by a multiple rearrangement (*cf.* [29–31]) via the dioxolenium ions of α-D-mannopyranose and α-D-altropyranose [28b,c]. In agreement with the results of NMR spectroscopy, under the action of aqueous sodium acetate solution, with *cis* ring opening [19, 24a,b, 29, 30], the salt gives only the two tetraacetylidoses *(246)* and *(247)* and with ethanol it yields the idose ortho ester *(248)*, from which *(245)* is readily regenerated with $SbCl_5$ [28c]. However, the idose acetoxonium ion *(245)* must still be in equilibrium with the expected *glucose* dioxolenium ion *(244)*: with acetic anhydride or glacial acetic acid the salt gives α-pentaacetylglucose *(243)*, probably via the orthodiacetate *(249)*; *cf. (225)* [19, 24a,b] and subsequent *cis* ring opening [28c]:

[Structures (244), α-D-gluco-; α-D-manno-; α-D-altro-; (245) α-D-ido-]

[Ac₂O/HOAc ↓]   [H₂O ↙]   [SbCl₅ ↕ C₂H₅OH]

[Structures (249); (246); (247); (248)]

The assumption of an equilibrium* between the four acetoxonium ions** is confirmed by the analogous behaviour of the 1,3-dioxolenium ion derived from α-D-xylopyranose *(251)* [28a,b,d]. A crystalline salt can also be obtained from β-acetylxylosyl chloride *(250)* with SbCl₅, but no uniform product is obtained [28d]. Hydrolysis of the salt with aqueous sodium acetate solution gives xylose (45%), lyxose (15%), and arabinose (40%) which indicates an equilibrium* of the three acetoxonium ions *(251)–(253)* [28b,d]:

α-D-xylo-   α-D-lyxo-   α-D-arabino-

[Structures (250); (251); (252); (253)]

---

* An equilibrium of this type was also confirmed by recent NMR investigations of the acetoxonium ions (A) and (B). The NMR signals of the acetyl- and acetoxonium-methyl groups coalesce with rising temperature [28e]:

(A) H₂C–CH–CH₂ ⇌ H₂C–CH–CH₂ (B)   (C)

Similar results were obtained for the acetoxonium ion (C) [28e].

** With the *tetrafluoroborate* of *(244)*—obtained with BF₃ from β-acetylglucosyl fluoride —no rearrangement to the idose acetoxonium salt has been found [28b,c].

In all the cases discussed so far in which 1,3-dioxolenium ions are formed preparatively or arise as intermediates by neighboring-group participation, the carbonyl *oxygen* of the ester carbonyl displaces a substituent in the *trans* position from the β-C atom (formation route A, "internal alkylation").

In the methods of preparation for 1,3-dioxolenium ions (Section 4.2.2.2) we have again assumed this mechanism if the substituent eliminated is an OR grouping, for which a proton acts as acceptor:

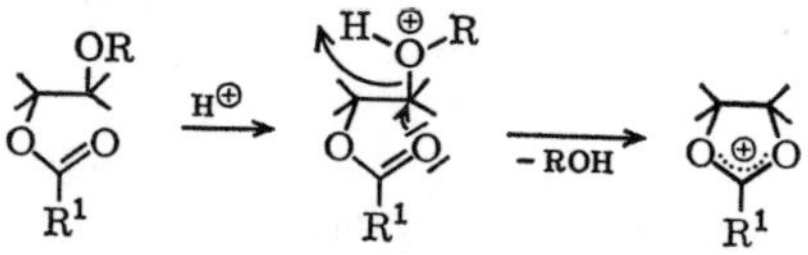

Attack of the carbonyl *carbon* atom on the ether oxygen of the OR group is also conceivable. Here an ortho-ester derivative should form as an intermediate, from which the 1,3-dioxolenium ion can again be produced in the subsequent step (formation route B). This will be possible particularly when the carbonyl group is activated by the addition of a proton. Another necessary condition is that (in contrast to formation route A) the acyloxy and OR groups must be orientated in the *syn* position with respect to one another:

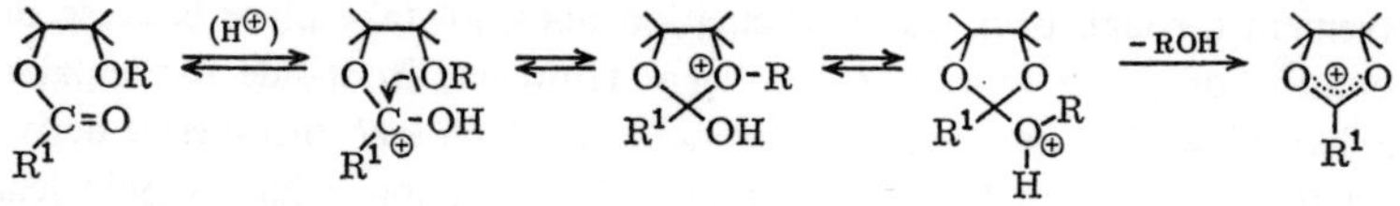

where $R = H, (CO)R^1$.

Winstein has introduced the term "front-side participation" for this type of neighboring-group participation [32]. It is to be expected, in particular with 1,2-diacyloxy- and 1,2-acyloxyhydroxyalkanes ($R = -(CO)R^1$, H) in a highly acid medium.

A distinction between back-side and front-side participation can be found very easily in the production of bicyclic 1,3-dioxolenium ions from cyclohexane derivatives.*

The first example of a front-side participation was found by Boschan and Winstein [32]: with hydrogen chloride, *cis*-1,2-diacetoxycyclohexane *(226)* smoothly gives 2-chlorocyclohexanol *(255)* or its acetate *(158)*. The same result is obtained with the *cis*-diol *(254)* with the addition of glacial acetic acid; in the absence of glacial acetic acid *no* chlorohydrin is produced from this compound. *Trans*-1,2-cyclohexanediol *(256)* and its diacetate *(244)* give no chlorocyclohexanol at all.

The following mechanism describes satisfactorily the course of the reaction; here the chloride ion cleaves the dioxolenium ring in the usual manner:

---

* The isomerization *(233)* ⇌ *(234)* with strong acids is probably also based on such front-side participation.

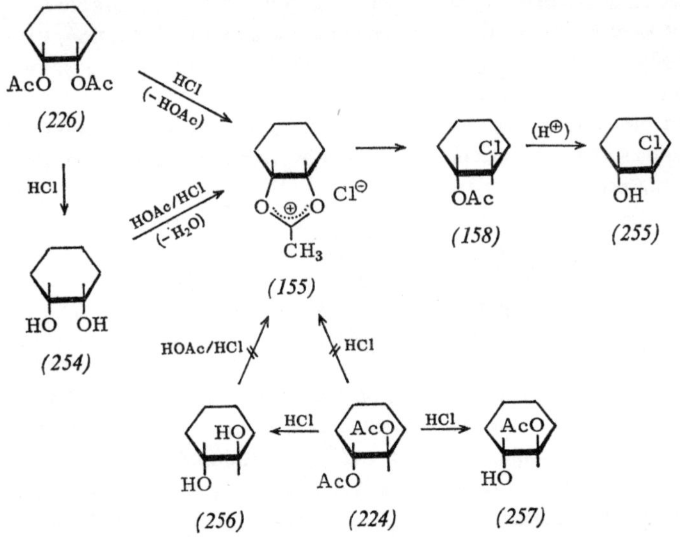

Consequently the intermediate formation of the 1,3-dioxolenium ion *(155)* from the *trans* derivatives in a strongly acid medium is impossible. Neighboring-group involvement through back-side participation does not take place because of the protonation of the ester carbonyl oxygen. However, front-side participation is more favorable with the *cis*-1,2 arrangement, in which a *cis* ortho-ester derivative can be formed, than in the case of *trans*-1,2 orientation, which should lead to more highly ring-strained *trans* ortho esters [32]:

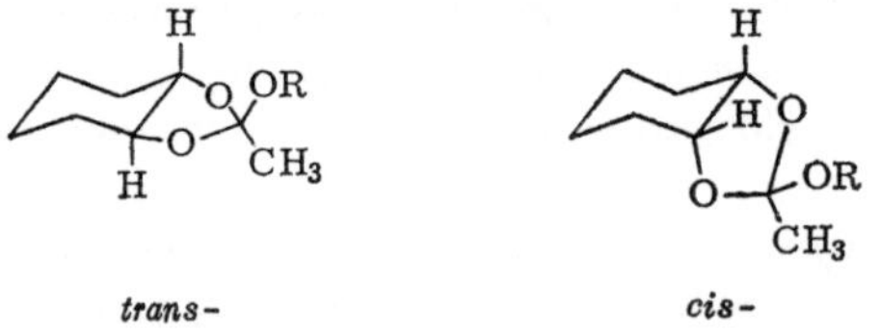

trans-   *cis-*

ortho ester

Recently it has also been proved by NMR spectroscopy that only *cis*-1,2-diacetoxycyclohexane *(226)* is converted smoothly in anhydrous hydrogen fluoride into the dioxolenium ion *(155)*, while the *trans*-diacetate *(224)* is stable in HF for days [33]. The formation of *(155)* has also been shown preparatively: From a solution of *(226)* in HF, BF_3 can precipitate the known tetrafluoroborate of *(155)* [24a] in crystalline form [33].

The explanation of the epimerization of cyclitol esters [29] or polyhydroxytetrahydropyran esters [30] in hydrofluoric acid is connected with these results.

In fact, Walden inversion is always found when three acyloxy groups are present in adjacent positions in *cis–trans* sequence: The epimerization always takes place at the *central* C-atom whose acyloxy group is the common constituent of a *cis* and a *trans* pair. Instead of the esters, the free alcohols themselves can be rearranged in the presence of glacial acetic acid and *p*-toluenesulfonic acid (or H_2SO_4), with the same result [31].

Since no epimerization is found with all-*trans* compounds (*cf.* [33]), here again we may assume the primary formation of an acyloxonium ion *(260)* by front-side participation [32]: The *cis*-diester pair [29, 30, 33] or a *cis*-acyloxyhydroxy grouping [31] of *(258)* can do this in a strongly acid medium, probably through the intermediate stages shown in the scheme below [31].* The ester grouping in the *trans* position then attacks the dioxolenium ion *(260)* from the back side with the formation of a new cation *(261)*; the two ions are in equilibrium with one another. The action of water leads as usual, with intermediate formation of the ortho-ester derivative *(259)* or *(262)*, to the *cis* ring opening [19, 24a,b] of *(259)* to *(263)* and of *(262)* to *(264)*. Here *(264)* is the epimer of the initial compound *(258)*, while *(263)* shows the original configuration. The epimerization is thus finally to be ascribed to the two 1,3-dioxolenium ions present in equilibrium with one another, and the mechanism is therefore similar to that in the case of the acetoxonium salts salts of sugar acetates *(245)*, *(251)* isolated as such [28a,b,c,d,e]:

Reaction Scheme

where R' = OH, OAc and R = H, Ac.

122   Tertiary Oxonium Ions as Intermediates in Chemical Reactions

The course of such reactions can be predicted on the basis of the stereochemistry of the starting materials. For example 2,3,4,6-tetra-*O*-acetyl-1,5-anhydro-D-glucitol *(265)* should not be epimerized with HF, since only *trans* ester pairs are present. In fact, *only* deacylation is found in practice [30]:

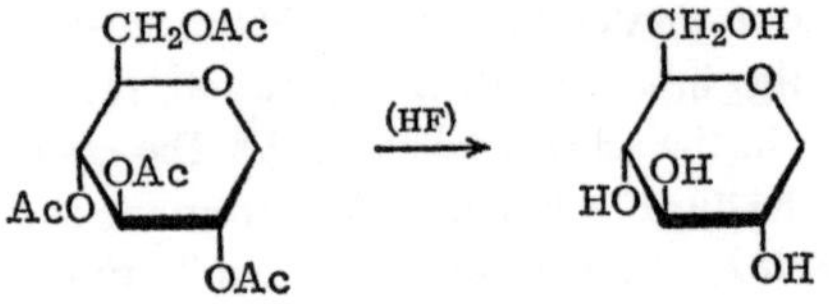

(265) → (with HF)

In contrast, the corresponding 1,5-anhydro-D-galactitol tetraester *(266)* undergoes not only deacylation but also inversion, since there is a *cis–trans* sequence of neighboring groups [30]:

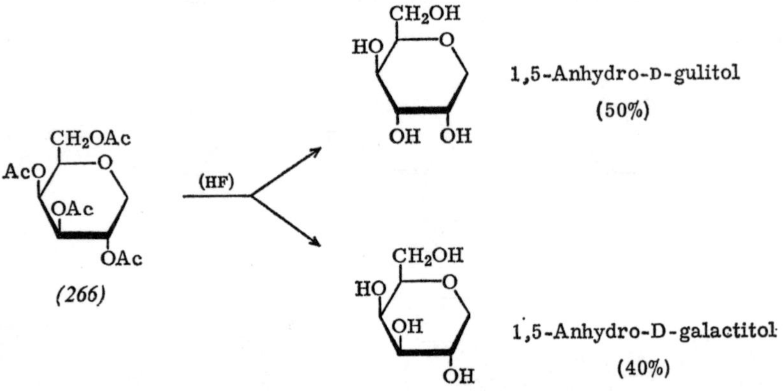

(266) → (HF) →
1,5-Anhydro-D-gulitol (50%)
1,5-Anhydro-D-galactitol (40%)

1,5-Anhydro-D-arabinitol tetraesters *(267)* can indeed be epimerized because of the *cis–trans* sequence that is present, but because of the particular molecular geometry even inversion leads again to arabinitol. The result of the reaction therefore appears as that of a simple deacylation [30]:

---

* The formation of the seven-membered ring discussed previously [29, 30] is unlikely here [33], cf. p. 121.

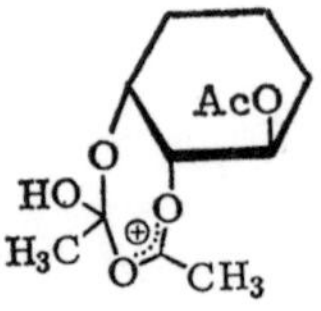

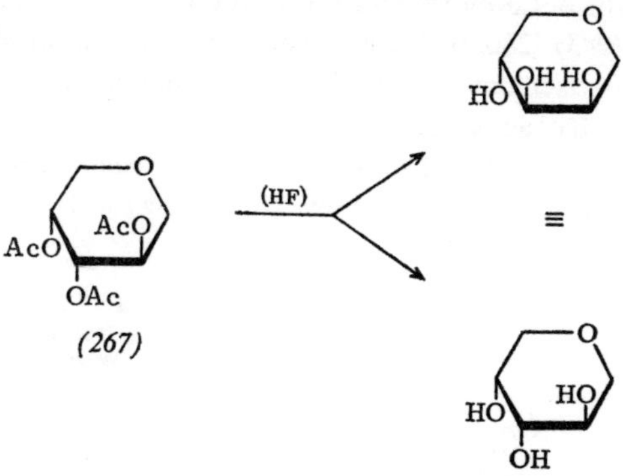

(267)

Sugar esters can also be epimerized in HF in a similar manner [34a,b].

6.3.2.2. 1,3-Dioxenium Ions [39]

The result of the reaction of the cholestane derivative *(268)* with tosyl chloride permitted the assumption of an intermediate formation of a 1,3-dioxenium ion [35a,b,c]:

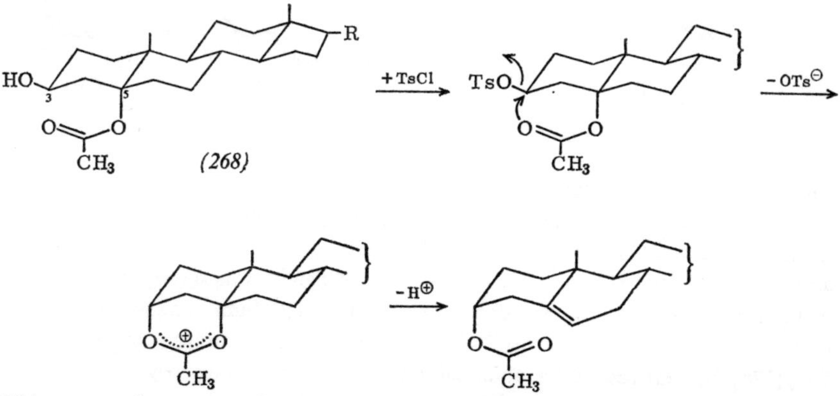

(268)

This assumption was confirmed by the preparation of 3α,5α-bridged acetoxonium perchlorates *(269)* from compounds corresponding to *(268)* [36a,b]:

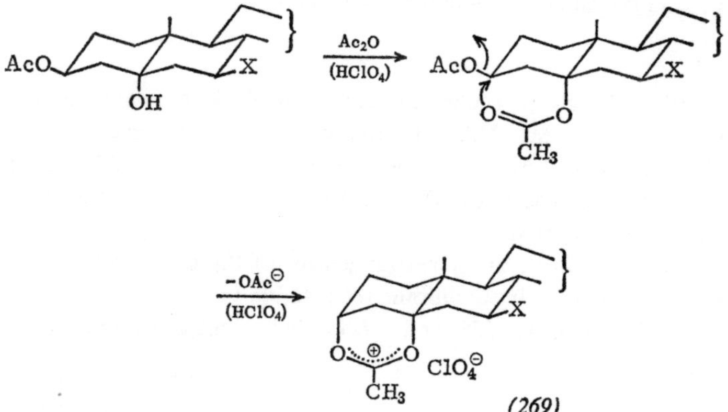

(269)

where X = Cl, F, OAc.

Tertiary Oxonium Ions as Intermediates in Chemical Reactions

Another example of a 1,3-dioxenium ion is found in the formation of the idose acetoxonium ion *(245)* [28b,d]. Ester-carbonyl interactions with the intermediate formations of 1,3-dioxenium ions have been investigated in detail by Schneider and Láng [23] and Kovács et al. [37a,b]. The latter authors again showed preparatively the direct access to 1,3-dioxenium ions by neighboring-group participation. The following reaction scheme shows the analogy with the results in the 1,3-dioxolenium series [37b]:

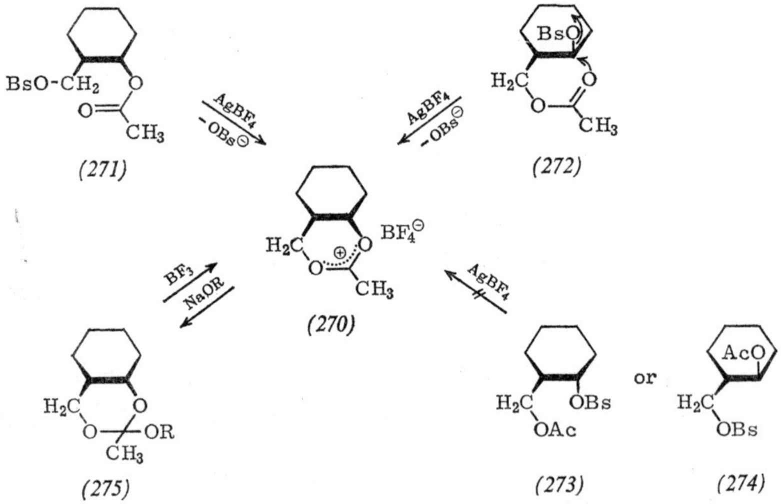

Thus, 2-methyl-*cis*-4,5-tetramethylene-1,3-dioxenium tetrafluoroborate can be obtained readily from both *cis*-2-brosyloxymethylcyclohexyl acetate *(271)* and from *trans*-2-acetoxymethylcyclohexyl brosylate *(272)*, but *not* from the two acetate–brosylates *(273)* and *(274)* in which the neighboring-group participation of the acetate carbonyl is excluded. The action of a sodium alkoxide on the salt *(270)* leads to the cyclic ortho ester *(275)* from which *(270)* can be regenerated with BF_3 [37b]. Similar results are obtained with the corresponding tosylate–benzoates [23, 37a]. The role of the cation *(270)* in the acetolysis of tosylates *(276)* and *(277)* has been demonstrated, while *(278)* and *(279)* can undergo only direct acetolysis without the intermediate formation of *(270)* [37b].

Consequently, with glacial acetic acid–acetic anhydride–potassium acetate, *(278)* reacts with retention and *(279)* with inversion to give the *trans*-diacetate *(282)*; the *cis*-diacetate is not formed. On the other hand, *(276)* and *(277)* give mixtures of the *cis*- and *trans*-diacetates. The *cis*-diacetate is most predominant; it can in fact be formed by direct substitution from *(276)* or *(277)*, by ring opening of the 1,3-dioxenium ion *(270)*, or by rearrangement of the orthodiacetate *(280)* that must be assumed to exist in equilibrium (cf. [24a,b]). In contrast, the *trans*-diacetate *(282)* can be formed from *(276)* and *(277)* only as a subsequent product of *(270)*.

Reaction Scheme

Neighboring-group participation can also be demonstrated by kinetic measurements: *(277)* is solvolyzed about 10^2 times faster than *(278)* [37a]. (In addition to the diacetates, small amounts of olefins are formed in the case of *(276)–(278)* and large amounts in the case of *(279)* [37b]). In the case of the *trans*-acetate–tosylate *(276)*, the greater part of solvolysis takes place without neighboring-group participation; solvolysis of the corresponding *trans*-acetate–brosylate *(272)* gives rise to the same conclusion [38a,b].

References

1. a) S. Winstein, E. Allred, R. Heck, and R. Glick, *Tetrahedron* **3**, 1 (1958); b) E. Allred and S. Winstein, *J. Am. Chem. Soc.* **89**, 3991 (1967); c) E. Allred and S. Winstein, *J. Am. Chem. Soc.* **89**, 4012 (1967); d) E. R. Novak and D. S. Tarbell, *J. Am. Chem. Soc.* **89**, 73 (1967).
2. S. Oae, *J. Am. Chem. Soc.* **78**, 4030 (1956).
3. D. J. Pasto and M. P. Serve, *J. Am. Chem. Soc.* **87**, 1515 (1965).
4. S. Winstein and R. E. Buckles, *J. Am. Chem. Soc.* **64**, 2780 (1942).
5. a) W. Lwoski, *Angew. Chem.* **70**, 483 (1958); b) B. Capon, *Quart. Rev. (London)* **18**, 45 (1964).
6. a) H. R. Ward and P. D. Sherman, Jr., *J. Am. Chem. Soc.* **89**, 4222 (1967); b) *J. Am. Chem Soc.* **90**, 3814 (1968); c) Y. Leroux, *Bull. Soc. Chim. France* p. 352 (1968); d) H. Schäfer and A. Alazrek, *Angew. Chem.* **80**, 485 (1968); *Angew. Chem. Intern. Ed. English* **7**, 474 (1968).

7. a) A. Kirrmann and N. Hamaide, *Bull. Soc. Chim. France* p. 789 (1957); b) A. Kirrmann and L. Wartski, *Bull. Soc. Chim. France* p. 3077 (1965); c) A. Kirrmann and L. Wartski, *Compt. Rend.* **259**, 2857 (1964); *Chem. Abstr.* **62**, 2748 (1965).
8. a) A. Kirrmann and L. Wartski, *Bull. Soc. Chim. France* p. 3825 (1966); b) see also H. Meerwein *in* "Methoden der organischen Chemie (Houben-Weyl)" (E. Müller, ed.), Bd. VI/3: Sauerstoffverbindungen I, p. 336. Thieme, Stuttgart, 1965.
9. a) S. Winstein and L. L. Ingraham, *J. Am. Chem. Soc.* **74**, 1160 (1952); b) S. Winstein, C. R. Lindegren, and L. L. Ingraham, *J. Am. Chem. Soc.* **75**, 155 (1953).
10. J. B. Lambert and D. H. Johnson, *J. Am. Chem. Soc.* **90**, 1349 (1968).
11. P. G. Gassmann and J. L. Marshall, *Tetrahedron Letters* p. 2429 (1968); see also J. R. Hazen, *J. Org. Chem.* **35**, 973 (1970).
12. G. Baddeley, E. K. Baylis, B. G. Heaton, and J. W. Rasburn, *Proc. Chem. Soc.* p. 451 (1961).
13. J. Weinstock, *J. Am. Chem. Soc.* **78**, 4967 (1956).
14. D. B. Denney and J. Ciacin, *Tetrahedron* **20**, 1377 (1964).
15. W. W. Epstein and A. C. Sonntag, *J. Org. Chem.* **32**, 3390 (1967).
16. a) H. Meerwein, P. Borner, O. Fuchs, H. J. Sasse, and J. Spille, *Chem. Ber.* **89**, 2060 (1956); b) K. Wunderlich, Ph.D. Thesis, University of Marburg, 1957.
17. a) V. Prelog and S. Heinbach-Juhasz, *Ber. Deut. Chem. Ges.* **74**, 1702 (1942); b) K. B. Wiberg, *J. Am. Chem. Soc.* **74**, 3957 (1952).
18. D. S. Noyce and H. J. Weingarten, *J. Am. Chem. Soc.* **79**, 3093 (1957).
19. S. Winstein, N. V. Hess, and R. E. Buckles, *J. Am. Chem. Soc.* **64**, 2796 (1942).
20. G. A. Olah and J. M. Bollinger, *J. Am. Chem. Soc.* **89**, 4744 (1967).
21. H. Meerwein, K. Bodenbenner, P. Borner, F. Kunert, and K. Wunderlich, *Ann. Chem. Liebigs* **632**, 38 (1960).
22. a) S. Winstein, E. Grunwald, and L. L. Ingraham, *J. Am. Chem. Soc.* **70**, 821 (1948); b) K. B. Gash and G. U. Yuen, *J. Org. Chem.* **31**, 4234 (1966).
23. G. Schneider and L. K. Láng, *Chem. Commun.* p. 13 (1967).
24. a) C. B. Anderson, E. C. Friedrich, and S. Winstein, *Tetrahedron Letters* p. 2037 (1963); b) R. M. Roberts, J. Corse, R. Boschan, D. Seymour, and S. Winstein, *J. Am. Chem. Soc.* **80**, 1247 (1958).
25. a) J. F. King and D. A. Allbutt, *Chem. Commun.* p. 14 (1966); b) *Can. J. Chem.* **47**, 1445 (1969).
26. J. F. King and D. A. Allbutt, *Tetrahedron Letters* p. 49 (1967).
27. a) J. Staněk, M. Černý, J. Kocourek, and J. Pacák, "The Monosaccharides," p. 296, Academic Press, New York and Czech. Acad. Sci., Prague, 1963; b) see also M. Schulz, H.-F. Boeden, and P. Berlin, *Ann. Chem. Liebigs* **703**, 190 (1967).
28. a) H. Paulsen, W.-P. Trautwein, F. Garrido Espinosa, and K. Heyns, *Tetrahedron Letters*, p. 4131 (1966); b) *Tetrahedron Letters* p. 4137 (1966); c) *Chem. Ber.* **100**, 2822 (1967); d) H. Paulsen, F. Garrido Espinosa, W.-P. Trautwein, and K. Heyns, *Chem. Ber.* **101**, 179 (1968); e) H. Paulsen and H. Behre, *Angew. Chem.* **81**, 905 (1969); *Angew. Chem. Intern. Ed. English* **8**, 886 (1969).
29. E. J. Hedgley and H. G. Fletcher, Jr., *J. Am. Chem. Soc.* **84**, 3726 (1962).
30. E. J. Hedgley and H. G. Fletcher, Jr., *J. Am. Chem. Soc.* **85**, 1615 (1963).
31. S. J. Angyal, P. A. J. Gorin, and M. E. Pitman, *J. Chem. Soc.* p. 1807 (1965).
32. R. Boschan and S. Winstein, *J. Am. Chem. Soc.* **78**, 4921 (1956).
33. C. Pedersen, *Tetrahedron Letters* p. 511 (1967).
34. a) C. Pedersen, *Acta Chem. Scand.* **17**, 1269 (1963); b) *Acta Chem. Scand.* **20**, 963 (1966).
35. a) P. A. Plattner and W. Lang, *Helv. Chim. Acta* **27**, 1872 (1944); b) P. A. Plattner, A. Fürst, F. Koller, and W. Lang, *Helv. Chim. Acta* **33**, 1455 (1948); c) see also A. D. Cross, E. Denot, R. Acevedo, R. Uquisa, and A. Bowers, *J. Org. Chem.* **29**, 2195 (1966).

36. a) J. W. Blunt, M. P. Hartshorn, and D. N. Kirk, *Chem. Ind. (London)* p. 1955 (1963); b) M. J. Coppen, M. P. Hartshorn, and D. N. Kirk, *J. Chem. Soc.* C, 576 (1966).
37. a) Ö. Kovács, G. Schneider, and L. K. Láng, *Proc. Chem. Soc.* p. 374 (1963); b) Ö. K. J. Kovács, G. Schneider, L. K. Láng, and J. Apjok, *Tetrahedron* **23**, 4181 (1967).
38. a) L. J. Dolby and M. J. Schwarz, *J. Org. Chem.* **30**, 3581 (1965); b) see also L. J. Dolby, C. N. Lieske, D. R. Rosencrantz, and M. J. Schwarz, *J. Am. Chem. Soc.* **85**, 47 (1963).
39. For a recent review see: H. Paulsen, H. Behre, and C.-P. Herold, *Fortschr. Chem. Forsch.* **14**, 473 (1970).

7. Preparative Application of Oxonium Salts

7.1. Alkylation with Trialkyloxonium Salts

7.1.1. Activation of C–X Multiple Bonds

When alkyl groups are transferred from trialkyloxonium salts to the nucleophilic centers Y of electrically neutral compounds R′–Y:, very reactive cationic intermediates *(136)* can frequently be found as primary products and these can be used advantageously in syntheses:

$$\text{R'-Y:} + R_3O^{\oplus}\ BF_4^{\ominus} \longrightarrow \text{R'-}\overset{\oplus}{Y}\text{-R}\ BF_4^{\ominus} + R_2O$$
$$\qquad (7) \qquad\qquad\qquad\qquad (136)$$

Of the large number of such cations *(136)* in Table 8 (see Chapter 5), we shall discuss here only those in which the positive charge is distributed over a heteroatom X (O, S, N) and a carbon atom; that is, the alkylation products of compounds with carbonyl, thiocarbonyl, imine, or nitrile functions:

$$\underset{R^1}{\overset{R^2}{>}}C=X: \xrightarrow[-R_2O]{+R_3O^{\oplus}} \underset{R^1}{\overset{R^2}{>}}C=\overset{\oplus}{X}\text{-R} \longleftrightarrow \underset{R^1}{\overset{R^2}{>}}\overset{\oplus}{C}\text{-}\overline{X}\text{-R} \qquad (X = O, S, NR')$$

$$R^1\text{-}C\equiv X: \xrightarrow[-R_2O]{+R_3O^{\oplus}} R^1\text{-}C\equiv\overset{\oplus}{X}\text{-R} \longleftrightarrow R^1\text{-}\overset{\oplus}{C}=X\text{-R} \qquad (X = N)$$

Such carboxonium, carbosulfonium, immonium, or nitrilium ions are ambident cations that may be attacked by (sufficiently strong) nucleophiles at the carbonium center (*cf.* Section 5.4.2); e.g.:

$$\underset{R^1}{\overset{R^2}{>}}\overset{\oplus}{C}\overset{\cdot\cdot}{=}X\text{-R} \xrightarrow{+:Y^{\ominus}} \underset{R^1}{\overset{R^2}{>}}\underset{X\text{-R}}{\overset{Y}{C}} \qquad (X = O, S, NR')$$

By the addition of an alkyl cation to the heteroatom X, therefore, the C–X multiple bond of the starting material is activated for nucleophilic attack. Indeed, such activation is also possible by protonation:

$$\underset{R^1}{\overset{R^2}{>}}C=X: \xrightarrow{+H^{\oplus}} \underset{R^1}{\overset{R^2}{>}}\overset{\oplus}{C}\overset{\cdot\cdot}{=}X\text{-H}$$

The cations arising in this way are, however, strong acids and readily transfer their proton to the nucleophile. Consequently, in the case of proton activation the nucleophile is frequently inactive, or at least its reactivity is weakened; *cf.* Section 5.1.3. On the other hand, the C–X bond can be activated by alkylation (normally) without the nucleophile being deactivated, since alkyl groups cannot be detached so easily as protons [1].

7.1.1.1. Activation of Ketones

By alkylation with trialkyloxonium salts it is possible to activate the carbonyl group of only a very limited number of aldehydes and ketones, namely, those in which a tertiary alkyl group or a vinyl group is present on the carbonyl carbon atom [2]. Most aliphatic ketones (or aldehydes) with primary alkyl groups undergo polymerization under the alkylation conditions because of the formation of vinyl ethers. On the other hand, the less nucleophilic carbonyl oxygen of the aromatic aldehydes and ketones cannot be alkylated by trialkyloxonium salts [3a].*

The O-alkylation of ketones is occasionally important in the preparation of ketals that are otherwise difficult to obtain, such as camphor diethyl ketal *(283)*, which is produced from camphor and ortho formate in moderate yields and from the carboxonium salt and ethoxide in up to 87% yield [4]:

$$\text{(176)} \xrightarrow{(C_2H_5)_3O^{\oplus} BF_4^{\ominus}} \xrightarrow{+C_2H_5O^{\ominus}} \text{(283)}$$

In some cases O-alkylation is the only way to obtain functional derivatives of the ketones concerned. Thus the condensation of tropone with reactive methylene components generally takes place only via the alkoxytropylium salt *(46)* [4, 5a] to give heptafulvene derivatives; e.g. *(284)*:

$$\xrightarrow{+R_3O^{\oplus} BF_4^{\ominus}} \text{(46)} \xrightarrow[-ROH,-HBF_4]{+H_2C(CN)_2} \text{(284)}$$
$$(N(C_2H_5)_3)$$

where $R = C_2H_5$. A recent investigation also deals with 2-chlorotropone [5b].

Such reactions take place in the same manner in the case of diphenylcyclopropenone *(285)* with the alkoxycyclopropenylium salt *(45)* [6a,b,c]; the reactions with secondary amines [7], enamines [6a,b,c], and dimethylaniline [8] to *(287)–(289)* also take place smoothly with the cation *(45)*:

---

* *Cf.* Section 7.3.1 [3a,b].

where R = C_2H_5 and X, Y = $(CO)R_1$, CO_2CH_3, CN, for exemple.

Similarly, pentatriafulvenes are available from di-*n*-propylcyclopropenone after activation with a triethyloxonium salt [9]; e.g.:

where R = C_2H_5.

The reactions recall the condensations of reactive methylene compounds with pyrones, which also take place very easily when the carbonyl group is activated by alkylation to alkoxypyrylium salts [10]; however, in the case of γ-pyrones, O-alkylation can be carried out even with dimethyl sulfate; cf. Section 4.2.3.1.

7.1.1.2. Conversion of Carboxylic Amide Functions

The alkylation of the carbonyl group of carboxylic esters cannot be achieved with trialkyloxonium salts (Sections 4.2.2.1), although that of carbonic esters and lactones takes place readily.* However, the O-alkylation of carboxylic amide groups in amides, lactams, urethanes, and ureas has acquired preparative importance.

Carboxylic Acid Amides

Carboxylic acid amides can be alkylated very readily with trialkyloxonium salts at the carbonyl oxygen atom [12a,b]:

$$R^1-C\overset{O}{\underset{N-R''}{\diagdown R'}} + R_3O^\oplus BF_4^\ominus \longrightarrow R^1-C\overset{OR}{\underset{N-R''}{\diagdown R'}}{}^\oplus BF_4^\ominus + R_2O$$

(290)

where R = alkyl and and R′,R″ = H, alkyl

The preparation of a mercury–carbene complex *(291)* from the bis(dialkylcarbamoyl)mercury compound is based on such alkylation [13]:

$$\underset{\underset{R''}{R'-N}}{\overset{O}{\diagdown}}C-Hg-C\overset{O}{\underset{\underset{R''}{N-R'}}{\diagdown}} \xrightarrow{+2\,R_3O^\oplus BF_4^\ominus} \left[\underset{\underset{R''}{R'-N}}{\overset{RO}{\diagdown}}C-Hg-C\overset{OR}{\underset{\underset{R''}{N-R'}}{\diagdown}}\right]^{2\oplus} 2\,BF_4^\ominus$$

(291)

Alkoxy(amino)carbonium ions *(290)*, which are derived from primary (R′, R″ = H) and secondary (R′ = alkyl, R″ = H) carboxylic amides, are transformed smoothly by alkali–metal carbonate solutions (or tertiary amines) into the corresponding imidic esters. Thus, for example, the following transformation of the bis(amide) into a bis(imidic ester), which is the intermediate product of a corrin synthesis, takes place in this way [14]:

---

* The O-alkylation of carboxylic ester groups was possible in the case of alkoxycarbonyl-alkylidene-triphenylphosphoranes with formation of vinyl phosphonium ions [11]; e.g.:

$$\underset{(C_6H_5)_3P}{\overset{H}{\diagdown}}C=C\overset{O}{\underset{OR'}{\diagdown}} \xrightarrow{+R_3O^\oplus BF_4^\ominus} \left[\underset{(C_6H_5)_3P^\oplus}{\overset{H}{\diagdown}}C=C\overset{OR}{\underset{OR'}{\diagdown}}\right]BF_4^\ominus + \left[\underset{(C_6H_5)_3P^\oplus}{\overset{H}{\diagdown}}C=C\overset{OR'}{\underset{OR}{\diagdown}}\right]BF_4^\ominus$$

where R′ = CH₃ and R = C₂H₅.

Preparative Application of Oxonium Salts

$$\text{structures: diamide} \xrightarrow{2\,(C_2H_5)_3O^\oplus\,BF_4^\ominus} \text{bis-ethoxy cation} \xrightarrow[-2\,HBF_4]{(K_2CO_3)} \text{bis-imidate}$$

The cations *(290)* are easily attacked by nucleophiles at the carbonium center. Then the alkoxy group is often expelled in a subsequent reaction:

$$R^1\text{-}\underset{\underset{R''}{\overset{|}{N\text{-}R'}}}{\overset{\overset{OR}{|}}{C}}\text{-}OR \qquad R^1\text{-}CH_2\text{-}N\underset{R''}{\overset{R'}{\diagup}}$$

(292)   (293)

R', R'' = alkyl   R' = H, alkyl; R'' = alkyl

central cation (290)

$$R^1\text{-}\underset{O}{\overset{OR}{\diagdown}}C \;+\; HN\underset{R''}{\overset{R'}{\diagup}} \longrightarrow \left[R^1\text{-}\underset{\underset{R''}{\overset{|}{N\text{-}R'}}}{\overset{\overset{OR}{|}}{C}}\text{-}OH\right]$$

$$R^1\text{-}C\underset{\underset{R''}{\overset{|}{N\text{-}R'}}}{\overset{\overset{N\text{-}R'''\,H}{|}}{\oplus}} \xrightarrow{-H^\oplus} R^1\text{-}C\underset{\underset{R''}{\overset{|}{N\text{-}R'}}}{\overset{\parallel N\text{-}R'''}{}}$$

(294)   (295)

R', R'', R''' = H, alkyl

The primary addition products, amide acetals *(292)*, can be isolated by the action of an alkoxide [12b]. The reaction of *(290)* with water leads via the hydroxy derivative corresponding to *(292)* to cleavage into carboxylic ester and amine. A carboxylic amide can be hydrolyzed under mild conditions in this way after transformation into the cation *(290)*, and use is made of this in the synthesis of a tetracycline derivative [15]:

$$\text{tetracycline amide} \xrightarrow[3\%\,CH_3CO_2H]{(a)\,(C_2H_5)_3O^\oplus\,BF_4^\ominus \;\; (b)\,\text{Dioxane-}H_2O} \text{tetracycline amine}$$

On the other hand, reduction of the carboxylic amide function is possible under the action of sodium tetrahydroborate with *(290)*; this method gives very good yields of chain-extended amines *(293)* [16].

Compound *(290)* can also be converted into amidines *(295)* via the stage of the amidinium ions *(294)* by reaction with ammonia or a primary amine [17]. The ready replacement of the carbonyl group by the imino group, which takes place in this way with carboxylic amides, has also been used to convert the polyamide Nylon into the corresponding polyamidine [18]:

$$\cdots-\overset{O}{\underset{\|}{C}}-(CH_2)_4-\overset{O}{\underset{\|}{C}}-\overset{H}{\underset{|}{N}}-(CH_2)_6-\overset{H}{\underset{|}{N}}-\overset{O}{\underset{\|}{C}}-(CH_2)_4-\overset{O}{\underset{\|}{C}}-\overset{H}{\underset{|}{N}}-(CH_2)_6-\overset{H}{\underset{|}{N}}-\cdots$$

$$\downarrow \begin{array}{l} (a)\ (C_2H_5)_3O^\oplus\ BF_4^\ominus \\ (b)\ NH_2R' \end{array}$$

$$\cdots-\overset{R'N}{\underset{\|}{C}}-(CH_2)_4-\overset{R'N}{\underset{\|}{C}}-\overset{H}{\underset{|}{N}}-(CH_2)_6-\overset{H}{\underset{|}{N}}-\overset{NR'}{\underset{\|}{C}}-(CH_2)_4-\overset{R'N}{\underset{\|}{C}}-\overset{H}{\underset{|}{N}}-(CH_2)_6-\overset{H}{\underset{|}{N}}-\cdots$$

It may be mentioned that the *methosulfates* of *(290)* can also be obtained by O-alkylation of the carboxylic amides with dimethyl sulfate [19a,b]. Although these are frequently not so well crystallized as the tetrafluoroborates, they are equivalent to them in reactivity. From the methosulfates, amidinium salts have been prepared with primary or secondary amines in a manner analogous to that described above [19a,b].

As vinylogous amides, the enaminoketones *(296)* can be alkylated with trialkyloxonium salts [12b, 20a] (or with dimethyl sulfate [20b]) at the carbonyl group; the reactions of the alkylation products with primary amines yield vinylogous amidinium salts *(297)* [20a,b]:

(296) → (with $(C_2H_5)_3O^\oplus\ BF_4^\ominus$) → intermediate $BF_4^\ominus$ → (with RNH_2) → *(297)* $BF_4^\ominus$

Functional derivatives of the carbonyl group of carboxylic amides may also be obtained via the amide acetals *(292)*, which in turn are readily available from *(290)*. The reactions with semicarbazide, *p*-nitrophenylhydrazine, arylamines, or active methylene components and also trans acetalization (e.g., with glycol) can be understood if weak ionization of *(292)* by a reversal of its formation reaction is postulated.* Actually, a low electrical conductivity has been found for amide acetals [12b]:

---

* In addition to the subsequent products of the kinetically controlled addition of the nucleophile to the cation *(290)* formed as an intermediate, the thermodynamically controlled path of the reaction is also realized (Sections 5.3.2 and 5.3.3.1); for example, in the action of carboxylic acids (or phenols) on amide acetals [21a,b],

$$(H_3C)_2N-C(OR)_2H + R^1CO_2H \xrightarrow{-HOR} (H_3C)_2N=C^\oplus(OR)(H) + R^1CO_2^\ominus \rightarrow (H_3C)_2N-C(=O)H + R^1CO_2R$$

Carboxyl groups of peptides can be esterified in this way [21a].

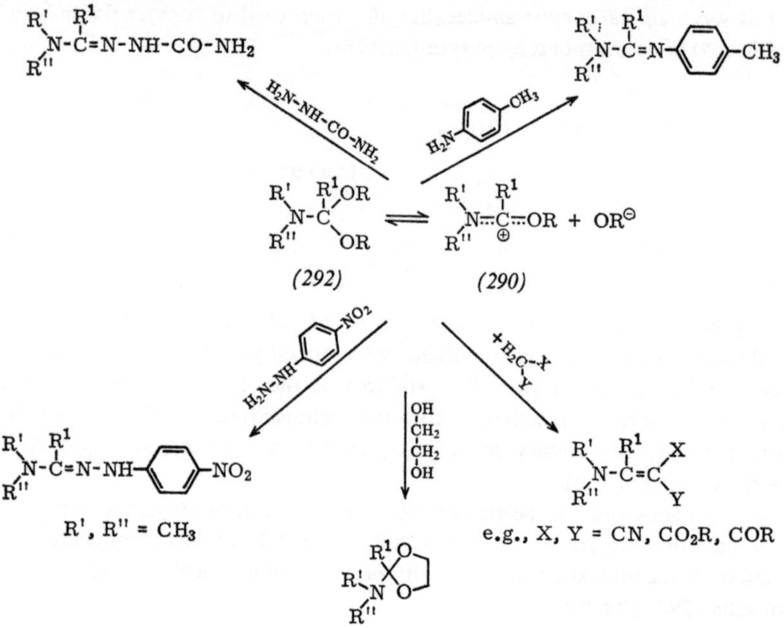

N,N-Dimethylformamide diethyl acetal has also been used with advantage for the preparation of cyanine dyes from heterocyclic quaternary salts [12b]; e.g.:

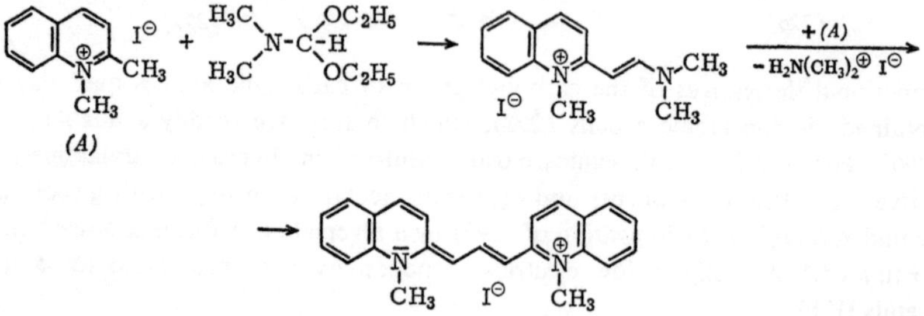

If an N,N-dialkylamide acetal *(292)* bears a methyl or methylene grouping as the substituent R^1, it is in equilibrium not only with the cation *(290)* but also with the ketene O,N-acetal *(298)*. The representatives of this class of compound, which is difficult to obtain by other means, can be readily obtained by heating suitable amide acetals with sodium [12b] (or calcium [19c]):

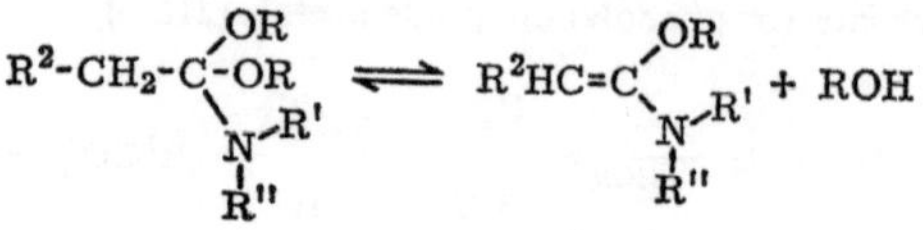

$$R^2\text{-}CH_2\text{-}C(OR)(OR)(NR'R'') \rightleftharpoons R^2HC=C(OR)(NR'R'') + ROH$$

(298)

where $R', R'' = CH_3$ and $R^2 = H$, alkyl.

Lactams

As cyclic amides, lactams behave analogously to the acyclic derivatives. They are readily alkylated at the carbonyl oxygen and then display the reactivity of the alkoxy(amino)carbonium ions [12b]. The possibility of the preparation of cyclic imidic esters is utilized very frequently when a hydrogen-bearing nitrogen atom is present in the lactam [22a,b,23]; e.g.:

[22a]

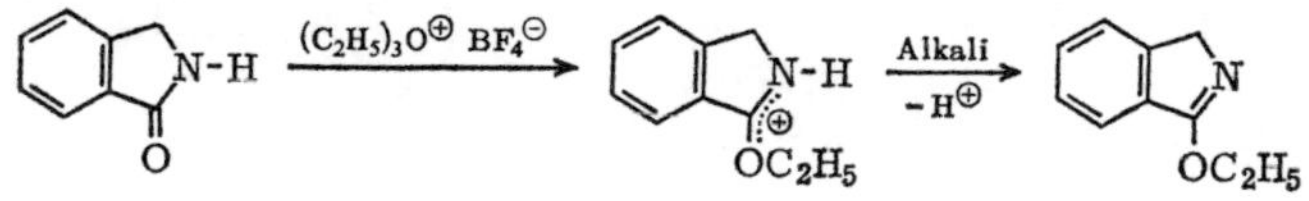

[22b]

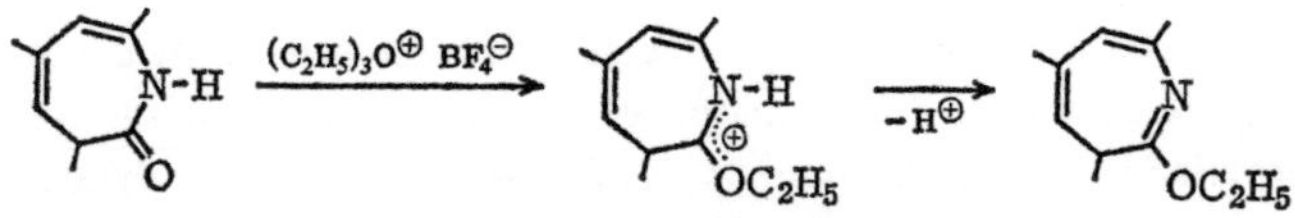

This transformation of lactams has been used successfully for the synthesis of intermediates in the building up of the corrin system [14, 24a,b]:

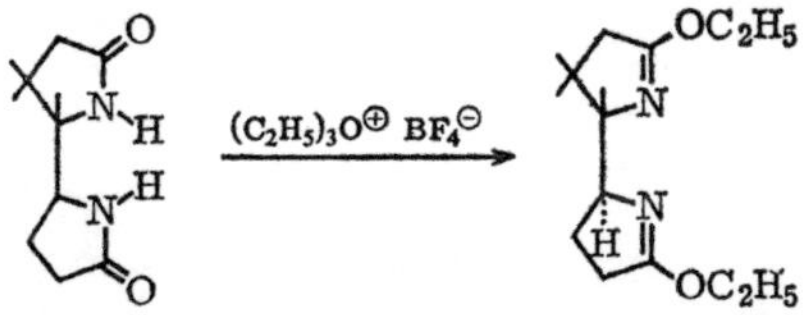

This elegant process for the introduction of C=N double bonds into heterocycles is also made use of in the synthesis of alkoxy derivatives of azacyclooctatetraene (*299*) [25] or azabullvalene (*300*) [26a,b], which are obtainable by the same principle, using chlorosulfonyl isocyanate:

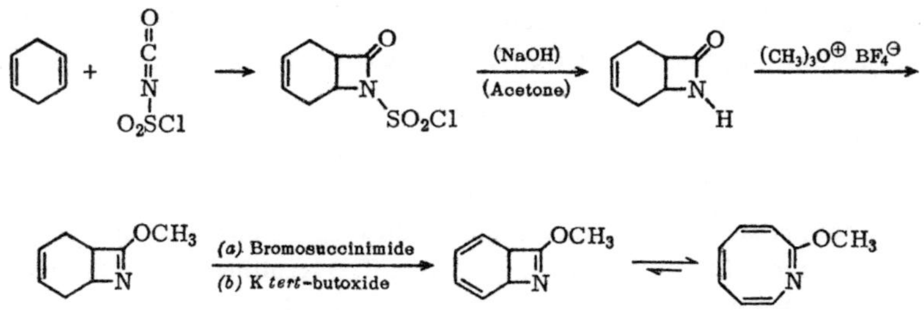

(*299*)

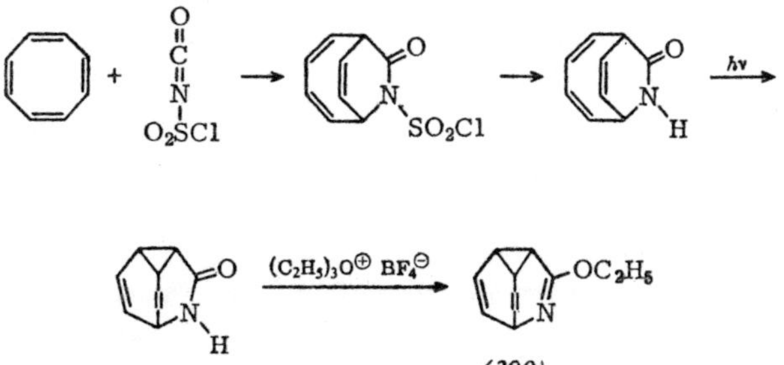

(*300*)

Preparative Application of Oxonium Salts

The O-alkylation of N-alkyllactams leads to salts that can be converted with alkoxides into the corresponding "lactam acetals" *(301)*. Like amide acetals *(292)*, these are also capable of undergoing condensations with semicarbazide, and especially with active methylene compounds [12b]:

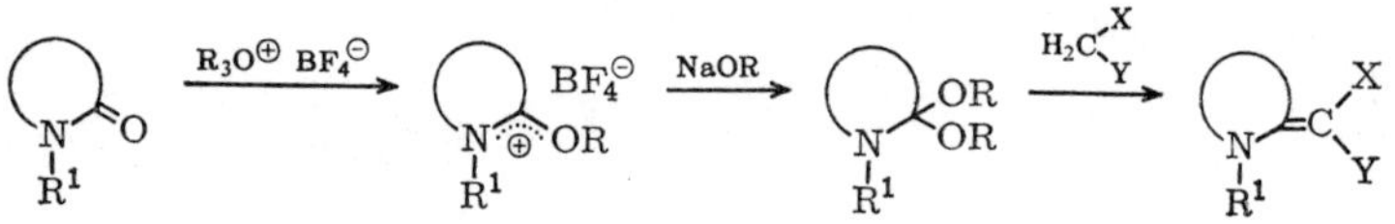

(301)

Here again the weak ionization of the acetals by a reversal of the formation reaction may be assumed. The reaction of lactam acetals with suitable quaternized heterocyclic nitrogen bases permits the formation of cyanine dyestuffs; e.g. *(302)* without additional condensation agents [12b]:

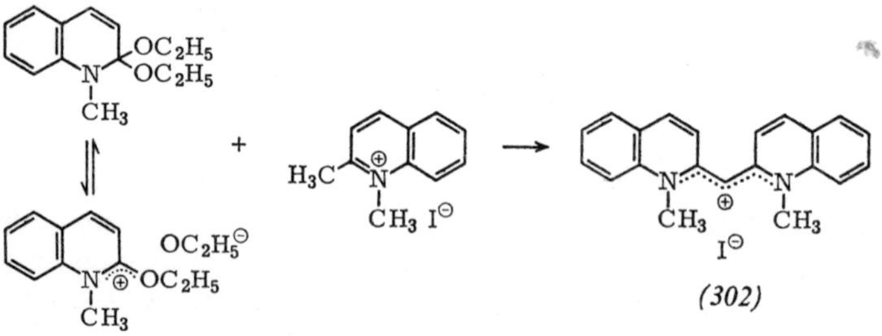

(302)

An analogy with the O-alkylated lactams is shown by the S-alkylated thiolactams [22a, 27], which in some cases have likewise been obtained with trialkyloxonium salts; *cf.* Table 8 in Chapter 5.

Urethane and Urea Derivatives

Ethyl-N,N-dimethylurethane and tetramethylurea can be alkylated smoothly with trialkyloxonium salts* at the carbonyl oxygen. The resulting alkylation products can be converted with alkoxides into the ortho-ester derivatives *(303)* and *(304)* [12b]. In the urea acetal *(304)*, an amino group can be readily replaced by an alkoxy group by the action of an alcohol [12b, 28]:

---

* Dialkyl sulfates have also been used successfully for alkylating urea derivatives [28].

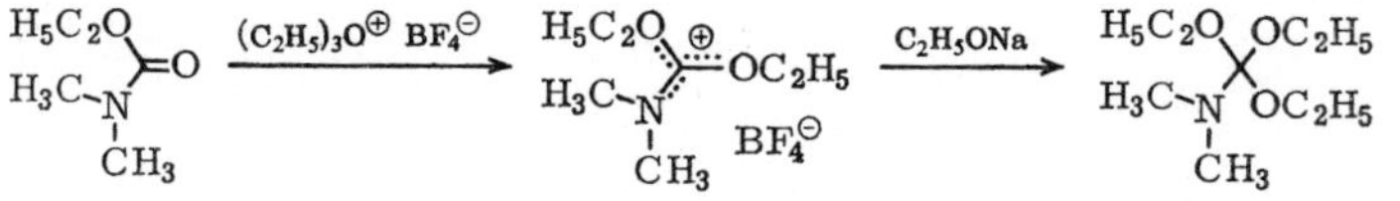

Alkylation with Trialkyloxonium Salts

(303)

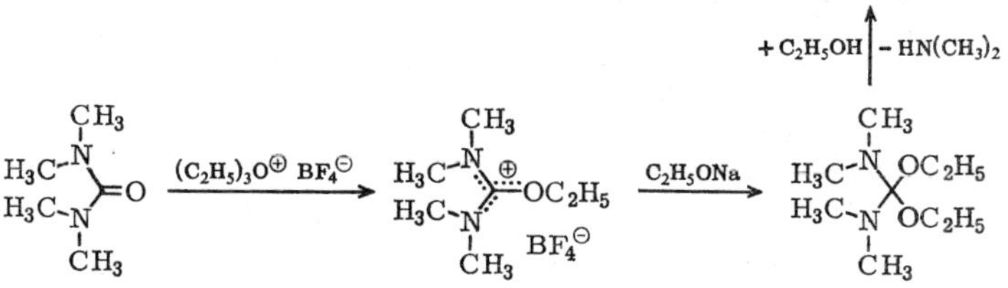

(304)

The alkylation product of N,N-dimethylbenzimidazolone is particularly stable and is not hydrolyzed even on heating with water. On the other hand, the urea acetal is readily obtainable by the action of an alkoxide, and this (like amide acetals) can be smoothly condensed with active methylene (or methyl) components* [12b]:

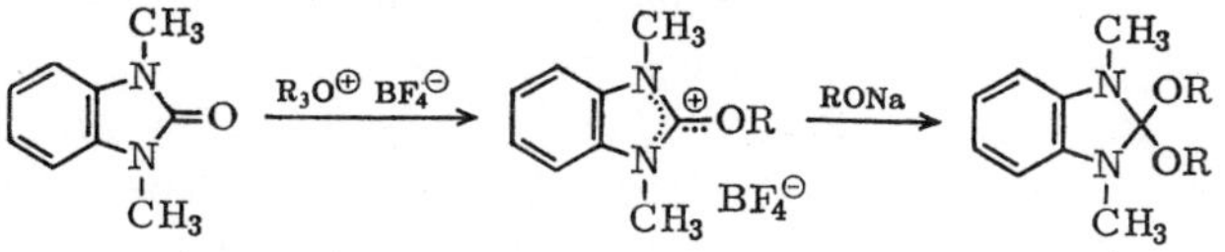

where $R = C_2H_5$.

7.1.1.3. Nitrile Alkylation

The reactions of nitrilium salts are related to the reactions of O-alkylated amides. Meerwein and associates obtained the first N-alkylnitrilium salts by the alkylation of nitriles with trialkyloxonium salts [29]**. Here the carbon atom of the nitrile $C \equiv N$ bond is activated for nucleophilic attack. Thus the action of water or hydroxyl ions on N-alkylnitrilium salts gives rise to amides; alcohols give imidic esters; esters; ammonia or primary amines yield amidines [29]; and, with sodium borohydride and alcohol, reduction gives secondary amines [30]. In the last-mentioned reaction the formation of imidic esters can be detected at low temperatures [30].

Possibilities of convenient transformations of the nitrile function via N-alkylnitrilium salts are shown by the following scheme:

---

* Acyclic urea acetals do not react uniformly with active methylene compounds, since, besides the alkoxygroups, an amino group is readily replaced [12b].
** Nitrile alkylation also takes place with dialkoxy carbonium hexachloroantimonates [3a,b,c]. In addition, nitrilium salts have meanwhile become available in the following ways: by the reaction of nitrile–$SbCl_5$ adducts with alkyl halides or of N-substituted imidochlorides with $SbCl_5$, and by the N-arylation of nitriles with aryldiazonium tetrafluoroborates [29].

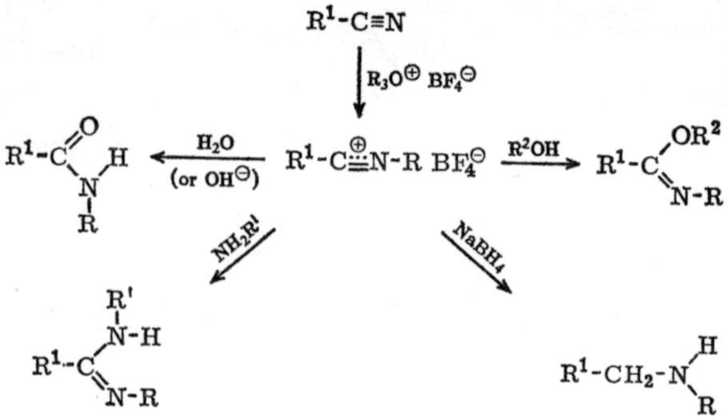

7.1.2. Trialkyloxonium Salts as Anion Acceptors

7.1.2.1. Alkoxide Transfer

The alkyl cations provided by trialkyloxonium salts can take up anionically expelled groups from suitable compounds; see Section 5.2.1. If the leaving groups are alkoxide residues the transfer of ions must be formulated via trialkoxonium ions occurring as intermediates. Such reactions have been used by Meerwein et al., especially for the production of alkoxycarbonium and dialkoxycarbonium ions from acetals and ortho esters [1]. Cf. also Table 5 in Chapter 3, e.g.:

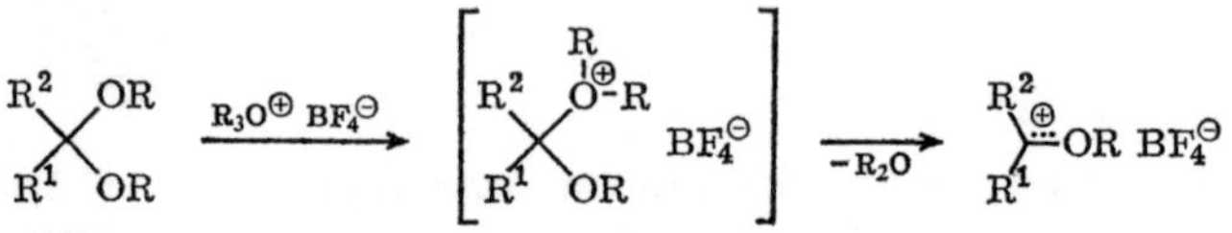

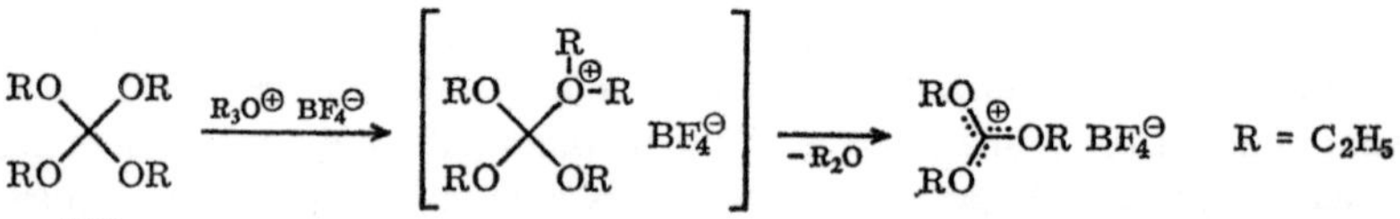

Ethers can also be split in this manner, provided stable carbonium ions are formed, as in the production of the 1,3-dioxolenium salt (66) from β-alkoxyethyl acetate [31a]:

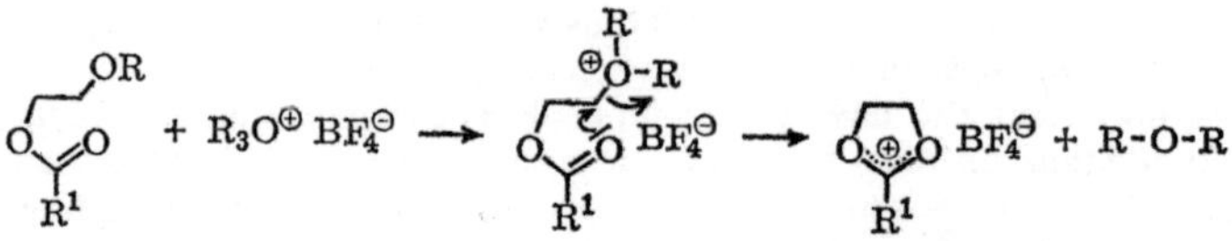

where $R = C_2H_5$ and $R_2 = CH_3$.

In the following newer example a 1,4-chlorine shift is enabled by a solvolytic reaction, in which dimethyl ether derived from an intermediate oxonium ion serves as the leaving group [31d]:

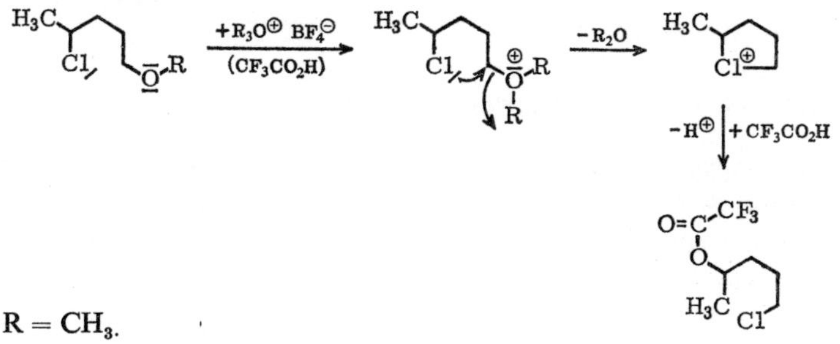

where R = CH_3.

7.1.2.2. Hydride Transfers

The reaction of 2-phenyl-1,3-dioxolane with trialkyloxonium salts has in fact been formulated as a direct transfer of a hydride ion to an alkyl cation [31b,c]:

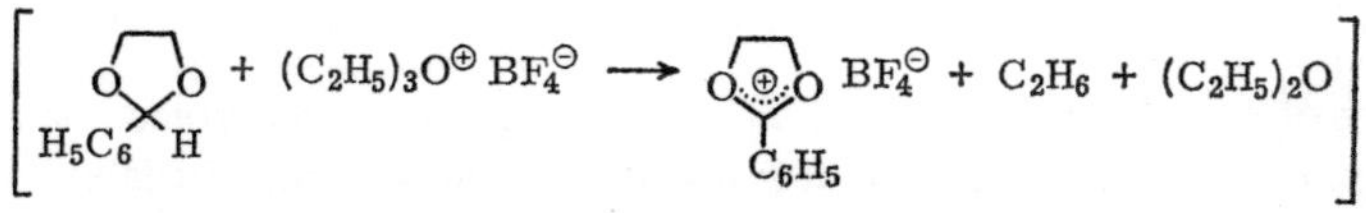

The ethane expected according to this equation appeared to have been detected by combustion analyses [31b], but this could not be confirmed by gas chromatography [3c]*. The active hydride acceptor is probably an aldehydium ion formed by the O-alkylation of the dioxolane and its ring opening [3b,c]:

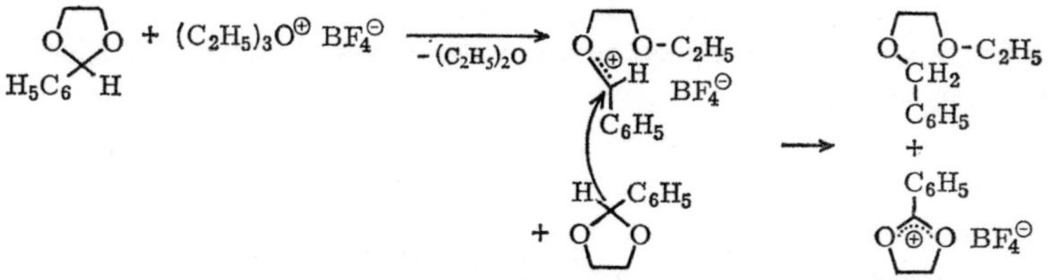

The benzyl ether formulated here is also incapable of detection under the reaction conditions, but this disproportionation mechanism is supported by the fact that the yields of dioxolenium salt never exceed 50% [3b,c, 31b]. On the other hand, the preparation of the boronylide *(305)* with a triethyloxonium salt can hardly be explained otherwise than by direct hydride transfer [32a]:

---

* *Note added in proof (Jan. 1970)*: In contrast to this result ethane has been detected in a recent investigation on the reaction of 1,3-dioxolane with triethyloxonium tetrafluoroborate [32b]. Therefore direct hydride transfer may be possible, too.

$$[(C_4H_9)_3\overset{\ominus}{B}-CH_2-C_3H_7]\,Li^{\oplus} + (C_2H_5)_3O^{\oplus}\,BF_4^{\ominus} \xrightarrow[-(C_2H_5)_2O]{-LiBF_4} (C_4H_9)_3\overset{\ominus}{B}-\overset{\overset{\oplus}{O}\diagdown H}{\underset{C_3H_7}{C}\diagup} + C_2H_6$$

(305)

7.1.3. Catalytic Action of Trialkyloxonium Salts

Like protons, alkyl cations can act as acidic catalysts: Thus, for example, the rearrangement of 2,2-dialkoxytetrahydrofuran into γ-alkoxybutyric esters (Section 5.3.2), which is catalyzed by protons or Lewis acids, can be brought about smoothly by catalytic amounts of trialkyloxonium salts [1,2].

where $R = C_2H_5$.

The butyrolactonium salt *(54)* can also be obtained from butyrolactone and trialkyloxonium salts. Consequently it is possible in the presence of trialkyloxonium salts to bring about the addition of ether to butyrolactone via the cations *(54)* and *(306)* [1]. The overall reaction so achieved,

does not take place with catalysis by protons.

The formation of glycol dialkyl ethers by the action of ethers to epoxides takes place analogously [33]:

where $R = C_2H_5, n\text{-}C_3H_7$.

With BF_3 etherate, epichlorohydrin can form the internal oxonium salt *(68)* as an intermediate—as well as the trialkyloxonium salt $R_3O^{\oplus}BF_4^{\ominus}$ itself—see Section 4.2.1. Consequently the formation of glycol ethers can also be achieved with a reaction mixture consisting of epichlorohydrin, the ether, and catalytic amounts of BF_3 etherate [33], which can be shown schematically in the following way:

Alkylation with Trialkyloxonium Salts

[Reaction scheme showing epichlorohydrin + R$_2$O→BF$_3$ leading via cyclic oxonium ion (307) to various products]

Besides the glycol dialkyl ether, ethers of higher molecular weight are always observed as by-products in these reactions, being derived from the cyclic oxonium ion *(307)* by the addition of epichlorohydrin [33]*:

[Reaction scheme showing (307) reacting with epichlorohydrin to give extended ether products + R$_3$O$^⊕$]

This reaction is only a special case of polymerizations of cyclic ethers under the influence of trialkyloxonium salts. The polymerization of tetrahydrofuran has been studied very thoroughly [34a,b]. The cyclic oxonium ion arising in the primary step reacts with another tetrahydrofuran molecule with ring opening to give a new tetrahydrofuranium ion, which in turn can add to another tetrahydrofuran molecule:

[Reaction scheme showing THF polymerization with R$_3$O$^⊕$ BF$_4^⊖$]

---

* *Note added in proof (June 1970):* A similar example of this type is the alkoxylation of alcohols with alkylene oxides catalyzed by trialkyloxonium salts [34d].

142   Preparative Application of Oxonium Salts

The reaction of the long-chain oxonium ion forming in this way with nucleophilic anions interrupts the growth of the chain. The polyether chains therefore become particularly long when stable complex anions such as BF_4^- and $SbCl_6^-$ are used.

Not only trialkyloxonium salts can start such a polymerization. Quite generally, all Lewis acids A^+ capable of addition to the oxygen atom of the cyclic ether can be used if a complex anion X^- is added simultaneously*:

$$A^\oplus \ X^\ominus + O\overset{}{\underset{}{\bigcirc}} \longrightarrow A-\overset{\oplus}{O}\overset{}{\underset{}{\bigcirc}} X^\ominus \xrightarrow{+O\overset{}{\underset{}{\bigcirc}}} A-O-(CH_2)_4-\overset{\oplus}{O}\overset{}{\underset{}{\bigcirc}} \text{ etc.}$$

Generally, however, trialkyloxonium salts are used as polymerization catalysts. Examples of this are, in addition to the polymerization of tetrahydrofuran [34a,b,c], copolymerizations in which tetrahydrofuran [35] or trioxane act as components [36].

7.2. Acyclic Alkoxycarbonium Salts

7.2.1. Use as Alkylating Agents

The weakly nucleophilic carbonyl functions of aryl aldehydes and aryl ketones cannot be alkylated satisfactorily with trialkyloxonium salts, but they can be with dialkoxycarbonium salts; see Section 7.1.1.1. Consequently it must be assumed that the aryl aldehydium and ketonium ions [3a] themselves can also act as alkylating agents. In fact, the action of the O-ethylbenzophenonium salts *(308)* on diethyl ether yields the trialkyloxonium salt [3b]:

$$\underset{R^1}{\overset{H_5C_6}{\diagdown}}C\overset{\oplus}{=}OC_2H_5 \ SbCl_6^\ominus + (C_2H_5)_2O \longrightarrow \underset{R^1}{\overset{H_5C_6}{\diagdown}}C=O + (C_2H_5)_3O^\oplus \ SbCl_6^\ominus$$

(308)

where $R^1 = C_6H_5$.

Aryl ketonium salts can alkylate other nucleophilic, electrically neutral molecules in this way. The preparative importance of this reaction is still slight. Table 10 gives a review of the alkylations carried out. Nitriles cannot be alkylated by this method [3b].

---

* The following act in this sense: Alkoxycarbonium and dialkoxycarbonium salts (generally produced in situ by the action of BF_3 or $SbCl_5$ on acetals or orthocarboxylic esters), acylium salts (also generally prepared in situ by the reaction of acetic anhydride with HCl, BF_3, or $AlCl_3$, or of acyl halides with BF_3, $AlCl_3$, or $FeCl_3$); *cf.* Section 5.2.3. Even complex protonic acids (e.g., $HClO_4$, HBF_4) or mixtures of two Lewis acids (e.g., $FeCl_3$ and $AlCl_3$) can be used [34a].

TABLE 10. ALKYLATION WITH PHENYL KETONIUM SALTS [3b]

$$\underset{R^1}{\overset{H_5C_6}{>}}\overset{\oplus}{C}\text{-OR SbCl}_6^{\ominus} + R'\text{-Y:} \longrightarrow \underset{R^1}{\overset{H_5C_6}{>}}C=\overset{\oplus}{O} + R'-\overset{\oplus}{Y}-R \text{ SbCl}_6^{\ominus}$$

Ketonium salt		Nucleophile	Alkylation products	
R^1	R	R'–Y:	$R'-\overset{\oplus}{Y}-R$ SbCl$_6^{\ominus}$	% yield
C_6H_5	C_2H_5	$(C_6H_5)_2S=O$:	$(C_6H_5)_2\overset{\oplus}{S}-O-C_2H_5$	80
C_6H_5	C_2H_5	$(C_2H_5)_2O$:	$(C_2H_5)_3O^{\oplus}$	81
C_6H_5	C_2H_5	⟨O:⟩	⟨$\overset{\oplus}{O}$-C$_2$H$_5$⟩	90
C_6H_5	C_2H_5	$H_5C_6-C\underset{N(C_2H_5)_2}{\overset{O:}{\diagup}}$	$H_5C_6-C\underset{N(C_2H_5)_2}{\overset{OC_2H_5}{\diagup\oplus}}$	43
CH_3	C_2H_5	$(C_2H_5)_2O$:	$(C_2H_5)_3O^{\oplus}$	65
CH_3	C_2H_5	⟨O:⟩	⟨$\overset{\oplus}{O}$-C$_2$H$_5$⟩	72

7.2.2. Use as Anion Acceptors

In the case of alkoxycarbonium ions, the kinetically controlled nucleophilic addition to the carbonium center is in general preferred to dealkylation; see Section 5.3.3.1. We have already become acquainted with a series of such addition reactions in connection with carbonyl activation by means of trialkyloxonium salts; see Section 7.1.1.1. Alkoxycarbonium salts should therefore also be suitable for taking up anionically expelled groups of uncharged molecules with the formation of new cations. Such anion transfers have recently been investigated, particularly with aryl ketonium salts [3b,c]. Thus, O-ethylacetophenonium salts are capable of taking up alkoxide, hydride, or carbeniate ions from dioxolane derivatives:

$$\underset{R^1}{\overset{H_5C_6}{>}}\overset{\oplus}{C}\text{-OC}_2H_5 \text{ SbCl}_6^{\ominus} + \underset{H_5C_6\ \ Y}{\langle O\diagdown O\rangle} \longrightarrow \underset{R^1}{\overset{H_5C_6}{>}}\underset{OC_2H_5}{\overset{Y}{C}} + \underset{C_6H_5}{\langle O\overset{\oplus}{\diagdown} O\rangle} \text{ SbCl}_6^{\ominus}$$

where Y = OR, H, CH$_3$.

The splitting out of hydride or alkylide ions is of particular preparative importance. Table 11 gives a review of anion transfers of this type. Occasionally, alkoxycarbonium ions arising in situ may also act as hydride acceptors [37a,b,c]. Particularly in the case of cyclic acetals, hydride transfers with ring opening occur at high temperatures (350–400 °C) in the presence of acidic catalysts.

Here a hydride ion from the neighboring hydroxymethyl group is probably transferred to the alkoxycarbonium ion, which can be shown formally in the following way [37b]:

TABLE 11. ANION TRANSFERS TO PHENYL KETONIUM SALTS [3b,c]

$$R'-Y + \underset{R^1}{\overset{H_5C_6}{\diagdown}}\!\!\!\underset{OR}{\overset{\oplus}{\diagup}} SbCl_6^{\ominus} \longrightarrow R'^{\oplus} SbCl_6^{\ominus} + \underset{R^1}{\overset{H_5C_6}{\diagdown}}\!\!\!\underset{OR}{\overset{Y}{\diagup}}$$

Transferred anion $Y^{\ominus}$	Anion-donor R'–Y:	Anion-acceptor R^1	R	Salt isolated R$^{\oplus}$ SbCl$_6^{\ominus}$	% yield	Addition product isolated $\underset{R^1}{\overset{H_5C_6}{\diagdown}}\!\!\!\underset{OR}{\overset{Y}{\diagup}}$	% yield
H	[structure: dioxolane with C$_6$H$_5$, H]	C$_6$H$_5$	C$_2$H$_5$	[cyclopentadienyl-C$_6$H$_5$]	67		
	[cycloheptatriene]	C$_6$H$_5$	C$_2$H$_5$	[tropylium]	98	H$_5$C$_6$–C(OC$_2$H$_5$)(OC$_2$H$_5$)–C$_6$H$_5$ ... H$_5$C$_2$ ... OC$_2$H$_5$	71
	same	CH$_3$	C$_2$H$_5$	same	60	H$_5$C$_6$ H [a] / H$_3$C OR (Cl)	~57
	[xanthene]	C$_6$H$_5$	n-C$_3$H$_7$	[xanthylium]	80		
		CH$_3$	C$_2$H$_5$		51		
	Ar$_3$CH[b]	C$_6$H$_5$	n-C$_3$H$_7$	Ar$_3$C$^{\oplus}$	60		
		CH$_3$	C$_2$H$_5$	Same	~60		
	Ar$_2$(C$_6$H$_5$)CH[b]	C$_6$H$_5$	n-C$_3$H$_7$	Ar$_2$(C$_6$H$_5$)C$^{\oplus}$[b]	41		
	Ar(C$_6$H$_5$)$_2$CH[b]	C$_6$H$_5$	n-C$_3$H$_7$	Ar(C$_6$H$_5$)$_2$C$^{\oplus}$[b]	38		
	N(CH$_2$C$_6$H$_5$)$_3$	C$_6$H$_5$	n-C$_3$H$_7$	(C$_6$H$_5$–CH$_2$)$_2$N$^{\oplus}$=CH(C$_6$H$_5$)	75		
CH$_3$	[dioxolane C$_6$H$_5$/CH$_3$]	CH$_3$	C$_2$H$_5$	[cyclopentadienyl-C$_6$H$_5$]	55	$\left(\underset{H_3C}{\overset{H_5C_6}{\diagdown}}C{=}CH_2 \right)_2$ [c]	
C$_2$H$_5$O	[dioxolane C$_6$H$_5$/OC$_2$H$_5$]	CH$_3$	C$_2$H$_5$	[cyclopentadienyl-C$_6$H$_5$]	95		
	[dioxolane H/OC$_2$H$_5$]	C$_6$H$_5$	C$_2$H$_5$	[dioxolanylium with H]	100	H$_5$C$_6$ H / H$_5$C$_6$ OR	51
		CH$_3$	C$_2$H$_5$	same	32		

[a] Under the reaction conditions the ether is partially converted into the chloride.

[b] Ar = –C$_6$H$_4$–OCH$_3$

[c] Under the reaction conditions the ether formed first is converted into an alkene; yield not determined.

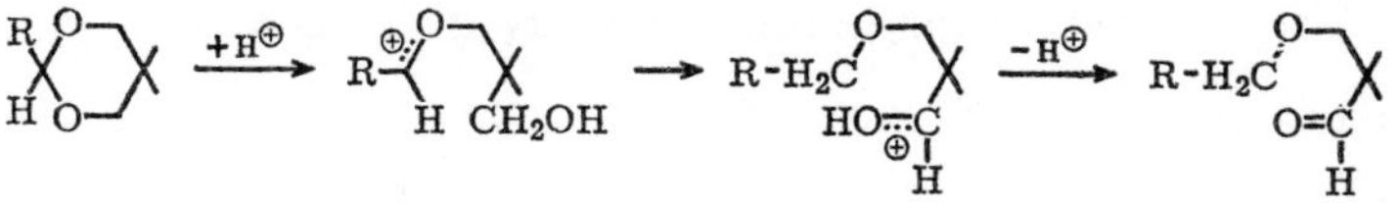

Intermolecular hydride transfer is also known, as in the action of 30% sulfuric acid on 1-hydroxyisochromane [37c]:

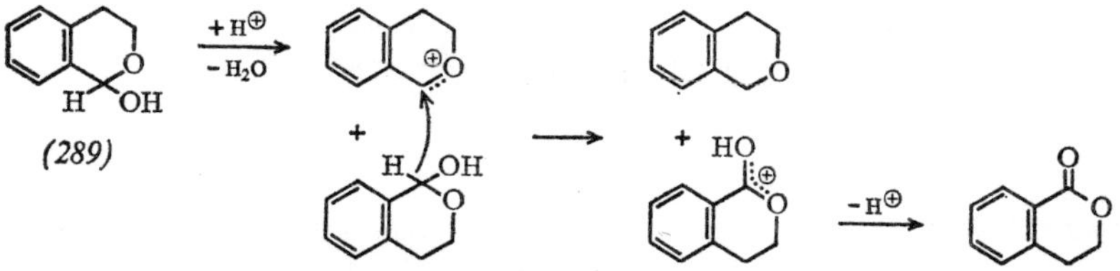

Reference may be made to the connection with the disproportionation of cyclic acetals or ketals; see Sections 4.2.2.2 and 7.1.2.2.

The stereochemistry of the addition of hydride to the alkoxycarbonium salts of camphor and norcamphor has also been observed in a recent investigation [37d]. The action of lithium tetrahydroborate on the two carbonium salts leads to reduction in each case, preferentially from the sterically less-inhibited side. Consequently, the camphor derivative (R = CH$_3$) gives predominantly the *exo*-ether (*a*) and the norcamphor derivative (R = H) preferentially the *endo*-ether (*b*) [37d]:

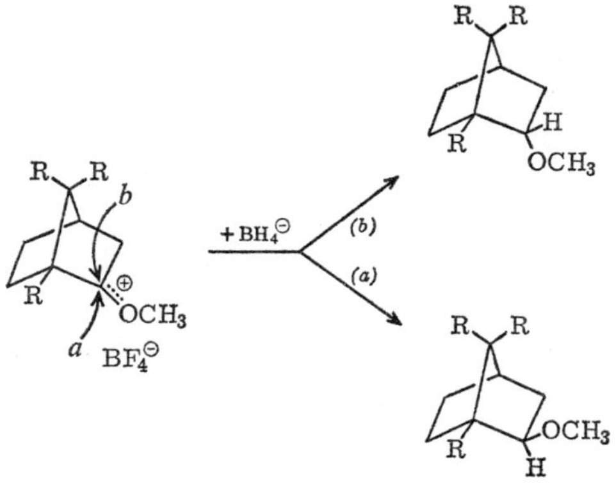

	R = CH$_3$	H
(a)	13	1
(b)	1	20

7.3. Preparative Use of Dialkoxycarbonium Salts

The monoalkoxycarbonium ions treated in the preceding section resemble carbonium ions in chemical behavior and (as the latter do) accept nucleophiles predominantly at the carbonium center. However, the dialkoxycarbonium ions *(9)*, which have lower energy, are more like oxonium ions, and like the latter they easily transfer alkyl cations to suitable nucleophiles; see Section 5.3.2.

7.3.1. Use of Acyclic Dialkoxycarbonium Salts as Alkylating Agents

The general preparative importance of acyclic dialkoxycarbonium salts as alkylating agents has been recognized only recently. The dialkoxycarbonium salts *(9)*, which are readily available from orthocarboxylic esters with BF_3 or $SbCl_5$, are dealkylated by numerous nucleophilic, uncharged molecules R'–Y to carboxylic esters; schematically:

$$R^1-C(OR)(OR)^{\oplus} \; SbCl_6^{\ominus} + R'-Y: \longrightarrow R^1-C(=O)(OR) + R'-\overset{\oplus}{Y}-R \; SbCl_6^{\ominus}$$

(9)

Acyclic dialkoxycarbonium salts are actually stronger alkylating agents than are trialkyloxonium salts. This is shown by the fact that they can smoothly alkylate ethers to trialkyloxonium salts, while the reverse reaction (alkylation of carboxylic esters with trialkyloxonium salts) is impossible [3a]. The very weakly nucleophilic carbonyl oxygen atoms of acetophenone, benzaldehyde, and even higher carboxylic esters can be alkylated with *(9)*, which again is impossible with trialkyloxonium salts. Table 12 gives a review of the more important alkylations.

It is worth noting that di-*n*-butoxycarbonium salts also have an alkylating action on aromatics. However, toluene gives no *n*-butyltoluene derivatives but does give the *sec*-butyl-substituted product. Again, in the decomposition of di-*n*-butoxycarbonium hexachloroantimonate the formation of *sec*-butyl chloride is observed [3b]:

$$H-C^{\oplus}(O-CH_2-CH_2-CH_2-CH_3)_2 \; SbCl_6^{\ominus} \longrightarrow H-C(=O \rightarrow SbCl_5)(O-(n\text{-Bu})) + H_3C-\overset{Cl}{\underset{|}{C}}H-CH_2-CH_3$$

where *n*-Bu = *n*-butyl.

In contrast, the *n*-butyl cation is transferred to aryl ketones without isomerization [3b]:

$$Ar-C(=O)(R^1) + H-C^{\oplus}(O-(n\text{-Bu}))_2 \; SbCl_6^{\ominus} \longrightarrow H-C(=O)(O-(n\text{-Bu})) + Ar-C(\overset{\oplus}{O}\text{-}(n\text{-Bu}))(R^1) \; SbCl_6^{\ominus}$$

where Ar = aryl.

The different behaviors are explained by different reaction mechanisms: In the cases where no isomerization of the alkyl group is observed an S_N2 mechanism may be inferred, and where isomerization does occur, an S_N1 mechanism [3b].

7.3.2. Use of Dialkoxycarbonium Salts as Anion–Acceptors

The carbonium ion character of dialkoxycarbonium ions is weakly expressed; in general, only very strong nucleophiles are added at the carbonium center of *(9)*. *Cf.* Table 9. However, some cases are known in which anionically expelled groups

TABLE 12. ALKYLATIONS WITH ACYCLIC DIALKOXYCARBONIUM SALTS (9) $R^1-\overset{OR}{\underset{OR}{C}}\cdot X^\ominus + R^1\text{-}Y\text{:} \longrightarrow R'\text{-}\overset{\oplus}{Y}\text{-}R\ X^\ominus + R^1-\overset{O}{\underset{}{C}}\diagdown_{OR}$

Dialkoxycarbonium salt (9)			Nucleophile R'-Y:	Alkylation product R'-Y-R X⊖	% yield	Reference
R^1	R	$X^\ominus$				
H	CH_3	$BF_4^\ominus$	(indole structure)	(N-methylindolium)	~95	[38a]
H	C_2H_5	$SbCl_6^\ominus$	$C_6H_5-C\equiv N$:	$C_6H_5-\overset{\oplus}{C}\equiv N-C_2H_5$	90	[3a]
CH_3	C_2H_5	$SbCl_6^\ominus$			61	[3a]
H	C_2H_5	$SbCl_6^\ominus$	$(C_6H_5)_2S=O$:	$(C_6H_5)_2\overset{\oplus}{S}-OC_2H_5$	92	[3a]
C_6H_5	CH_3	$BF_4^\ominus$	(pyranone)	(methoxypyrylium)	~30	[38b]
2,4,6-$(CH_3)_3C_6H_2$	CH_3	$BF_4^\ominus$			~60	[38b]
H	C_2H_5	$SbCl_6^\ominus$	$(C_6H_5)_2C=O$:	$(C_6H_5)_2\overset{\oplus}{C}=OC_2H_5$	90	[3a]
H	C_2H_5	$SbCl_6^\ominus$	$\underset{H_3C}{\overset{H_3C}{>}}C=O:$	$\underset{H_3C}{\overset{H_3C}{>}}\overset{\oplus}{C}-OC_2H_5$	91	[3a]
H	C_2H_5	$SbCl_6^\ominus$	$\underset{H_3C}{\overset{H}{>}}C=O:$	$\underset{H_3C}{\overset{H}{>}}\overset{\oplus}{C}-OC_2H_5$	91	[3a]
H	C_2H_5	$SbCl_6^\ominus$	$\underset{(CH_3)_2N}{\overset{H_5C_6}{>}}C=O:$	$\underset{(CH_3)_2N}{\overset{H_5C_6}{>}}\overset{\oplus}{C}-OC_2H_5$	95	[3a]
CH_3	C_2H_5	$SbCl_6^\ominus$			91	[3a]
H	C_2H_5	$SbCl_6^\ominus$	$\underset{(C_6H_5)_2N}{\overset{H_3C}{>}}C=O:$	$\underset{(C_6H_5)_2N}{\overset{H_3C}{>}}\overset{\oplus}{C}-OC_2H_5$	97	[3a]
H	C_2H_5	$SbCl_6^\ominus$	$C_6H_5-CH_2-\overset{}{\underset{}{C}}\diagdown_{OC_2H_5}^{O:}$	$C_6H_5-CH_2-\overset{\oplus}{C}\diagdown_{OC_2H_5}^{OC_2H_5}$	92	[3a]
H	C_2H_5	$SbCl_6^\ominus$	$(CH_3)_2CH-\overset{}{\underset{}{C}}\diagdown_{OC_2H_5}^{O:}$	$(CH_3)_2CH-\overset{\oplus}{C}\diagdown_{OC_2H_5}^{OC_2H_5}$	62	[3a]
H	C_2H_5	$BF_4^\ominus$	$(H_3C_2O)_3P$:	$(H_3C_2O)_3\overset{\oplus}{P}(C_2H_5)$	(a)	[38c]
2,4,6-$(CH_3)_3C_6H_2$	CH_3	$BF_4^\ominus$	(catechol)	(methoxyphenol, b)	83	[38b]

[a] Yield not reported. [b] Subsequent product of R'-Y-R.

Y of electrically neutral molecules R′–Y are transferred to dialkoxycarbonium salts; schematically:

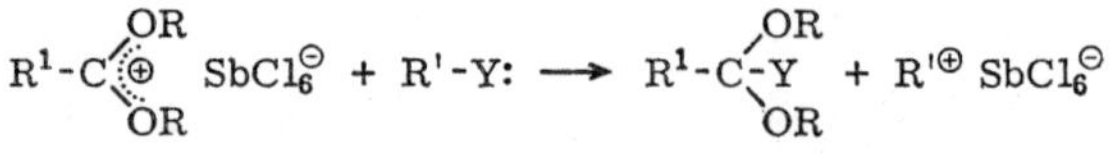

$$R^1-C(OR)_2^{\oplus}\ SbCl_6^{\ominus} + R'-Y: \longrightarrow R^1-C(OR)_2-Y + R'^{\oplus}\ SbCl_6^{\ominus}$$

(9)

Nevertheless the use of dialkoxycarbonium salts offers no advantage as compared with other anion–acceptors with an equivalent action; cf. Section 4.1.3.

The taking up of *hydride* ions (Y = H) occurs smoothly only when tri(p-methoxyphenyl)methane and xanthene are used as hydride donors; triarylcarbonium or xanthylium salts are then produced [3b]. If in tri(p-methoxyphenyl)methane one methoxyphenyl residue is replaced by a phenyl group, the corresponding diarylcarbonium salt is obtained in only low yield (<10%) [3b].

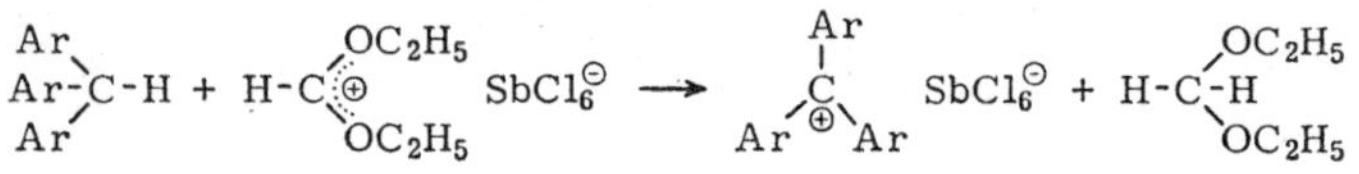

$$Ar_3C-H + H-C(OC_2H_5)_2^{\oplus}\ SbCl_6^{\ominus} \longrightarrow Ar_3C^{\oplus}\ SbCl_6^{\ominus} + H-C(OC_2H_5)_2-H$$

(Ar = aryl)

correspondingly:

xanthene $\longrightarrow$ xanthylium $SbCl_6^{\ominus}$

The transfer of *alkoxide* groups (Y = OR) to acyclic dialkoxycarbonium salts is observed only in isolated instances. Here the alkoxide donors are orthocarboxylic esters or amide acetals [31c]:

cyclic orthoester $+ H-C(OC_2H_5)_2^{\oplus}\ BF_4^{\ominus} \longrightarrow$ [cyclic carbonate]$-CH_3\ BF_4^{\ominus} + H-C(OC_2H_5)_3$

(73%)

correspondingly:

$C(OC_2H_5)_4 \longrightarrow (H_5C_2O)_3C^{\oplus}-OC_2H_5\ BF_4^{\ominus}$

(88%)

$HC(OC_2H_5)_2N(CH_3)_2 \longrightarrow (H_5C_2O)_2C^{\oplus}-N(CH_3)_2\ BF_4^{\ominus}$

(92%)

The reaction between an ortho ester and the dialkoxycarbonium ions *(9)* takes place in this sense, i.e., kinetically controlled (by route B) only when the salt of the newly formed cation *(9a)* is sparingly soluble and precipitates. Otherwise the thermodynamically stable products are produced (by route A) with dealkylation of *(9)*, which finally leads to the formation of trialkyloxonium salts *(7)* [3b]:

$$R^1-\overset{OR}{\underset{OR}{C^\oplus}} + \underset{R^2}{\overset{R'O}{\underset{}{\times}}}\overset{OR'}{\underset{OR'}{}} \underset{\text{route B}}{\rightleftharpoons} \underset{R^1}{\overset{R'O}{\underset{}{\times}}}\overset{OR}{\underset{OR}{}} + R^2-\overset{OR'}{\underset{OR'}{C^\oplus}}$$

(9) route A *(9a)*

$$R^1-\overset{O}{\underset{OR}{C}} + \left[\underset{R^2}{\overset{R'O}{\underset{}{\times}}}\overset{OR'}{\underset{\overset{\oplus}{O}-R'}{}} \rightarrow \underset{R}{\overset{R'}{\underset{}{O}}} + R^2-\overset{OR'}{\underset{OR'}{C^\oplus}} \rightarrow \right] R'-\overset{\oplus}{\underset{R}{O}}\overset{R'}{\underset{}{}} + R^2-\overset{O}{\underset{OR'}{C}}$$

(9a) *(7)*

It may also be mentioned that the transfer of the cyano group to *(9)* has been observed; see Table 7 [31c] in Chapter 4.

Reactions of cyclic dialkoxycarbonium salts have been treated in detail in Sections 5.3.2 and 6.3.

7.4. Reactions of Pyrylium Salts

Among the cyclic alkoxycarbonium salts, only the pyrylium salts have acquired preparative importance. These are important key compounds for the preparation of heterocyclic and isocyclic aromatics, which can be obtained by ring-closure reactions from the initially formed addition products of nucleophiles with pyrylium ions.

7.4.1. Primary Reactions

7.4.1.1. Review of Reaction Possibilities

Nucleophiles $Y^\ominus$ can add to the pyrylium ion at C atoms 2, 6, or 4. The resulting 2*H*- or 4*H*-pyrans *(85)* and *(90)* correspond to the products of the kinetically controlled nucleophilic addition, in the case of carboxonium ions, by route B; *cf.* Section 5.4.1. Dealkylation of the oxonium oxygen with the direct formation of the pentadienone *(84)*, route A, is indeed conceivable, but *(84)* is normally the subsequent product from the 2*H*-pyran *(85)* with which it is valence-isomeric [39a,b]. The conversion of *(90)* into *(85)* can be carried out by photolysis in special cases, and is occasionally assumed as an intermediate reaction step [40]:

In addition, the nucleophile may also take up protons by route C from suitable proton-bearing substituents X–H at C_2 (or C_6) or C_4 of the pyrylium ion. This gives α- or γ-derivatives *(87)* or *(88)* of the pyrones (X=O) or methylenepyrans ("dehydropyrans") (X=CR$_2$):

where X=O or CR$_2$.

7.4.1.2. Isolable Primary Products from Pyrylium Salts

Nucleophilic Addition to C_2

Provided no particular steric effects are operating, nucleophilic addition normally takes place preferentially in position 2 (or 6) of the pyrylium ion [41]. However, the 2H-pyrans so formed *(85)* can be isolated only rarely. The 2H-pyran structure *(309)* appears certain in the reaction product from the 2,4,6-triphenylpyrylium salt and the anion of acetonedicarboxylic ester [42], while for the compound produced, with acetylacetonate, it has not yet been definitively elucidated whether it is present in the form of the 2H-pyran *(310)* or as the dienone *(311)* [43]:

Reactions of Pyrylium Salts 151

[Structures 309, 310, 311 with reaction scheme]

The dienone structures are assumed for the products that can be obtained with aryllithium or aryl Grignard compounds and trimethylpyrylium salt, but the valence isomers cannot be excluded [44a,b]:

[Reaction scheme of trimethylpyrylium perchlorate with Ar-Li]

The reaction product from pentaphenylpyrylium salts and benzylmagnesium chloride is also formulated as a dienone *(312)* [40]:

[Reaction scheme yielding (312)]

Normally, 2H-pyran derivatives readily undergo ring opening so that only ketones of type *(84)* can be isolated. This has been known for a long time in the case of the action of hydroxyl ions (or water) on pyrylium salts [45]. In this way, 2,3,5,6-tetraphenylpyrylium salt does not yield pyranol but enol ketone *(313)*, which rearranges thermally into the diketone *(314)* [46]:

[Reaction scheme yielding (313) and (314)]

In fact, isolation of the enol ketone is an exception; in general, only diketones can be obtained, for example, from 2,4,6-triphenyl- [47a], 2,3,4,6-tetraphenyl-, and 3-methyl-2,4,6-triphenylpyrylium salts [48]. In the last two examples the isomerizations of the initially formed *cis*-diketones *(315)* and *(317)* into the *trans*-diketones *(316)* and *(319)* have been studied extensively [48]:

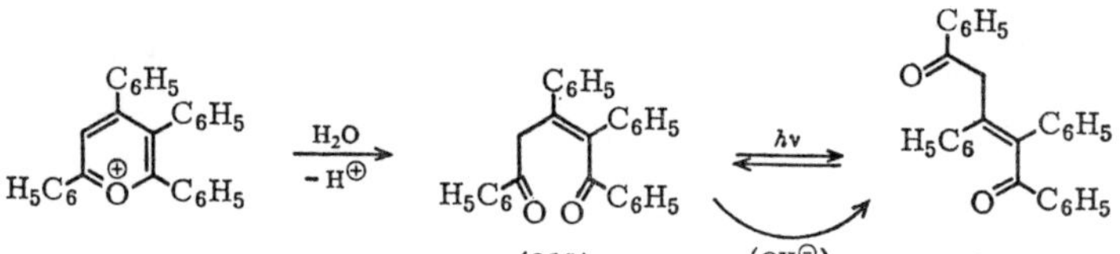

(315) *(316)*

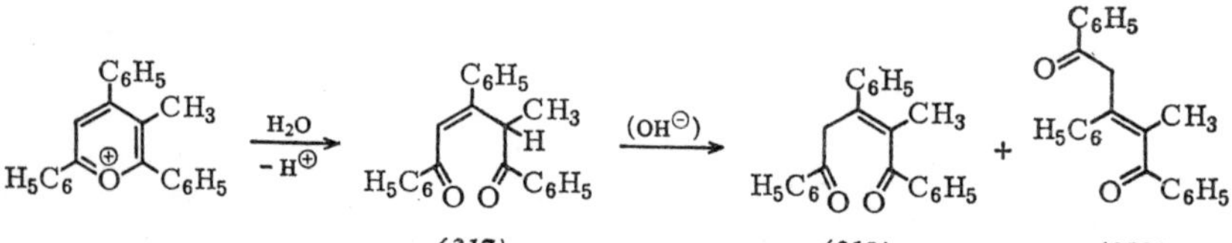

(317) *(318)* *(319)*

Recently the first compound with the structure of a 2-hydroxy-2*H*-pyran could also be isolated from the alkaline hydrolysis of 2,4,6-triphenyl pyrylium perchlorate. The pyranol was found to be the main reaction product at high pH values [47b]:

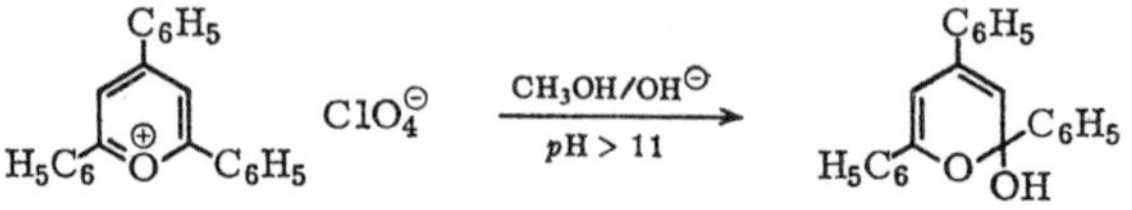

In the reaction of pyrylium salts with ammonia, which leads to pyridine derivatives (Section 7.4.2.2), an enol imine *(320)* can be obtained as the primary product of ring opening [49]; e.g.:

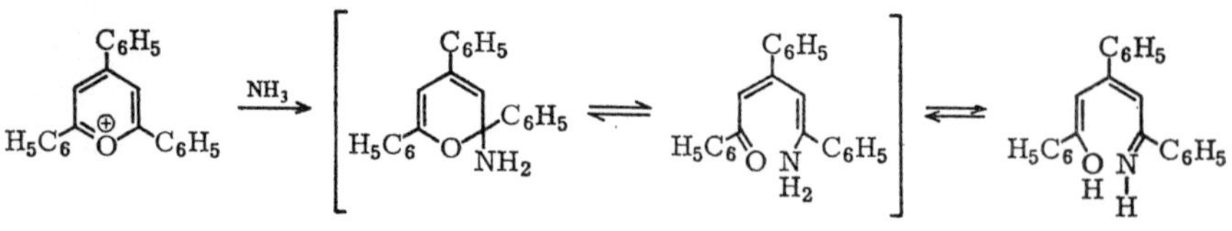

(320)

Correspondingly, amino pentadienones *(321)* are generally obtained from primary amines [50,51] in poor yields, but from secondary amines [51] in good yields:

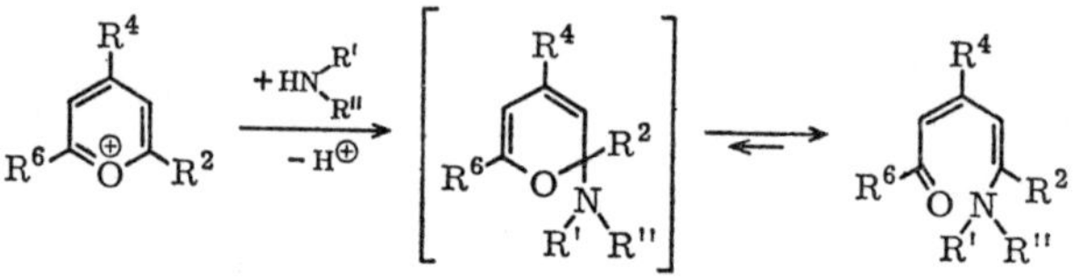

(321)

The products *(322)* that can be obtained from triphenylpyrylium salt and hydroxylamine [52a] or phenylhydrazine [52a, 53, 54] fall within this scheme:

where Z = OH.

With phenylhydrazine, in addition to the compound *(322)*, (Z = NHC$_6$H$_5$), termed an "α-pyranolphenylhydrazide," an isomeric "β-pyranolphenylhydrazide" is obtained to which the *trans* configuration *(323)* was previously ascribed [54]. In fact the isomer is the pyrazoline derivative *(324)* [52a], which is also obtained from *(322)* by ring closure*:

The isoxazoline *(325)* is formed similarly from the oxime *(322)*.

In the case of the reaction product from 2,4,6-trialkylpyrylium salts and cyanide [55a,b], *cis–trans* isomerism has been detected and the steric position of the substituents has been elucidated [55b]:

where R = CH$_3$, C$_2$H$_5$.

---

* In contrast to this the action of hydrazine on 2,4,6-triphenylpyrylium salt yields an unstable monohydrazone that dehydrates readily to give 3,5,7-triphenyl-4H-1,2-diazepine 52b]:

Nucleophilic Addition to C_4

Pyrylium ions with an unsubstituted position 4 are attacked by nucleophiles almost exclusively at C_4. Thus, for example, 2,6-diphenylpyrylium salts give with reactive methylene components (such as acetylacetone and nitromethane) in the presence of potassium *tert*-butoxide [10, 57a,b] or with Grignard compounds in ether [10] good yields of 4H-pyrans *(326)*:

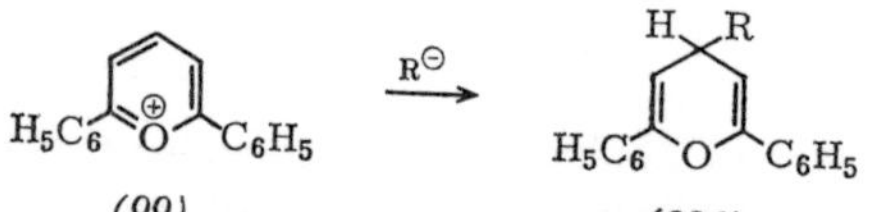

$$(99) \xrightarrow{R^{\ominus}} (326)$$

where $R = CH(COCH_3)_2$, CH_2NO_2, CH_3, $CH_2C_6H_5$, C_6H_5.

Besides other products, 4H-pyrans can also be obtained by the action of sodium tetrahydroborate on 2,4,6-trisubstituted pyrylium salts by hydride addition in position 4 [56]. Obviously, 2,4,6-trisubstituted pyrylium salts react with benzylmagnesium chloride (in contrast to other alkyl Grignard compounds or metal aryls) preferentially in position 4 [40],* even if a bulky substituent is present in this position [40, 58]:

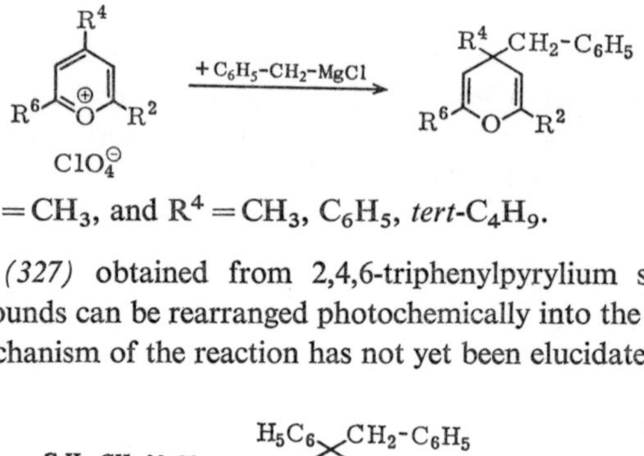

where $R^2 = R^6 = CH_3$, and $R^4 = CH_3$, C_6H_5, *tert*-C_4H_9.

The 4H-pyran *(327)* obtained from 2,4,6-triphenylpyrylium salt and benzyl Grignard compounds can be rearranged photochemically into the 2H-pyran *(328)* [40, 58]. The mechanism of the reaction has not yet been elucidated:

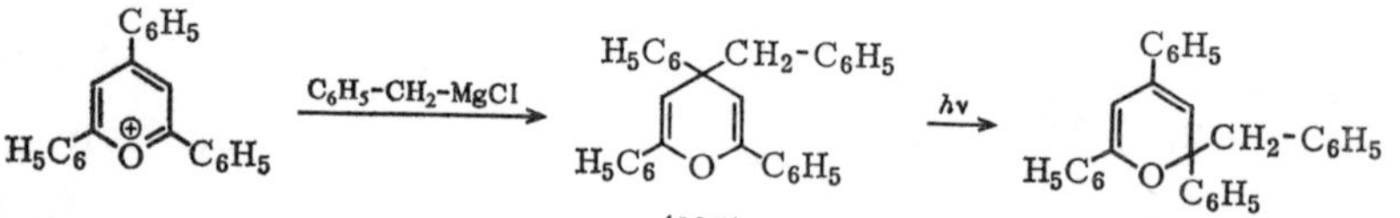

(327)   *(328)*

Occasionally the anions of active methylene compounds also react with 2,4,6-trisubstituted pyrylium salts in position 4; e.g. [59]:

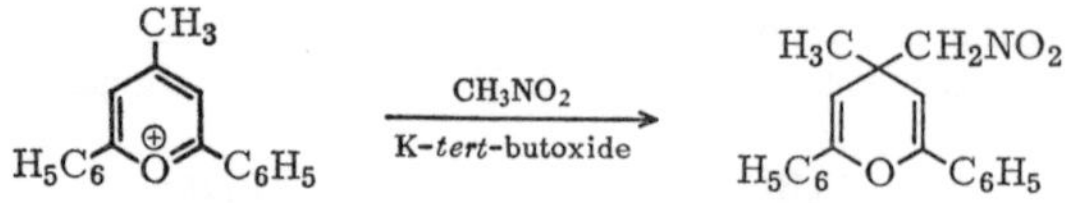

---

* The 2,3,4,6-tetrasubstituted pyrylium salts that have been investigated so far react similarly [40]. The dienone *(312)* that can be obtained from the 2,3,4,5,6-pentaphenyl-pyrylium salt with a benzyl Grignard compound appears to be an exception.

In this special case the steric differences between the substituents could be responsible for the formation of the 4H-pyran [59].

Deprotonation

In a reversal of the formation reaction, 4-hydroxypyrylium ions can be deprotonated to γ-pyrones in the presence of bases [60]:

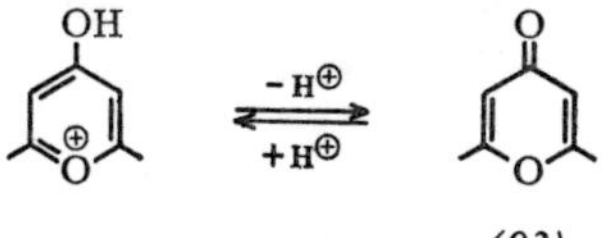

(93)

The corresponding deprotonation of pyrylium ions bearing a CH acidic residue in position 4 is more important, since in this process 4-methylene pyrans (4H-4-dehydropyrans) arise as reactive intermediates [10, 61a,b,c]:

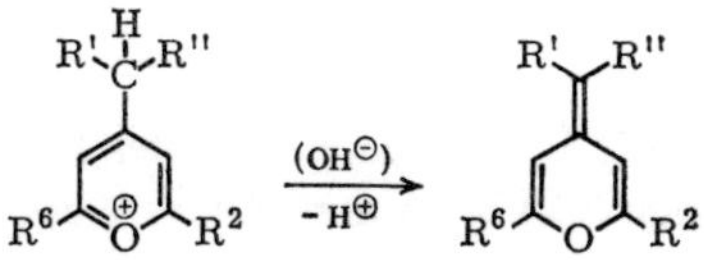

where $R^2 = R^6 = CH_3, C_6H_5$. This reaction takes place with hydroxyl ions preferentially before the aforementioned ring opening of the pyrylium ion; in general, electron-withdrawing groups R' and R" (CN, COOR, CO–R) are necessary for this purpose, but in the presence of very strong bases this side reaction must be taken into account even with 4-alkylpyrylium salts [59]; e.g.:

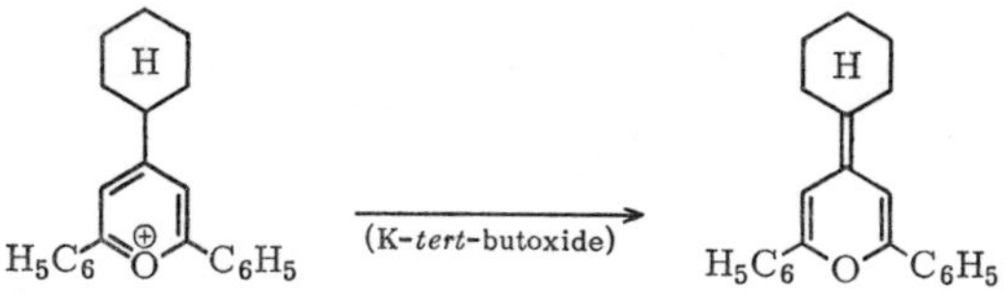

The analogous deprotonation to a 2-methylenepyran derivative *(329)* is also known [62]:

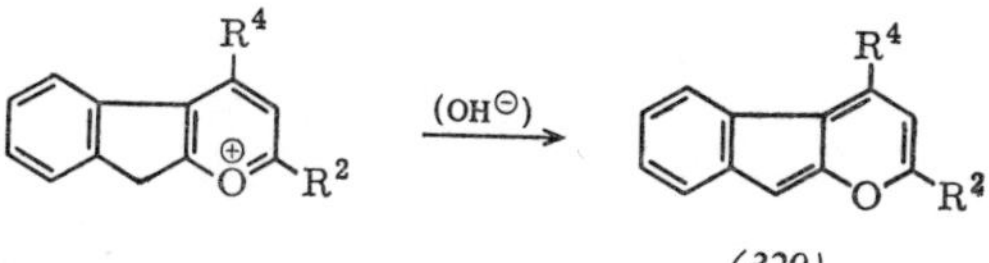

(329)

7.4.2. Secondary Reactions of Nucleophilic Addition at C_2

7.4.2.1. Review

The isolation of primary products of nucleophilic addition at C_2—namely, 2H-pyran derivatives *(85)* or their ring-opened valence isomers *(84)*—is limited to only a few cases. In general, the only subsequent products that are isolated are those formed by ring-closure reactions from the ketone derivatives *(84)*. In these reactions the C_2-atom of the original pyrylium ion can be linked through an atom of the added nucleophilic compound Y with the carbon atom C_6 ($C_2 \rightarrow C_6$), C_4 ($C_2 \rightarrow C_4$), or C_5 ($C_2 \rightarrow C_5$); schematically:

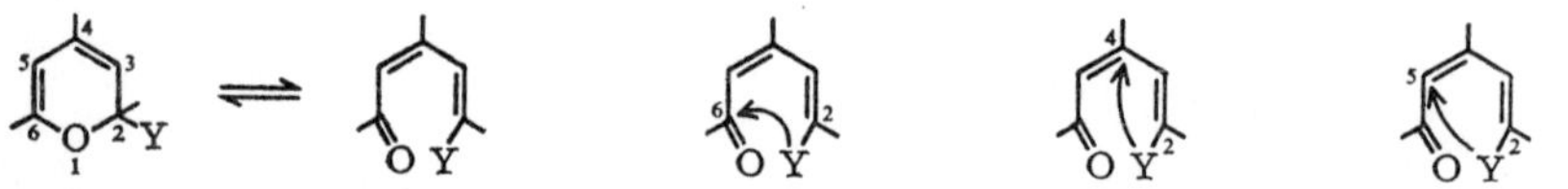

(85) *(84)* ($C_2 \rightarrow C_6$) ($C_2 \rightarrow C_4$) ($C_2 \rightarrow C_5$)

$C_2 \rightarrow C_6$ Linkage

Ring-closure reactions in which a nucleophilic center Z on C_2 attacks the carbon atom of the carbonyl group (C_6) by means of its free-electron pair are by far the most important. After elimination of water the intermediate product *(330)* irreversibly gives the compound *(331)*, in which Z has formally replaced the oxygen atom of the original pyrylium ion [63]. If one takes into account the fact that in the 2H-pyran *(85)*, ($Y = ZH_2$), the center Z may be neutral ($n = 0$) or singly positively charged ($n = +1$), the following scheme is obtained:

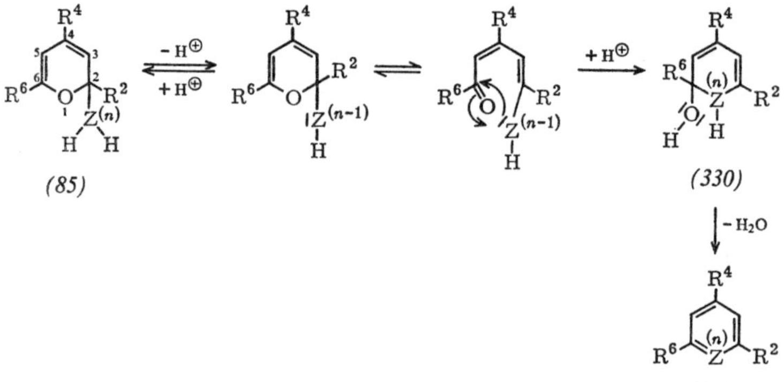

(85) *(330)* *(331)*

The general reaction may be illustrated by three examples:

Z = S [scheme] (332)

Z = N-R [scheme] (333)

Z = C-R [scheme] (334)

$C_2 \rightarrow C_4$ Linkage

Position C_4, which is a vinylogous position with respect to the carbonyl carbon C_6, is occasionally involved in ring closure when a five-membered ring system can be formed; i.e., the reactive nucleophilic center may not be bound directly to C_2. Consequently the nucleophilic compound Y added to the pyrylium ion must possess two nucleophilic centers (Y = X–Z–H) or, schematically:

[scheme]

Apart from the pyrazoline or isoxazoline derivatives (324) and (325), no other compounds have so far become available from pyrylium ions [52a] by this reaction type.

$C_2 \rightarrow C_5$ Linkage

More important, again, are syntheses in which carbon atoms 2 and 5 of the pyran are linked; in this way, for example, a positive center X can attack C_5 electrophilically, as in the oxidation of pyrylium salts, with H_2O_2 [64]:

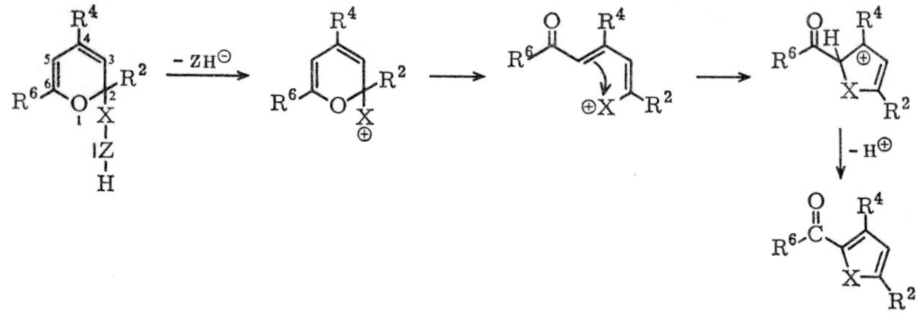

Preparative Application of Oxonium Salts

In other cases, electrocyclic reactions should preferably be formulated [65], particularly when Y = vinyl; e.g.:

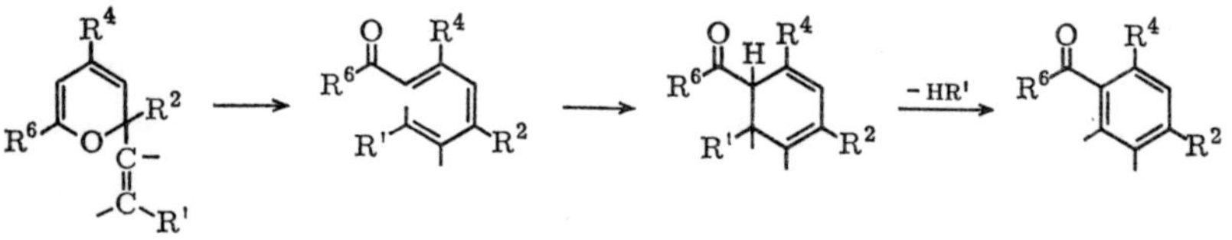

No general scheme can be drawn up as in the case of the $C_2 \rightarrow C_6$ and $C_2 \rightarrow C_4$ linkages.

7.4.2.2. Syntheses with $C_2 \rightarrow C_6$ Linkage

Formation of Heterocycles

The $C_2 \rightarrow C_6$ reaction scheme given above is followed in the longest-known heterocyclic syntheses from pyrylium salts [66a,b,c]; namely, their reaction with ammonium carbonate or ammonia in aqueous or alcoholic solution to give pyridine derivatives *(336)* [63, 66a,b,c, 67, 68a,b,c,d,e, 69]:

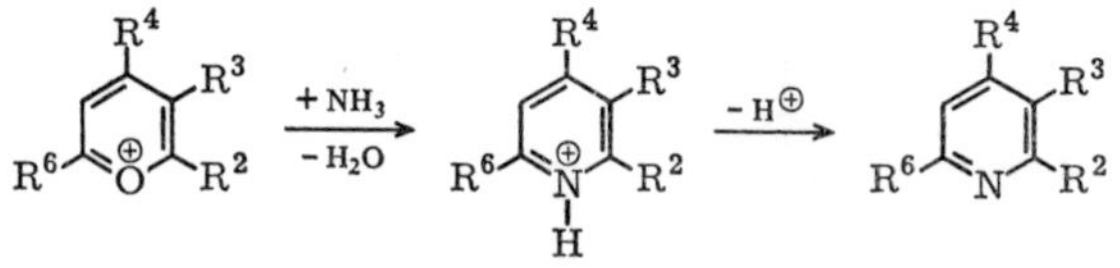

(335)   *(336)*

The reaction with compounds containing an NH_2 group takes place in a fully analogous manner; whether the N-substituted pyridinium salt *(333)* then gives neutral molecules (as betaines) depends on whether the substituent R can still act as a proton donor. Thus, with primary amines, *N*-alkyl- or *N*-arylpyridinium salts are produced [66a,b, 70a,b]; if R is a hydroxyaryl residue, with bases the highly solvatochromic *N*-(hydroxyphenyl)pyridinium, betaines *(337)* are produced [71a,b]. With glycine, the betaine *(338)* is obtained via the pyridinium *N*-carboxylic acid [72], with phenylhydrazine the *N,N*-betaine *(339)* [73] via the *N*-anilinopyridinium salt [53], and in favorable cases pyridine *N*-oxides *(340)* via the *N*-hydroxypyridinium salt [63, 74]:

Reactions of Pyrylium Salts

[Scheme showing pyrylium salt (reacting with H$_2$N-R, -H$_2$O) converting to pyridinium salt (333), which under various R groups gives products (337), (338), (339), (340)]

The reaction with phosphine, analogous to the action of ammonia on pyrylium salts, has recently led to the discovery of phosphorin (phosphabenzene) derivatives *(341)* [75, 76a,b,c, 77]:

[Scheme: pyrylium salt + PH$_3$ → phosphorin, with loss of H$^+$ and H$_2$O]

(341)

Although free phosphine cannot be used in this reaction, it is possible to prepare aryl- or alkyl-substituted phosphorins with phosphine obtained in situ from phosphonium iodide and *n*-butanol [75]. Particularly advantageous—although only for the synthesis of triaryl- or tri-*tert*-butylphosphorins—is the use of tris(hydroxymethyl)phosphine [76a,b,c], which reacts with the elimination of formaldehyde and the formation of a C=P linkage [78].

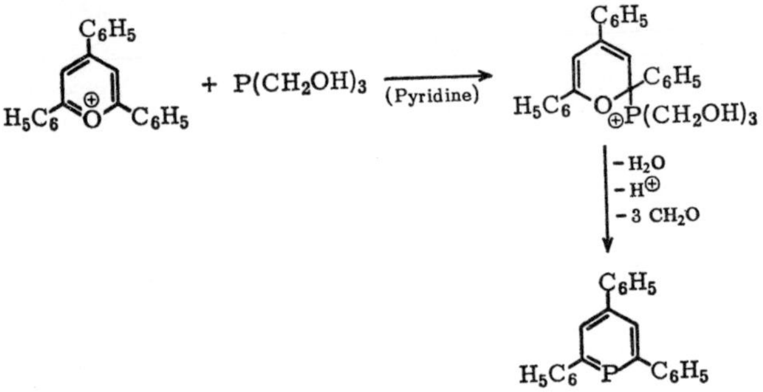

A possibility of not binding the oxygen atom of the pyrylium ion to protons as water but of transferring it to an organosilicon compound has been utilized in the case of tris (trimethylsilyl)phosphine and pyrylium iodides [77]:

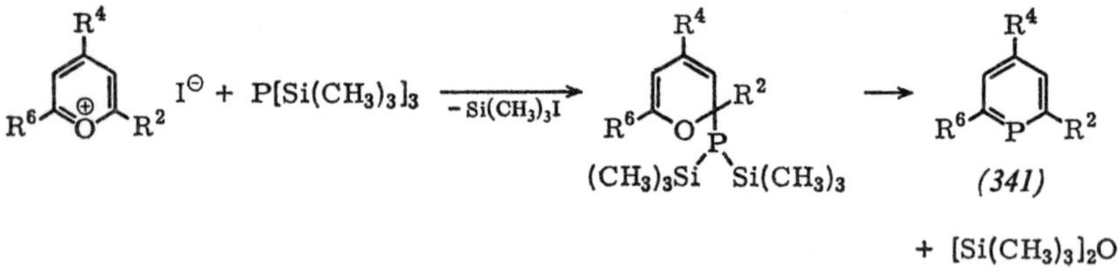

The approach to the thiopyrylium salt series *(332)* by the action of sodium sulfide on pyrylium salts may also be mentioned [79a,b]. However, the reaction cannot be extended to purely alkyl-substituted pyrylium salts:

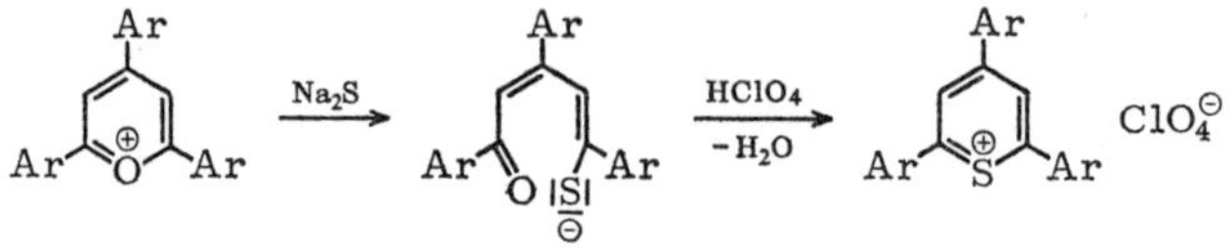

Formation of Benzene Derivatives

Isocyclic, six-membered ring systems are formed when 2H-pyrans produced as intermediates are substituted on C_2 by a CH_2R residue that can condense with the carbonyl group. The residue capable of condensation may be either the original pyrylium substituent in position 2 (where $R^2 = CH_3$, CH_2R) or a methylene group added at C_2 (where $Y = CH_2R$):

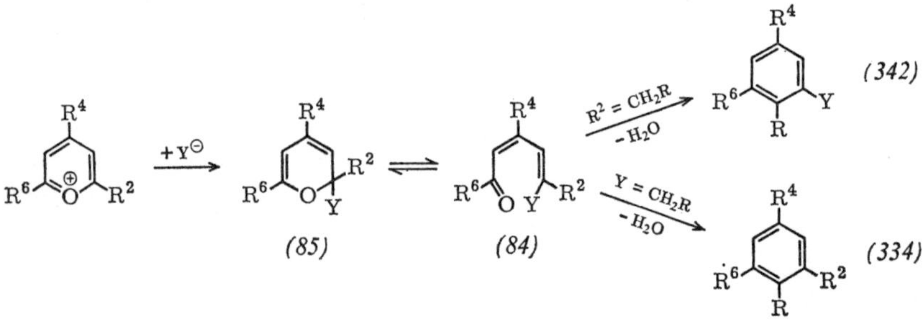

Reactions of Pyrylium Salts 161

The action of hydroxyl ions on pyrylium salts bearing a methyl group in position 2 to give phenols *(343)* takes place by the first method [63, 66b]. The *m*-disubstituted amines *(344)* are obtained in a completely similar manner with secondary amines [70b]; even primary amines give this reaction to a certain extent, in addition to the aforementioned formation of pyridinium salts *(333)* [50]. Finally, the reaction of 2-methyl-substituted pyrylium salts with Grignard compounds to give symmetrically trisubstituted benzenes also belongs here [80]:

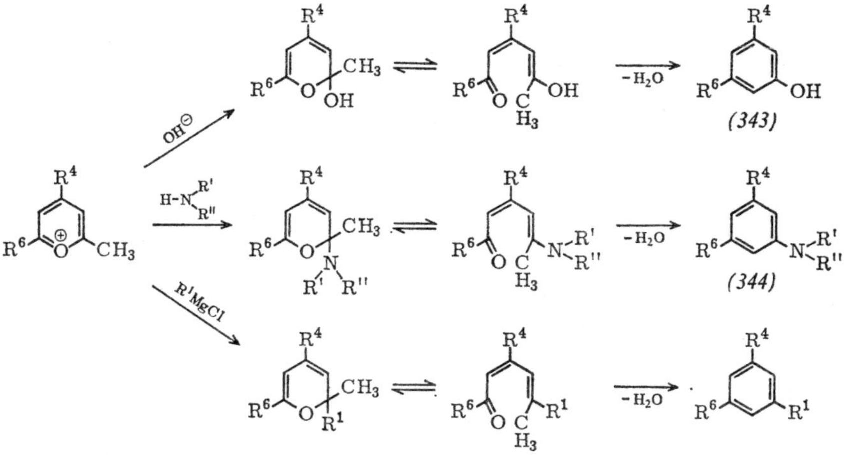

The preparative importance of the second method—the addition of nucleophiles with a CH_2R group to pyrylium ions—is greater.

The broadest field of application is in the conversion of 2,4,6-trisubstituted pyrylium salts with nitromethane into 2,4,6-trisubstituted nitrobenzene derivatives *(346)*, which can be obtained smoothly in the presence of bases (potassium *tert*-butoxide, triethylamine) [59, 81a,b,c]:

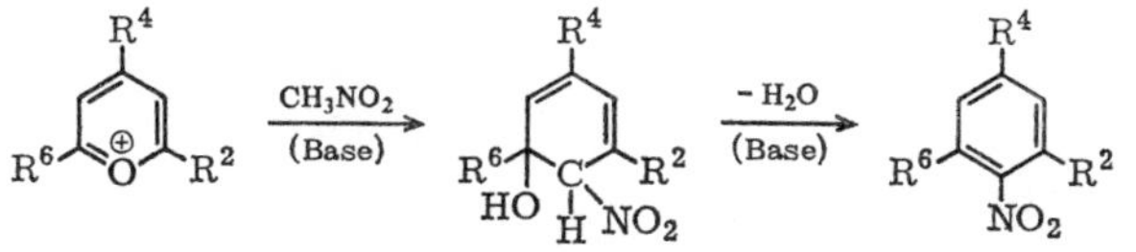

The reaction can also be applied to tetra- and pentasubstituted pyrylium salts. The nitromethane condensation has been investigated particularly in relation to 2,4,6-triphenylpyrylium salt which smoothly gives 2,4,6-triphenylnitrobenzene *(347)* with an excess of a base. When a smaller amount of base (triethylamine) is present, no detachment of the proton at C_1 from the (not isolated) intermediate product *(345)* takes place, but there is an allyl rearrangement of the nitro group into the nitrous ester *(348)*, which yields 2,4,6-triphenylphenol *(349)*. A second elimination leads, after anionic expulsion of the nitro group and with the migration of a phenyl group, to 2,3,5-triphenylphenol *(350)* [82]:

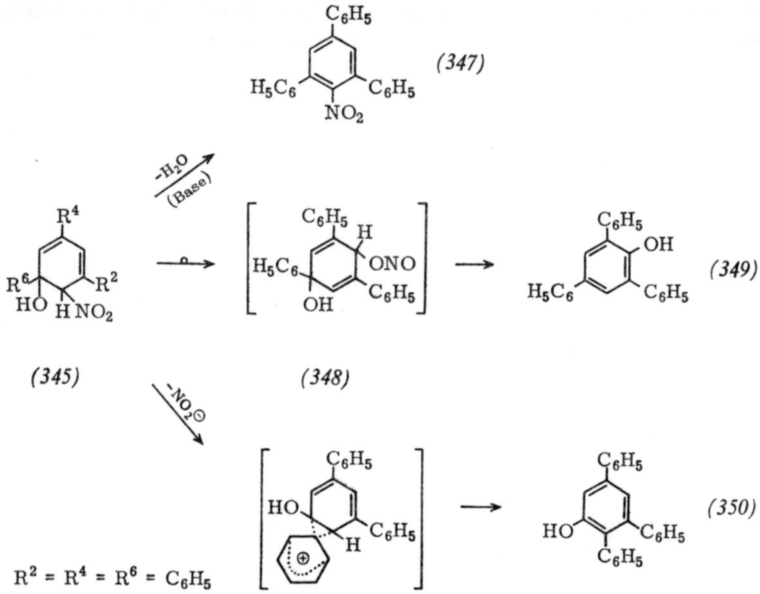

These reaction routes have also been confirmed in the case of tetraarylpyrylium salts [82, 83]. Phenylnitromethane reacts in the first step like nitromethane itself. However, the intermediate product *(351)* corresponding to *(345)*, whose isolation has been achieved in one case, can be stabilized only by rearrangement of the nitro group to form the aromatic compound [84]. In this case the tetraarylnitro compound *(353)* and the tetraarylphenol *(352)* are obtained side by side [42, 84]:

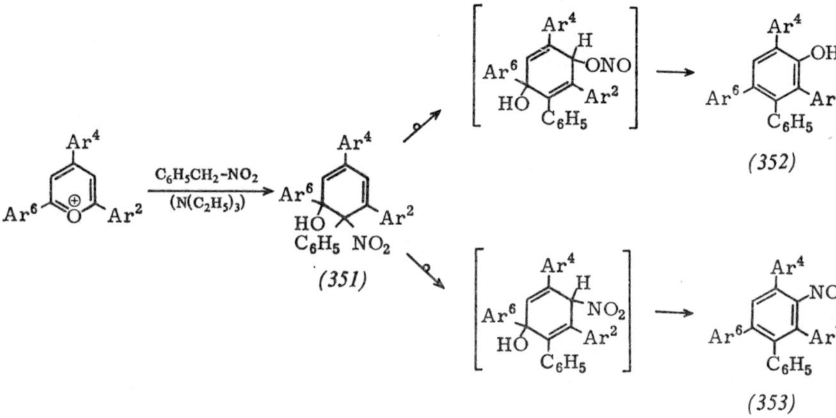

In the case of 2,4,6-triphenylpyrylium salt the condensation has also been studied with other active methylene compounds (in the presence of potassium *tert*-butoxide) [85]. When acetoacetic ester or acetylacetone was used, the acetyl group was found to split out; in the case of cyanoacetic ester the carboxy group was eliminated [85] and in the case of acetonedicarboxylic ester the elimination of malonic ester occurred [42]:

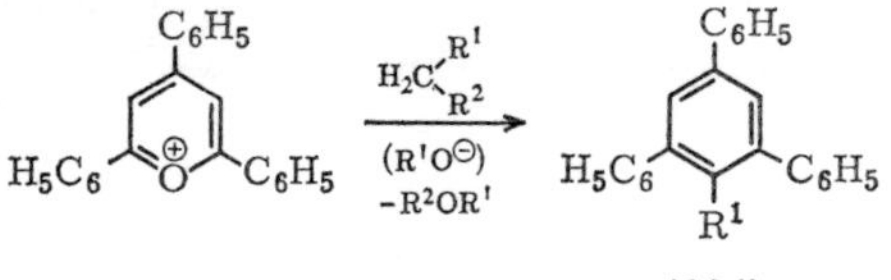

$R^1 = CO_2R,\ R^2 = COCH_3$
$R^1 = R^2 = COCH_3$
$R^1 = CN,\ R^2 = CO_2R$
$R^1 = CO_2R,\ R^2 = COCH_2CO_2R$

(334)

The 2H-pyran derivatives or their valence isomers [40, 44a] obtained by the action of alkyl (or benzyl) Grignard compounds can be converted by heating with strong bases (sodium in diethylene glycol) [40] or with calcium oxide [85] into aryl hydrocarbons; e.g.:

(312)

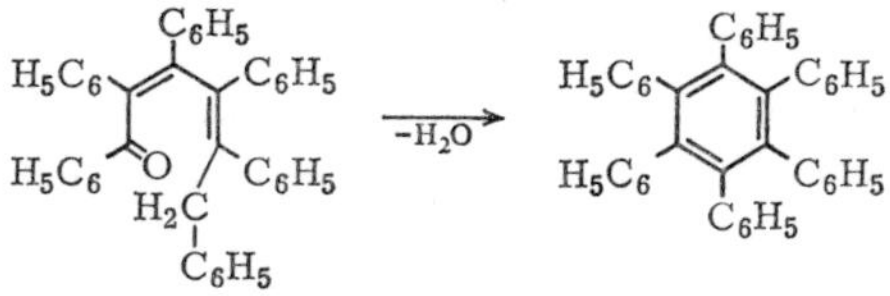

In many cases, compounds known to have the structure of 4H-pyrans also give rise to hydrocarbons [40]. It must therefore be assumed that cyclization is preceded in these compounds by the intermediate rearrangement of 4H- into 2H-pyrans [40, 58]; e.g.:

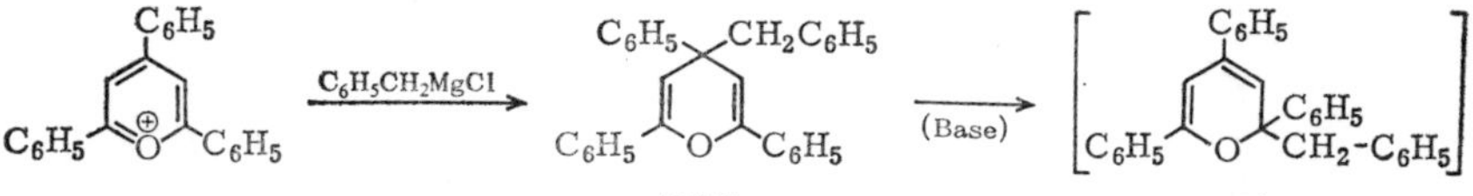

(327)   *(328)*

The action of methylenetriphenylphosphoranes on pyrylium ions may be given as another possibility for the synthesis of aryl hydrocarbons; here the benzene derivative is produced with the elimination of triphenylphosphine oxide [86]:

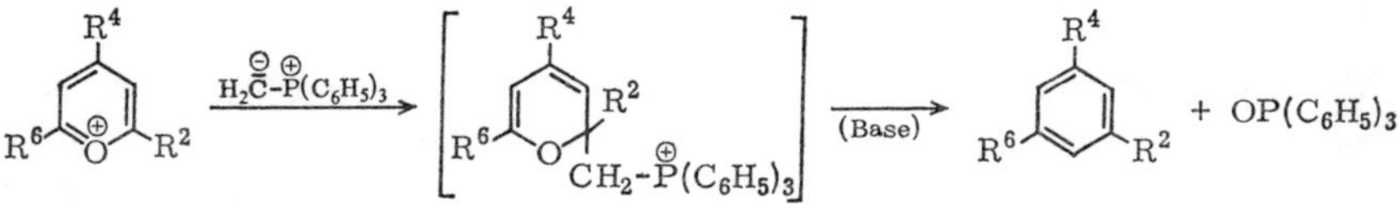

Formation of Azulene Derivatives

In the reaction of the sodium derivative of cyclopentadiene with 2,4,6-trimethylpyrylium perchlorate, carbon atoms C_2 and C_6 of the pyrylium ion are linked to one another by *two* carbons with the formation of 4,6,8-trimethylazulene *(354)* [87]:

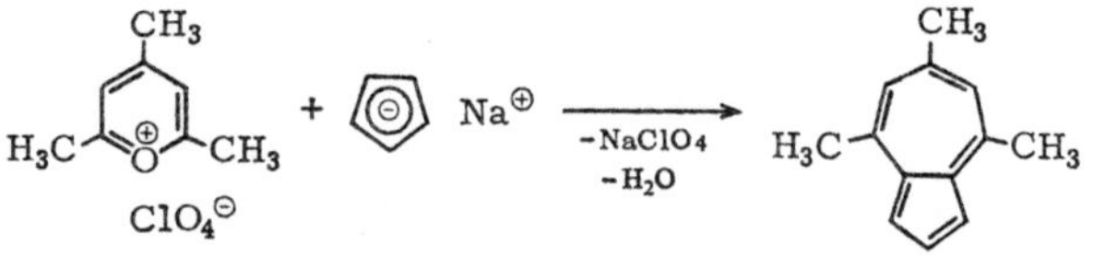

This elegant azulene synthesis can also be applied to other pyrylium salts, provided at least one methyl group is present in position 2 or 6.

7.4.2.3. Syntheses with $C_2 \rightarrow C_5$ Linkage

As already mentioned, syntheses with the linking of C_2 and C_4 are not important, whereas ring closures linking carbon atoms C_2 and C_5 to one another have preparative value.

Formation of Furan Derivatives

Besides the synthesis of 2-acylfurans *(356)* (*cf.* Section 7.4.2.1) by the action of hydrogen peroxide on pyrylium salts ($R^2 = R^6$) [64, 76c],

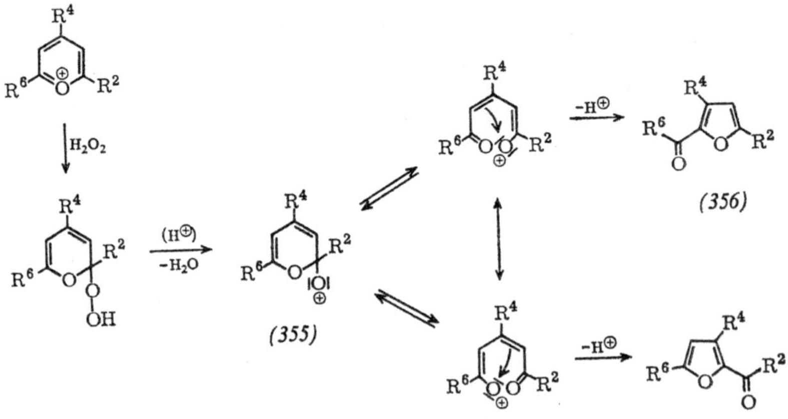

an analogous oxidation by means of hypobromite (alkaline hydrolysis of pyrylium perbromides) can be observed [88].

Formation of Benzene Derivatives

The reaction of malonic ester with pyrylium salts yields the phenolcarboxylic acid ester *(358)*; it might have been formed from the addition product *(357)* in the manner of an electrocyclic reaction. The analogous reaction of the malodinitrile adduct *(359)* to give the amino nitrile *(360)* is described as an attack of the electrophilic nitrile carbon on the C_5 center [85]:

An electrocyclic mechanism probably plays a part in the reaction of pyrylium salts with enamines [65]:

where $n = 5, 6, 8, 12$ and $X = CH_2$, O.

Here the elimination of amine leads to the aryl ketone *(361)* which is comparable with *(358)* and *(360)*: on the other hand, the formation of the hydrocarbon *(362)* with elimination of benzamides is surprising. The ratio of *(361)* to *(362)* is affected decisively by the enamine base [65].

Formation of Azulene Derivatives

The following azulene syntheses may be mentioned in connection with the linkage of carbon atoms C_2 and C_5 of pyrylium salts [89]:

In the action of methylenetriphenylphosphorane on 2,4,6-triphenylpyrylium salts, a bis(2H-pyran) must be required as an intermediate which yields an azulene derivative with the participation of atoms C_2 and C_5. The reaction takes place in competition with the aforementioned formation of aryl hydrocarbons by the same method [86]:

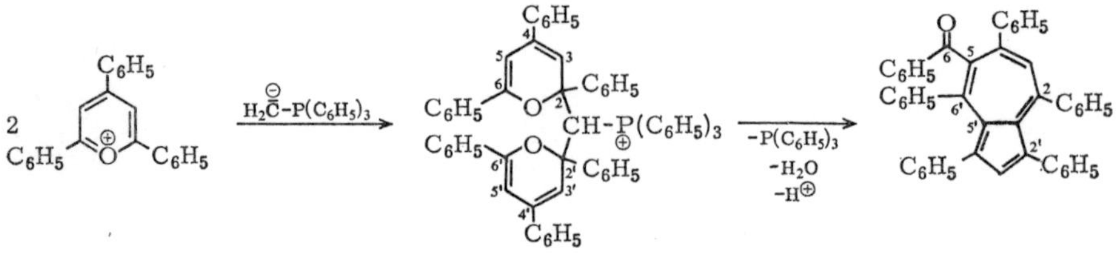

7.4.3. Secondary Reactions of Nucleophilic Addition at C_4

7.4.3.1. Review

The variety of the reactions of 2H-pyrans must be ascribed above all to their ring opening to give valence-isomeric ketones and the re-formation of rings. In the case of 4H-pyrans, where no analogous isomerization can take place, ring opening with the addition of water to form the diketone *(363)* is indeed possible, but exchange and elimination reactions at the tetrahedral center of the pyran *(90)* are more important. The anionic expulsion of Y or of R^4 leads either to the original pyrylium ion or to a new one; if R^4 contains a hydrogen atom in the α-position, with a suitable Y the formation of 4H-dehydropyrans with the elimination of H–Y is possible. The photochemical and intermediate thermal transformation of *(90)* into 2H-pyrans *(85)* has been mentioned above. The condensation of 4-benzyl-4H-pyrans with perchloric acid to give naphthalene derivatives *(364)* with the elimination of a methyl ketone R^2–CO–CH_3, discovered by Dimroth and Wolf [90], is preparatively important. The reaction scheme is

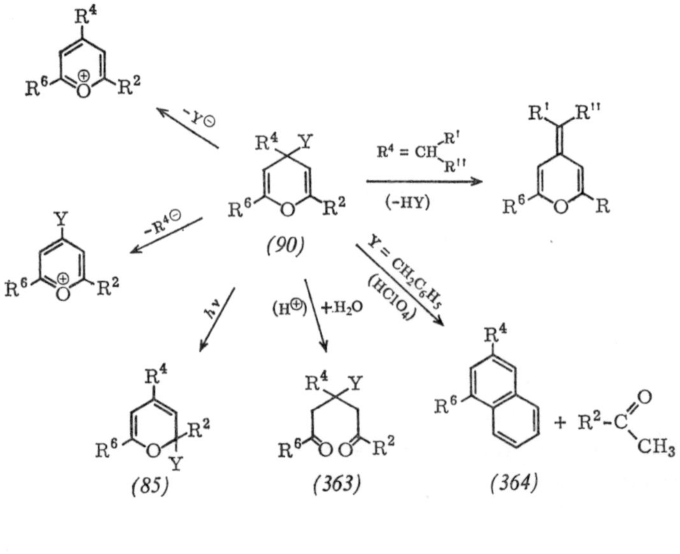

7.4.3.2. Exchange Reactions of Substituents in Position 4

Alkoxy (or alkylthio) substituents in 2,4,6-trisubstituted pyrylium salts can easily be displaced by the reaction of other nucleophiles via 4H-pyran derivatives as intermediates with the production of new pyrylium salts.

With secondary amines, 4-dialkylaminopyrylium salts are formed [91]; active methylene compounds yield methylene-4H-pyrans directly [61b,c]:

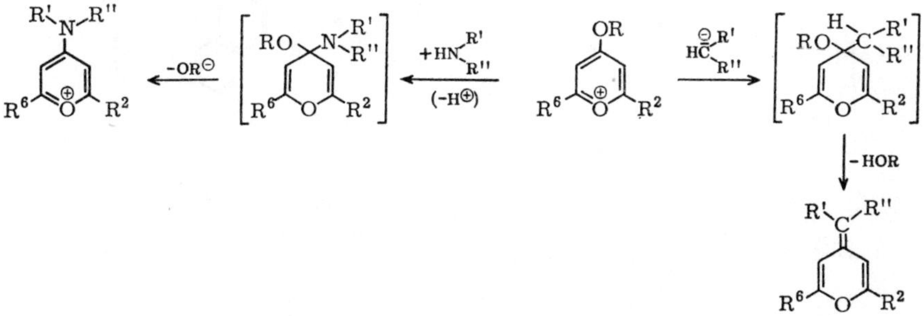

The splitting out of hydride ions in nucleophilic addition to 2,6-disubstituted pyrylium salts must be formulated quite analogously [10, 57a,b, 59]:

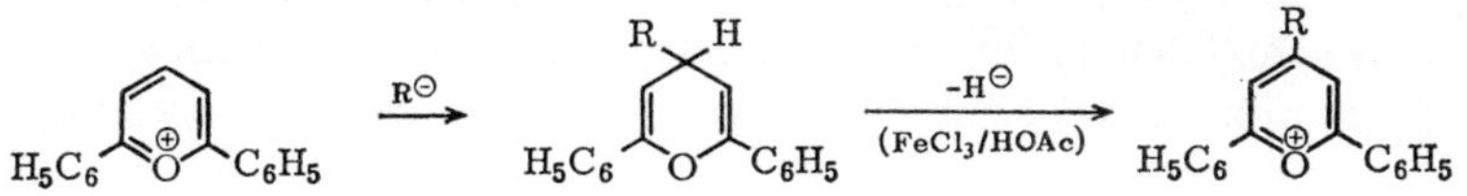

If a CH acidic residue enters as R, oxidizing agents (cyanoferrate(III) solution) give methylene-4H-pyrans [61a,b].

7.4.3.3. Preparation of Naphthalene Derivatives from 4-Benzyl-4H-Pyrans

The ring opening of 4H-pyrans to give diketones *(363)* can also be applied to 4-benzyl derivatives [90]. However, the reaction of the 4-benzyl-4H-pyrans takes place differently on heating with 70% perchloric acid; namely, it gives naphthalene derivatives *(364)* with the elimination of a methyl ketone [58, 90]. The initial reaction step is probably again ring opening to form the diketone *(365)* [92]. In the subsequent step the phenyl nucleus of the benzyl group condenses with one of the two acyl groups. Schematically:

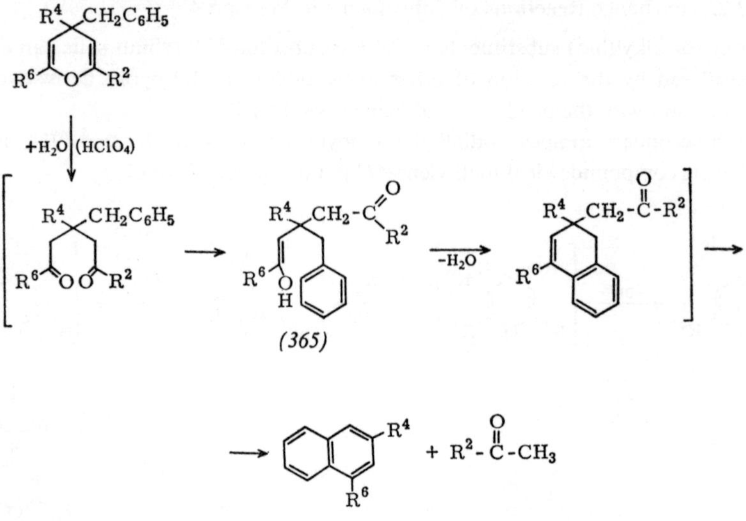

168 Preparative Application of Oxonium Salts

(365)

(364)

The reaction can be applied to numerous trisubstituted 4-benzyl-4H-pyrans, which are easily obtainable from pyrylium salts and benzyl Grignard compounds, and 4-(1-Naphthylmethyl)-4H-pyran *(366)* can be converted into the phenanthrene derivative *(367)* [90]. In principle, the mechanism of the reaction can probably be compared with the very similar reaction of the ketone *(368)* with perchloric acid [92]:

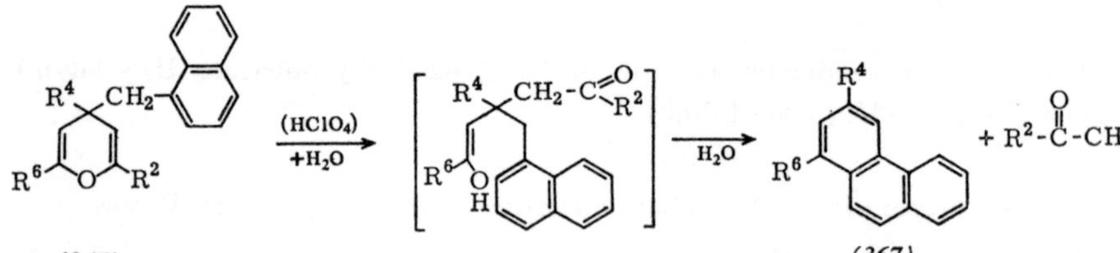

(366)   *(367)*

$R^4 = R^6 = C_6H_5$

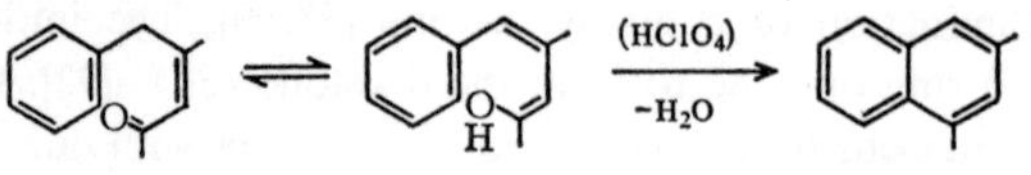

(368)

References

1. H. Meerwein, P. Borner, H. J. Sasse, H. Schrodt, and J. Spille, *Chem. Ber.* **89**, 2060 (1956).

2. H. Meerwein, in "Methoden der organischen Chemie (Houben-Weyl)" (E. Müller, ed.), Bd. VI/3: Sauerstoffverbindungen I, p. 359. Thieme, Stuttgart, 1965.
3. a) S. Kabuss, *Angew. Chem.* **78,** 714 (1966); *Angew. Chem. Intern. Ed. English* **5,** 675 (1966); b) S. Kabuss, Lecture to the German Chemical Society, Marburg, Dec. 12, 1968; c) S. Kabuss, personal communication, Feb. 1969, unpublished results.
4. H. Meerwein, see Reference 2, p. 362.
5. a) K. Hafner, H. Riedel, and M. Danielisz, *Angew. Chem.* **75,** 344 (1963); M. Danielisz, Ph.D. Thesis, University of Marburg, 1963; cf. however G. Seitz, *Angew. Chem.* **79,** 96 (1967); *Angew. Chem. Intern. Ed. English* **6,** 82 (1967); b) C. E. Forbes and R. H. Holm, *J. Am. Chem. Soc.* **90,** 6884 (1968).
6. a) T. Eicher and G. Frenzel, *Z. Naturforsch.* **20b,** 274 (1965); b) T. Eicher and A. Löschner, *Z. Naturforsch.* **21b,** 295, 899 (1966); c) T. Eicher and A. M. Hansen, *Chem. Ber.* **102,** 319 (1969).
7. R. Breslow, T. Eicher, A. Krebs, R. A. Peterson, and J. Posner, *J. Am. Chem. Soc.* **87,** 1320 (1965).
8. B. Föhlisch and P. Bürgle, *Ann. Chem. Liebigs* **701,** 67 (1968).
9. A. S. Kende, P. T. Izzo, and P. T. McGregor, *J. Am. Chem. Soc.* **88,** 3359 (1966).
10. K. Dimroth and K. H. Wolf in "Newer Methods of Preparative Organic Chemistry" (W. Foerst, ed.), Vol. 3, p. 369, 408. Academic Press, New York, 1964.
11. H. J. Bestmann, R. Saalfrank, and J. P. Snyder, *Angew. Chem.* **81,** 227 (1969); *Angew. Chem. Intern. Ed. English* **8,** 216 (1969).
12. a) H. Meerwein, E. Battenberg, H. Gold, E. Pfeil, and G. Willfang, *J. Prakt. Chem.* **154,** 83 (1939); b) H. Meerwein, W. Florian, N. Schön, and G. Stopp, *Ann. Chem. Liebigs* **641,** 1 (1961).
13. U. Schöllkopf and F. Gerhart, *Angew. Chem.* **79,** 990 (1967); *Angew. Chem. Intern. Ed. English* **6,** 970 (1967).
14. E. Bertele, H. Boos, J. D. Dunitz, F. Elsinger, A. Eschenmoser, I. Felner, H. P. Gribi, H. Gschwend, E. F. Meyer, M. Pesaro, and R. Scheffold, *Angew. Chem.* **76,** 393 (1964); *Angew. Chem. Intern. Ed. English* **3,** 490 (1964).
15. H. Muxfeldt and W. Rogalski, *J. Am. Chem. Soc.* **87,** 933 (1965).
16. R. F. Borch, *Tetrahedron Letters* p. 61 (1968).
17. L. Weintraub, S. R. Oles, and N. Kalish, *J. Org. Chem.* **33,** 1679 (1968).
18. French Patent 1,381,790 (Cl. C 08g D 06mp), Dec. 11, 1964, Ciba Ltd.; *Chem. Abstr.* **63,** 728 (1965).
19. a) H. Bredereck, F. Effenberger, and G. Simchen, *Angew. Chem.* **73,** 493 (1961); *Angew. Chem.* **74,** 353 (1962); *Chem. Ber.* **96,** 1350 (1963); b) H. Bredereck, F. Effenberger, and E. Henseleit, *Angew. Chem.* **75,** 790 (1963); *Chem. Ber.* **98,** 2754, 2887 (1965); c) H. Bredereck, F. Effenberger, and H. P. Beyerlin, *Chem. Ber.* **97,** 3076, 3081 (1964).
20. a) S. G. McGeachin, *Can. J. Chem.* **46,** 1903 (1968); b) H. Bredereck, F. Effenberger, and D. Zeyfang, *Angew. Chem.* **77,** 219 (1965); *Angew. Chem. Intern. Ed. English* **4,** 242 (1965).
21. a) H. Brechbühler, H. Büchi, E. Hatz, J. Schreiber, and A. Eschenmoser, *Angew. Chem.* **75,** 296 (1963); b) H. Vorbrüggen, *Angew. Chem.* **75,** 296 (1963).
22. a) S. Petersen and E. Tietze, *Ann. Chem. Liebigs* **623,** 133 (1959); b) L. A. Paquette, *J. Am. Chem. Soc.* **86,** 4096 (1964).
23. T. Oishi, M. Ochiai, M. Nagai, and Y. Ban, *Tetrahedron Letters* p. 497 (1968).
24. a) I. Felner, A. Fischli, A. Wick, M. Pesaro, D. Bormann, E. L. Winnacker, and A. Eschenmoser, *Angew. Chem.* **79,** 863 (1967); *Angew. Chem. intern. Ed. English* **6,** 864 (1967); b) Y. Yamada, D. Miljkovic, P. Wehrli, B. Golding, P. Löliger, R. Keese, K. Müller, and A. Eschenmoser, *Angew. Chem.* **81,** 301 (1969); *Angew. Chem. Intern. Ed. English* **8,** 343 (1969); cf. A. Eschenmoser, *Quart. Rev.* **34,** 366 (1970).
25. L. A. Paquette and T. Kakihana, *J. Am. Chem. Soc.* **90,** 3897 (1968).

26. a) L. A. Paquette and T. J. Barton, *J. Am. Chem. Soc.* **89,** 5480 (1967); b) P. Wegener, *Tetrahedron Letters* p. 4985 (1967).
27. Japanese Patent 9,420 ('65), May 14, 1965, Konishiroku Photo Ind. Co., Ltd.; *Chem. Abstr.* **63,** 18321 (1965).
28. H. Bredereck, F. Effenberger, and H. P. Beyerlin, *Chem. Ber.* **97,** 1834 (1964); see also H. Meerwein *et al.*, Reference 12b, p. 12, footnote 17.
29. H. Meerwein, P. Laasch, R. Mersch, and J. Spille, *Chem. Ber.* **89,** 209 (1956).
30. R. F. Borch, *Chem. Commun.* p. 442 (1968); *J. Org. Chem.* **34,** 627 (1969).
31. a) H. Meerwein, K. Bodenbenner, P. Borner, F. Kunert, and K. Wunderlich, *Ann. Chem. Liebigs* **632,** 38 (1960); b) H. Meerwein, V. Hederich, and K. Wunderlich, *Arch. Pharm.* 291/63, 547 (1958); c) H. Meerwein, V. Hederich, H. Morschel, and K. Wunderlich, *Ann. Chem. Liebigs* **635,** 1 (1960); d) P. E. Peterson and F. J. Slama, *J. Am. Chem. Soc.* **90,** 6516 (1968).
32. a) H. Jäger and G. Hesse, *Chem. Ber.* **95,** 345 (1962); see also W. Tochtermann, *Angew. Chem.* **78,** 366 (1966); *Angew. Chem. Intern. Ed. English* **5,** 351 (1966); b) F. R. Jones and P. H. Plesch, *Chem. Commun.* p. 1230, 1231 (1969).
33. H. Meerwein, U. Eisenmenger and H. Matthiae, *Ann. Chem. Liebigs* **566,** 150 (1950).
34. a) H. Meerwein, D. Delfs, and H. Morschel, *Angew. Chem.* **72,** 927 (1960); b) R. Wegler *in* "Methoden der organischen Chemie (Houben-Weyl)" (E. Müller, ed.), Bd. XIV, 2: Makromolekulare Stoffe II, p. 556. Thieme, Stuttgart, 1963; c) B. A. Rozenberg, O. M. Chekhuta, E. B. Lyudvig, A. R. Gantmakher, and S. S. Medvedev, *Vysokomolekul. Soedin.* **6,** 2030 (1964); *Chem. Abstr.* **62,** 5341 (1965); B. A. Rozenberg, E. B. Lyudvig, A. R. Gantmakher, and S. S. Medvedev, *Vysokomolekul. Soedin.* **6,** 2035 (1964); *Chem. Abstr.* **62,** 5342 (1965); E. B. Lyudvig, B. A. Rozenberg, T. M. Zvereva, A. R. Gantmakher, and S. S. Medvedev, *Vysokomolekul. Soedin.* **7,** 269 (1965); *Chem. Abstr.* **62,** 14830 (1965); N. V. Makletsova, I. V. Epel'baum, B. A. Rozenberg, and E. B. Lyudvig, *Vysokomolekul. Soedin.* **7,** 70 (1965); *Chem. Abstr.* **62,** 13255 (1965); d) W. Umbach and W. Stein, *Fette, Seifen, Anstrichmittel* **78,** 48 (1969).
35. A. Hilt, K. H. Reichert, and K. Hamann, *Makromol. Chem.* **101,** 246 (1967); *Chem. Abstr.* **66,** 95724 (1967).
36. U.S. Patent 3,207,726 (Cl. 260–67), Sept. 21, 1965, Fred Jaffe; *Chem. Abstr.* **63,** 18304 (1966); French Patent 1,409,957 (Cl. C 08g), Sept. 3, 1965, Toyo Rayon Co., Ltd.; *Chem. Abstr.* **65,** 4052 (1966); Dutch Patent 6,412,448 (Cl. C 08g), April 27, 1965, Farbenfabriken Bayer A.-G.; *Chem. Abstr.* **64,** 878 (1966); K. Weissermel, E. Fischer, K. Gutweiler, H. D. Hermann, and H. Cherdron, *Angew. Chem.* **79,** 512 (1967); *Angew. Chem. Intern. Ed. English* **6,** 526 (1967).
37. a) E. Schmitz and I. Eichhorn *in* "The Chemistry of the Ether Linkage" (S. Patai, ed.), p. 340. Wiley (Interscience), New York, 1967; b) C. S. Rondestvedt, Jr. and G. J. Mantell, *J. Am. Chem. Soc.* **82,** 6419 (1960); C. S. Rondestvedt, Jr., *J. Am. Chem. Soc.* **84,** 3319 (1962); c) A. Rieche and E. Schmitz, *Chem. Ber.* **90,** 531 (1957); d) T. G. Traylor and C. L. Perrin, *J. Am. Chem. Soc.* **88,** 4934 (1966).
38. a) N. Greif, Ph.D. Thesis, University of Marburg, 1967; b) K. Dimroth and P. Heinrich, *Angew. Chem.* **78,** 714 (1966); *Angew. Chem. Intern. Ed. English* **5,** 676 (1966); c) K. Dimroth and A. Nürrenbach, *Chem. Ber.* **93,** 1649 (1960).
39. a) P. Schiess, H. L. Chia, and C. Suter, *Tetrahedron Letters* p. 5747 (1968); b) E. N. Marvell, G. Caple, T. A. Gosink, and G. Zimmer, *J. Am. Chem. Soc.* **88,** 619 (1966).
40. K. Dimroth, K. H. Wolf, and H. Kroke, *Ann. Chem. Liebigs* **678,** 183 (1964).
41. K. Dimroth and K. H. Wolf, see Reference 10, p. 398.
42. H. Wache, Ph.D. Thesis, University of Marburg, 1964; K. Dimroth and K. H. Wolf, see Reference 10, p. 388.
43. W. Michel, Ph.D. Thesis, University of Marburg, 1961; K. Dimroth and K. H. Wolf, see Reference 10, p. 386.
44. a) G. Köbrich and D. Wunder, *Ann. Chem. Liebigs* **654,** 131 (1962); G. Köbrich, *Ann.*

Chem. Liebigs, **648**, 114 (1961); see also G. Köbrich and W. E. Breckoff, *Ann. Chem. Liebigs* **704**, 42 (1967); b) see however C. Decoret and J. Royer, *Compt. Rend.* **267**, 1614 (1968).
45. A. Baeyer and J. Piccard, *Ann. Chem. Liebigs* **384**, 213 (1911); **407**, 339 (1915).
46. J. J. Basselier, *Ann. Chim. (Paris)* **6**, 1131 (1961).
47. a) J. A. Berson, *J. Am. Chem. Soc.* **74**, 358 (1952); b) J.-P. Griot, J. Royer, and J. Dreux, *Tetrahedron Letters* p. 2195 (1969).
48. G. Rio and Y. Fellion, *Tetrahedron Letters* p. 1213 (1962).
49. A. T. Balaban and C. Toma, *Tetrahedron Suppl.* **7**, 1 (1966).
50. C. Toma and A. T. Balaban, *Tetrahedron Suppl.* **7**, 7 (1966).
51. R. Lombard and A. Kress, *Bull. Soc. Chim. France* p. 1528 (1960).
52. a) A. T. Balaban, *Tetrahedron* **24**, 5059 (1968); b) A. T. Balaban, *Tetrahedron* **26**, 739 (1970).
53. W. Schneider, *Ann. Chem. Liebigs* **438**, 115 (1924); W. Schneider and W. Müller, *Ann. Chem. Liebigs* **438**, 147 (1924); W. Schneider and K. Weiss, *Ber. Deut. Chem. Ges.* **61**, 2445 (1928); W. Schneider and W. Riedel, *Ber. Deut. Chem. Ges.* **74**, 1252 (1941).
54. A. T. Balaban, P. T. Frangopol, G. D. Mateescu, and C. D. Nenitzescu, *Bull. Soc. Chim. France* p. 298 (1962).
55. a) A. T. Balaban and C. D. Nenitzescu, *J. Chem. Soc.* p. 3566 (1961); b) A. T. Balaban, T. H. Crawford, and R. H. Wiley, *J. Org. Chem.* **30**, 879 (1965).
56. A. T. Balaban, G. Mihai, and C. D. Nenitzescu, *Tetrahedron* **18**, 257 (1962).
57. a) K. Dimroth and G. Neubauer, *Angew. Chem.* **69**, 720 (1957); b) F. Kröhnke and K. Dickoré, *Chem. Ber.* **92**, 46 (1959).
58. H. Kroke, Ph.D. Thesis, University of Marburg, 1962.
59. W. Krafft, Ph.D. Thesis, University of Marburg, 1961; K. Dimroth and K. H. Wolf, see Reference 10, p. 383.
60. J. N. Collie and T. Tickle, *J. Chem. Soc.* **75**, 710 (1899); on reversibility, see also J. Kendall, *J. Am. Chem. Soc.* **36**, 1222 (1914).
61. a) K. Dimroth and K. H. Wolf, *Angew. Chem.* **72**, 777 (1960); b) see also Reference 10, p. 383; M. Ohta and H. Kato, *Bull. Chem. Soc. Japan* **32**, 707 (1959); *Chem. Abstr.* **56**, 1422 (1962); c) J. A. Van Allan, G. A. Reynolds, and D. P. Meier, *J. Org. Chem.* **83**, 4418 (1968).
62. G. Fischer and W. Schroth, *Z. Chem.* **3**, 191 (1963).
63. A. T. Balaban and C. D. Nenitzescu, *Ann. Chem. Liebigs* **625**, 74 (1959).
64. A. T. Balaban and C. D. Nenitzescu, *Chem. Ber.* **93**, 599 (1960).
65. G. Märkl and H. Baier, *Tetrahedron Letters* p. 4379 (1968).
66. a) A. Baeyer, *Ber. Deut. Chem. Ges.* **43**, 2337 (1910); b) A. Baeyer and J. Piccard, *Ann. Chem. Liebigs* **384**, 208 (1911); *Ann. Chem. Liebigs* **407**, 332 (1914); c) see also A. B. Susan and A. T. Balaban, *Rev. Roumaine Chim.* **14**, 111 (1969).
67. Review see F. Brody and P. R. Ruby, in "The Chemistry of Heterocyclic Compounds" (A. Weissberger, ed.), p. 210. Wiley (Interscience), New York, 1960; K. Dimroth and K. H. Wolf, see Reference 10, p. 357.
68. a) W. Dilthey, *J. Prakt. Chem.* **94**, 53 (1916); **102**, 209 (1921); **104**, 28 (1922); b) C. Gastaldi, *Gazz. Chim. Ital.* **52**, 169 (1921); C. Gastaldi and G. L. Peyretti, *Gazz. Chim. Ital.* **53**, 11 (1922); c) F. Klages and H. Träger, *Chem. Ber.* **86**, 1330 (1953); d) R. M. Anker and A. H. Cook, *J. Chem. Soc.* p. 117 (1946); e) R. Wizinger, S. Losinger, and P. Ulrich, *Helv. Chim. Acta* **39**, 9 (1956).
69. G. Mutz, Ph.D. Thesis, University of Marburg, 1960; K. Dimroth and K. H. Wolf, see Reference 10, p. 358.
70. a) W. Dilthey, *Ber. Deut. Chem. Ges.* **55**, 59 (1922); W. Dilthey and W. Radmacher, *J. Prakt. Chem.* **111**, 153 (1925); b) O. Diels and K. Alder, *Ber. Deut. Chem. Ges.* **60**, 716 (1927).

71. a) W. Dilthey and H. Dierichs, *J. Prakt. Chem.* **144,** 1 (1936); W. Schneider, *Angew. Chem.* **39,** 412 (1926); W. Schneider, W. Döbling, and R. Cordua, *Ber. Deut. Chem. Ges.* **70,** 1645 (1937); b) K. Dimroth, C. Reichardt, T. Siepmann, and F. Bohlmann, *Ann. Chem. Liebigs* **661,** 1 (1962); K. Dimroth, C. Reichardt, and A. Schweig, *Ann. Chem. Liebigs* **669,** 95 (1963); C. Reichardt and K. Dimroth, *Fortschr. Chem. Forsch.* **11,** 1 (1968).
72. C. Toma and A. T. Balaban, *Tetrahedron Suppl.* **7,** 27 (1966).
73. K. Dimroth, G. Arnoldy, S. von Eicken, and G. Schiffler, *Ann. Chem. Liebigs* **604,** 221 (1957).
74. E. Schmitz, *Chem. Ber.* **91,** 1488 (1958).
75. G. Märkl, F. Lieb, and A. Merz, *Angew. Chem.* **76,** 947 (1967); *Angew. Chem. Intern. Ed. English* **6,** 944 (1967).
76. a) G. Märkl, *Angew. Chem.* **78,** 907 (1966); *Angew. Chem. Intern. Ed. English* **5,** 846 (1966); b) K. Dimroth, N. Greif, W. Städe, and F. W. Steuber, *Angew, Chem.* **79,** 725 (1967); *Angew. Chem Intern. Ed. English* **6,** 711 (1967); c) K. Dimroth and W. Mach, *Angew. Chem.* **80,** 489 (1968); *Angew. Chem. Intern. Ed. English* **7,** 460 (1968).
77. G. Märkl, F. Lieb, and A. Merz, *Angew. Chem.* **79,** 475 (1967); *Angew. Chem. Intern. Ed. English* **6,** 458 (1967).
78. K. Dimroth and P. Hoffmann, *Angew. Chem.* **76,** 433 (1964); *Angew. Chem. Intern. Ed. English* **3,** 384 (1964); *Chem. Ber.* **99,** 1325 (1966).
79. a) G. Suld and C. C. Price, *J. Am. Chem. Soc.* **83,** 1770 (1961); **84,** 2090, 2096 (1962); b) R. Wizinger and P. Ulrich, *Helv. Chim. Acta* **39,** 207, 217 (1956).
80. R. Gompper and O. Christmann, *Chem. Ber.* **94,** 1796 (1961).
81. a) K. Dimroth and G. Bräuniger, *Angew. Chem.* **68,** 519 (1956); K. Dimroth, G. Bräuniger, and G. Neubauer, *Chem. Ber.* **90,** 1634 (1957); K. Dimroth, G. Neubauer, H. Möllenkamp, and G. Oosterloo, *Chem. Ber.* **90,** 1668 (1957); K. Dimroth and K. H. Wolf, see Reference 10, p. 379; b) K. Dimroth, F. Kalk, and G. Neubauer, *Chem. Ber.* **90,** 2058 (1957); c) K. Dimroth, A. Berndt, and R. Volland, *Chem. Ber.* **99,** 3040 (1966).
82. K. Dimroth and G. Laubert, *Angew. Chem.* **81,** 392 (1969); *Angew. Chem. Intern. Ed. English* **8,** 370 (1969).
83. G. Laubert, Ph.D. Thesis, University of Marburg, 1968.
84. K. Dimroth and H. Wache, *Chem. Ber.* **99,** 399 (1966).
85. K. Dimroth and G. Neubauer, *Chem. Ber.* **92,** 2042 (1959).
86. G. Märkl, *Angew. Chem.* **74,** 696 (1962); K. Dimroth and K. H. Wolf, see Reference 10, p. 391.
87. K. Hafner, *Angew. Chem.* **69,** 393 (1957); K. Hafner and H. Kaiser, *Ann. Chem. Liebigs* **618,** 140 (1958); see also K. Hafner, *Angew. Chem.* **70,** 419 (1958).
88. F. Quint, R. Pütter, and W. Dilthey, *Ber. Deut. Chem. Ges.* **71,** 356 (1938); *Ber. Deut. Chem. Ges.* **55,** 1275 (1922).
89. K. Dimroth, K. H. Wolf, and H. Wache, *Angew. Chem.* **75,** 860 (1963).
90. K. Dimroth, H. Kroke, and K. H. Wolf, *Ann. Chem. Liebigs* **678,** 202 (1964).
91. L. King and F. Ozog, *J. Org. Chem.* **20,** 448 (1955).
92. A. T. Balaban and A. Barabas, *Chem. Ind. (London)* p. 404 (1967).

Appendix
The Structure of Oxonium Ions

A.1. Saturated Oxonium Ions

In analogy with amines and sulfonium ions, a pyramidal arrangement *(369)* of the substituents R, R', R" is also formulated for mono-, di-, and trialkoxonium ions [1–3].

The fact that a bicyclic oxonium ion *(370)* can be prepared shows at least that a pyramidal structure can be obtained forcibly at the oxonium oxygen on the basis of a particular steric situation [1]. The geometry of *(370)* is said to be similar to that of quinuclidine *(371)*.

(369)   *(370)*   *(371)*

However, no X-ray structural analysis of *(370)* has been reported.

Oxonium ions *(369)* can undergo inversion of their configuration in the same way as amines [3]. Only once has such an inversion been detected with certainty, by NMR spectroscopy [2]. At +40°C, the NMR spectrum of O-alkyloxiranium ion *(372)* shows a sharp singlet for the ring methylene protons, which broadens on cooling. At −70°C, different signals are found for the ring-methylene protons in the *cis* and *trans* positions with respect to the alkyl residue R,* and these coalesce to a single signal at −50°C [2]. Consequently, at a sufficiently low temperature, the inversion takes place slowly enough for the nonequivalence of the ring-methylene protons to be observed:

R = i-C$_3$H$_7$ *(372)*   R = CH$_3$ *(373)*

The activation energy for the inversion in the case of *(372)* has been determined as 10 ± 2 kcal/mole [2]. In the nitrogen analog N-methylaziridine *(373)*, 19 ± 3 kcal/mole, is found for the inversion barrier [4]; the figure calculated for N-alkylaziridines is 16 kcal/mole [3]. In six-membered cyclic oxonium ions, the tetrahydropyranium ions, even at −70°C, very rapid oxygen inversion still takes place: the NMR spectra are temperature-independent [2].

A.2. Aldehydium and Ketonium Ions

In the case of protonated carbonyl compounds *(8)*, two isomers can frequently be

---

* The spectrum observed at −70°C is ascribed by the authors to an AA'BB' system [2].

detected by low-temperature NMR measurements if there are two different substituents R^1 and R^2 on the carbonyl carbon [5a,b, 6a,b,c,d]. From this a considerable double-bond nature of the C–O bond has been deduced, and the occurrence of isomers has been ascribed to a *cis–trans* isomerism of the unsaturated oxonium ions [5a,b, 6a,b,c,d]:

$$\underset{(8a)}{\overset{R^2}{\underset{R^1}{>}}C=\overset{\oplus}{O}\diagdown H} \quad \rightleftharpoons \quad \underset{(8b)}{\overset{R^2}{\underset{R^1}{>}}C=\overset{\oplus}{O}\diagdown_H} \qquad \underset{(374)}{\overset{R^2}{\underset{R^1}{>}}C=\overset{\oplus}{O}\diagdown^H_{\odot}}$$

An sp^2 hybridization of the oxygen *(374)* is considered possible; in addition to the proton, an unbound sp^2 electron pair would then be present as a "substituent" on the oxygen [5b].

At a sufficiently high temperature the isomers *(8a)* and *(8b)* are transformed so rapidly into one another by rotation about the C–O bond (or the flipping over of the O–H bond) that they cannot be distinguished by NMR spectroscopy. At low temperatures both isomers can be detected together. Protonated acetaldehyde (in FSO_3H–SbF_5–SO_2 solution) at $-60\,°C$ may serve as an example [5a]; see Fig. A. 1. All resonance signals for the proton species α, β, and γ appear doubled. The particular chemical shifts for the protons β and γ are very similar for the

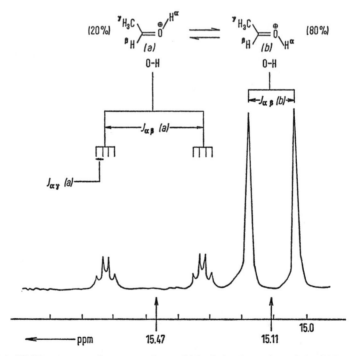

*Fig. A.*1. NMR spectrum of protonated acetaldehyde in the region of the OH resonance signals (in FSO_3H–SbF_5–SO_2 at $-60\,°C$); schematic, according to [5a].

isomers; therefore their signals partially overlap. However, the O–H signals, α, for *(a)* and *(b)* are clearly separated from one another.

A doublet is found at 15.11 ppm, the splitting of which is caused by the coupling of the O–H and the aldehyde proton ($J_{\alpha\beta} = 9.0$ Hz). In addition, there is a second doublet at 15.47 ppm ($J_{\alpha\beta} = 19.5$ Hz) in which each half of the doublet is split to form a quartet by the coupling of the O–H proton with the protons of the methyl group ($J_{\alpha\gamma} = 1.2$ Hz). The isomeric aldehydium ions are comparable with *cis–trans* isomeric alkene derivatives *(a′)* and *(b′)*. For these it is known that the coupling $J_{\alpha'\beta'}$ between *trans*-vinyl protons is greater than between *cis*-protons. Furthermore, the coupling constant $J_{\alpha'\gamma'}$ between a vinyl and a *cis*-allyl proton is in general found to be greater than with a *trans*-allyl proton [5a,b,7]:

$$J_{\alpha'\beta'}(a') > J_{\alpha'\beta'}(b')$$
$$J_{\alpha'\gamma'}(a') > J_{\alpha'\gamma'}(b')$$
$$J_{\alpha'\beta'} > J_{\alpha'\gamma'}$$

In the low-field doublet of the acetaldehydium ion an "allylic" coupling $J_{\alpha\gamma}$ can be observed, while in the high-field doublet $J_{\alpha\gamma}$ is not resolved ($J < 0.3$ Hz). The doublet at the low field also shows the larger of the two doublet couplings $J_{\alpha\beta}$; the signal at 15.47 ppm must therefore be ascribed to the cation *(8a)*.* It can also be understood that the aldehydium ion *(a)* is present in smaller amount than *(b)* because of the greater steric interaction between the O–H and methyl groups [5a].

Similarly, in the equilibrium of protonated ketones [5b, 6a,b,c,d] there is always a larger proportion of that isomer in which the O–H proton is in the *trans* position with respect to the larger substituent on the carbon atom; e.g., (in HSO_3F–SbF_5–SO_2 at $-60°C$) [5b]:

H_5C_2 ⊕ H H_3C =O	H_5C_2 ⊕ H_3C =O, H	$(CH_3)_3C$ ⊕ H H_3C =O	$(CH_3)_3C$ ⊕ H_3C =O, H
20%	80%	not observed	~100%

In protonated ketones *(8)* that bear two equal substituents ($R^1 = R^2$) on the carbonyl carbon atom, only one isomer is possible; correspondingly, only one O–H signal is found in the NMR spectrum. Nevertheless, at a sufficiently low temperature ($-60°C$) the alkyl groups are not equivalent so that in protonated di-*tert*-butyl ketone, for example, two signals occur for the *tert*-butyl groups (in a ratio of 1:1) [5b]:

---

* *Cis* allylic coupling constants in NMR spectra of some alkoxycarbonium ions were found to be smaller than *trans* coupling [12b]. This trend is a reversal of the results cited above; therefore assignment of isomers analogous to corresponding olefins may sometimes be uncertain.

$$(CH_3)_3C\!\!\!>\!\!\!=\!\!\!\overset{\oplus}{O}\!\!-\!\!H \;\rightleftharpoons\; (CH_3)_3C\!\!\!>\!\!\!=\!\!\!\overset{\oplus}{O}\cdots H$$
$(CH_3)_3C$ ... $(CH_3)_3C$

Cis–trans isomerism has also been observed with alkoxycarbonium ions. As an example, we may mention the chloro(methoxy)carbonium ion (in SbF_5–SO_2 at $-60\,°C$) [8]; the ratio of the isomers does not change between -90 and $-40\,°C$:

$$(19\%) \quad \underset{H}{\overset{Cl}{>}}\!\!=\!\!\overset{\oplus}{O}\!\!-\!\!CH_3 \;\rightleftharpoons\; \underset{H}{\overset{Cl}{>}}\!\!=\!\!\overset{\oplus}{O}\cdots CH_3 \quad (81\%)$$

Here again assignment of the isomers can be made on the basis of different "allylic" couplings [8].

A.3. Acidium Ions

In protonated carboxylic acids *(17)*, the C–O bonds (as in aldehydium and ketonium ions) have a partial double-bond character. Consequently, the following *cis–trans* isomers are to be expected [6a, 9a,b,c, 10]:

$$R^1\!-\!\overset{\overset{H}{|}}{\underset{\underset{H}{|}}{C^{\oplus}}}\!\!\!<\!\!\!\overset{O}{\underset{O}{}} \;\rightleftharpoons\; R^1\!-\!C^{\oplus}\!\!<\!\!\overset{O-H}{\underset{O-H}{}} \;\rightleftharpoons\; \cdots$$

(17a)	*(17b)*	*(17c)*	*(17d)*
cis–trans	*cis–cis*	*trans–cis*	*trans–trans*

The terms "*cis*" and "*trans*" relate to the orientation of the protons on the oxygen atoms relative to group R^1. Formulas *(17a)* and *(17c)* are identical, and the *trans–trans* isomer *(17d)* is sterically very unfavorable. For these reasons it can be understood that only two isomers are observed with low-temperature NMR measurements: the *cis–trans* isomer *(17a)* or *(17c)* and the *cis–cis* isomer *(17b)*. However, this is the case only with protonated formic acid [9a,b,c, 10] and protonated acetic acid [6a, 9c, 10]; with all higher carboxylic acids, protonation leads only to the *cis–trans* isomer [6a, 10]*. The figures in the accompanying table give the percentage distribution of the isomers in FSO_3H–SbF_5–SO_2 [6a, 10] HF–SbF_5 [9a,c], or HF–BF_3 [9a,b,c] between -60 and $-75\,°C$:

---

* *Note added in proof (June 1970):* The great preponderance of *(17a)* compared with *(17b)* indicates that *(17a)* may be stabilized by intramolecular hydrogen bonds [9d,e]:

$$R^1\!-\!C^{\oplus}\!\!<\!\!\overset{O\cdots H}{\underset{O}{}}\!\!\!H$$

Recently this has been proved by IR measurements of the carbonyl stretching frequencies of carboxylic acids dissolved in H_2SO_4 and D_2SO_4 [9e].

R^1	(17a) ≡ (17c) cis–trans, %	(17b) cis–cis, %	Reference
H	77	23	[9a,c]
	76	24	[6a]
CH_3	97	3	[6a, 10]
	96	4	[9c]
C_2H_5	100	–	[6a, 10]

The assignments were found by comparing the coupling constants [9a, 10], and in the case of the protonated formic acid also by deuterium-labeling [9a]. If *cis–trans* and *cis–cis* isomers are present together, three O–H signals are found in the NMR spectrum; two are due to the *cis–trans* isomer and one to the *cis–cis* isomer; e.g., in the case of protonated formic acid (in FSO_3H–SbF_5–HF) at $-60\,°C$ [10]:

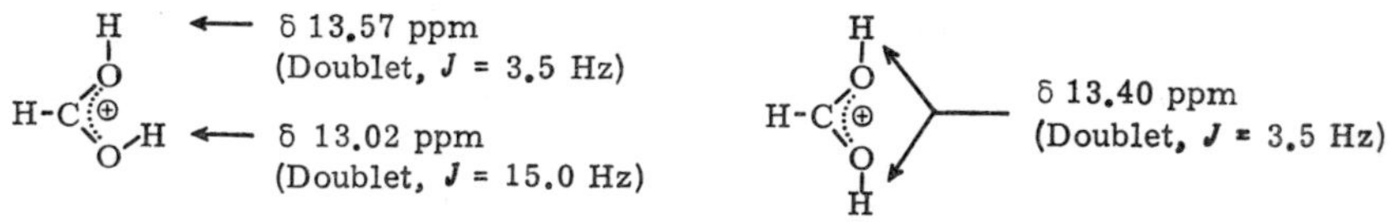

δ 13.57 ppm (Doublet, $J = 3.5$ Hz)

δ 13.02 ppm (Doublet, $J = 15.0$ Hz)

(17a)

δ 13.40 ppm (Doublet, $J = 3.5$ Hz)

(17b)

Recently, activation energies have been determined for the configurational equilibration of the cations $(17a) \rightleftharpoons (17b) \rightleftharpoons (17c)$: for protonated formic acid an activation barrier of 15.3 ± 0.9 was found and for protonated acetic acid, 11.2 ± 0.2 kcal/mole (in HF–SbF_5) [9c]. It is assumed that in the equilibration between (17a) and (17c) the cation (17b) is an intermediate [9c].

In the case of the O-alkylated esters of formic acid and acetic acid, the dialkoxycarbonium ions (9), only the *cis–trans* isomers $(9a) \equiv (9c)$* have been detected in the low-temperature NMR spectrum between -30 and $0\,°C$ [11a,b,c,d]:

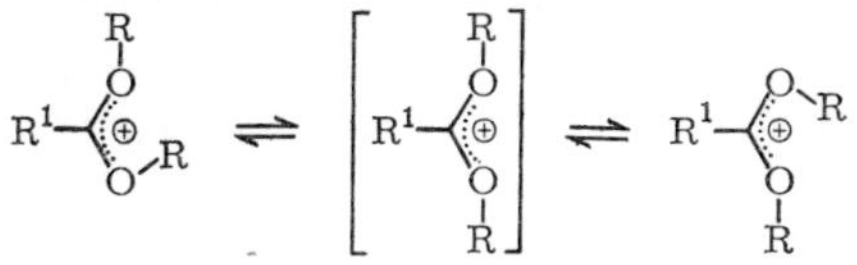

(9a) ⇌ (9b) ⇌ (9c)

where $R^1 = H$, CH_3 and $R = CH_3$, C_2H_5, n-C_3H_7.

Only *one* NMR signal is observed for R^1; at a sufficiently low temperature, on the other hand, the nonequivalence of the two alkyl groups R can be detected by doubling of the NMR signals (for all types of protons) from R. In the case of the dimethoxymethylcarbonium ion $(CH_3O)_2C(CH_3)^\oplus$, two methyl signals in a ratio

---

* In O-alkylated formates, there is probably 10% of a second isomer in the equilibrium [11b].

of 1:1 were found in trifluoroacetic acid below +14 °C. For this cation the activation energy for the hindered rotation about the C–O bonds* has also been determined and found to be 11 ± 4 kcal/mole [11a].

A higher energy barrier is assumed for the dimethoxycarbonium ion $(CH_3O)_2CH^+$, since even at room temperature two (broadened) methyl signals are still visible in the NMR spectrum of this cation [11a].

In the case of protonated esters *(18)*, four isomers are conceivable:

(18a)   *(18b)*   *(18c)*   *(18d)*

Between $-65°$ and $-90°C$, however, only in the case of protonated formates do two (or even three) isomers occur; in all other carboxylic esters only one cation can be observed [12a]. The *trans–trans* cation *(18d)*, like *(17d)*, is not to be expected because of its unfavorable steric structure. However, the assignment of the NMR data to the remaining isomers has not been definitively made [11b, 12a]. Thus it is concluded from the analogy with *(9)* that the *cis–trans* isomer *(18a)* should predominate in the case of the protonated formates and should be present exclusively in the case of the protonated esters of higher carboxylic acids [11b]. On the other hand, Olah and associates assumed that in the case of protonated ethyl and isopropyl formates, it is just the *cis–cis* isomer *(18b)* that is the main constituent of the equilibrium mixture (e.g., 82:12:6% for ethyl formate) [12a]. Neither do recent investigations permit final assignment [12b].

In the case of protonated carbonic acid, the sterically most favorable arrangement is *(375)*, R = H; in this all protons are equivalent. Even at $-80°C$ (in FSO_3H–SbF_5–SO_2) only one resonance signal for the O–H proton is observed, from which it is concluded that *(375)* is present as the only cation [13]. In the triethoxycarbonium ion, too, the steric strain is lowest when the ethyl groups are arranged as in *(375)*, R = C_2H_5. Down to $-60°C$ only one triplet and one quartet appear in the NMR spectrum for the methyl and methylene protons, respectively [11a]:

(375)   R = H, CH_3, C_2H_5

However, the signals broaden approximately twofold from room temperature to $-60°C$ [11a]. Again, for the trimethoxycarbonium ion only, the *(isomer 375)*, R = CH_3, has been detected [11a, d].

---

* It has not been established whether the transition from *(9a)* to *(9c)* takes place by rotation round the C–O bond or by inversion of the C–O–R bond. The corresponding remarks apply to *(17a)*⇌*(17c)* [9c].

References

1. F. Klages and H. A. Jung, *Chem. Ber.* **92**, 3757 (1965).
2. J. B. Lambert and D. H. Johnson, *J. Am. Chem. Soc.* **90**, 1349 (1968).
3. G. W. Koeppl, D. S. Sagatys, G. S. Krishnamurthy, and S. I. Miller, *J. Am. Chem. Soc.* **89**, 3396 (1967).
4. A. Loewenstein, J. F. Neumer, and J. D. Roberts, *J. Am. Chem. Soc.* **82**, 3599 (1960).
5. a) G. A. Olah, D. H. O'Brien, and M. Calin, *J. Am. Chem. Soc.* **89**, 3582 (1967); b) G. A. Olah, M. Calin, and D. H. O'Brien, *J. Am. Chem. Soc.* **89**, 3586 (1967).
6. a) M. Brookhart, G. C. Levy, and S. Winstein, *J. Am. Chem. Soc.* **89**, 1735 (1967); b) G. A. Olah and M. Calin, *J. Am. Chem. Soc.* **90**, 938 (1968); c) D. M. Brouwer, *Tetrahedron Letters* p. 453 (1968); d) D. M. Brouwer, *Rec. Trav. Chim.* **86**, 696 (1967).
7. E. B. Whipple, J. H. Goldstein, and L. Mandell, *J. Am. Chem. Soc.* **82**, 3010 (1962); A. A. Bothner-By and C. Naar-Colin, *J. Am. Chem. Soc.* **83**, 2318 (1961); A. A. Bothner-By, C. Naar-Colin, and H. Günther, *J. Am. Chem. Soc.* **84**, 2748 (1962).
8. G. A. Olah and J. M. Bollinger, *J. Am. Chem. Soc.* **89**, 2993 (1967).
9. a) H. Hogeveen, A. F. Bickel, C. W. Hilbers, E. L. Mackor, and C. MacLean, *Chem. Commun.* p. 898 (1966); *Rec. Trav. Chim.* **86**, 687 (1967); b) H. Hogeveen, *Rec. Trav. Chim.* **86**, 289 (1967); c) H. Hogeveen, *Rec. Trav. Chim.* **87**, 1312 (1968); d) T. Birchall and R. J. Gillespie, *Can. J. Chem.* **43**, 1045 (1965); e) C. J. Clemett, *Chem. Commun.* p. 211 (1970).
10. G. A. Olah and A. M. White, *J. Am. Chem. Soc.* **89**, 3591 (1967).
11. a) B. G. Ramsey and R. W. Taft, *J. Am. Chem. Soc.* **88**, 3058 (1966); b) R. F. Borch, *J. Am. Chem. Soc.* **90**, 5303 (1968); c) C. H. V. Dusseau, S. E. Schaafsma, H. Steinberg, and T. J. de Boer, *Tetrahedron Letters* p. 467 (1969); d) A. Gerlach, Ph.D. Thesis, University of Marburg, 1969.
12. a) G. A. Olah, D. H. O'Brien, and A. M. White, *J. Am. Chem. Soc.* **89**, 5694 (1967); b) A. M. White and G. A. Olah, *J. Am. Chem. Soc.* **91**, 2943 (1969).
13. G. A. Olah and A. M. White, *J. Am. Chem. Soc.* **90**, 1884 (1968).

Author Index

Numbers in brackets are reference numbers; numbers in italics show the page on which the complete reference is listed.

A

Abalyaeva, V. V. 70 [23, 26], *96* [23, 26]
Acevedo, R. 123 [35c], *126* [35c]
Ahmed, M. G. 26 [18c], *54* [18c]
Alazrek, A. 106 [6d], *125* [6d]
Alder, K. 158 [70b], 161 [70b], *171* [70b]
Alder, R. W. 26 [18c], *54* [18c]
Allbutt, D. A. 18 [37] *21* [37], 27 [19a], 29 [19a], 34 [19a], 40 [19a], *54* [19a], 114 [25a, 26], 115 [26], 116 [26], *126* [25a, 26]
Allendörfer, H. 28 [26a], 32 [26a], 34 [26a], 41 [26a], 44 [26a], *54* [26a]
Allred, E. 74 [56, 57], 75 [56], *98* [56, 57], 100 [1a], 101 [1a,b,c], 102 [1a,b], 103 [1c], *125* [1a,b,c]
Anderson, C. B. 3 [26], *4* [26], 9 [27], *10* [27], 24 [6b], 25 [6b], 29 [6b], 30 [6b], 31 [6b], 40 [6b], 41 [6b], *54* [6b], 84 [80], 90 [80], *98* [80], 112 [24a], 113 [24a], 117 [24a], 120 [24a], 121 [24a], 124 [24a], *126* [24a]
Angyal, S. J. 117 [31], 121 [31], *126* [31]
Anker, R. M. 47 [52a], *55* [52a], 158 [68d], *171* [68d]
Apjok, J. 24 [6c], 29 [6c], 30 [6c], 31 [6c], *54* [6c], 124 [37b], 125 [37b], *127* [37b]
Arens, J. F. 52 [81], *56* [81]
Arnett, E. M. 60 [10a], 61 [10a], *96* [10a]
Arnoldy, G. 158 [73], *172* [73]
Astle, M.J. 89[91], *99*[91]

B

Baddeley, G. 106 [12], *126* [12]
Bader, H. 49 [71], *56* [71]
Baeyer, A. 1 [11, 12, 14], 2 [11, 14, 18], *4* [11, 12, 14, 18], 16 [18a], 19 [42], *20* [18a], *21* [42], 26 [18a], 46 [47], 47 [18a], 48 [47], *54* [18a], *55* [47], 62 [9b], *96* [9b], 151 [45], 158 [66a,b], 161 [66a], *171* [45, 66a,b]
Baganz, H. 82 [77c], *98* [77c]
Baier, H. 158 [65], 165 [65], *171* [65]
Baker, E. B. 5 [4], 6 [4], *9* [4]
Balaban, A. T. 45 [41b, 43], 47 [43], 48 [43, 63], 49 [69, 70a], 50 [43], 51 [41b, 63, 77a, b,c,d,e], 53 [69], *55* [41b,43], *56* [63, 69, 70a, 77a,b,c,d,e], 152 [49, 50], 153 [52a, b, 54, 55a,b], 154 [56], 156 [63], 157 [52a, 64], 158 [63, 66c, 72], 161 [50, 63], 164 [64], 167 [92], 168 [92], *171* [49, 50, 52a,b, 54, 55a,b, 56, 63, 64, 66c, 72, 92].
Balli, H. 72 [49a], *97* [49a]
Ban, Y. 72 [47d], *97* [47d], 135 [23], *169* [23]
Barabas, E. 51 [77d], *56* [77d], 167 [92], 168 [92], *172* [92]
Barbulescu, N. S. 49 [69], 53 [69], *56* [69]
Barton, T. J. 72 [47c], *97* [47c], 135 [26a], *170* [26a]
Basselier, J. J. 151 [46], *171* [46]
Bastien, I. J. 5 [4], 6 [4], *9* [4]
Battenberg, E. 2 [24], *4* [24], 18 [29], 19 [29], *20* [29], 24 [8], 27 [8], 33 [8], 34 [8], 36 [8], 37 [8], 38 [8], *54* [8], 70 [22], 71 [22], 73 [22], 74 [22], *96* [22], 131 [12a], *169* [12a]
Bayer A.-G. 142 [36], *170* [36]
Baylis, E. K. 106 [12], *126* [12]
Beekmann, P. 28 [26a], 32 [26a], 34 [26a], 41 [26a], 44 [26a], *54* [26a]
Behre, H. 118 [28e], 121 [28e], *126* [28e], 110 [39], 123 [39], *127* [39]
Beringer, F. M. 29 [28d], 40 [28d], *54* [28d], 94 [100], *99* [100]
Berlin, P. 117 [27b], *126* [27b]
Berndt, A. 161 [81c], *172* [81c]
Bernhard, C. 72 [53a], *98* [53a]
Berson, J. A. 152 [47a], *171* [47a]
Bertele, E. 72 [47f], *97* [47f], 131 [14], 135 [14], *169* [14]
Besserer, K.18 [35c], *21* [35c], 69 [19], *96* [19]
Bestmann, H. J. 131 [11], *169* [11]
Beyer, E. 82 [77a], 84 [81], *98* [77a, 81]
Beyerlin, H. P. 134 [19c], 136 [28], *169* [19c], *170* [28]
Bickel, A. F. 3 [33], *4* [33], 19 [44b], *21* [44b], 176 [9a], 177 [9a], *179* [9a]
Birchall, T. 19 [45], *21* [45], 176 [9d], *179* [9d]
Blackett, B. N. 42 [11b], *54* [11b]
Blackwood, R. K. 58 [103], *99* [103]
Blunt, J. W. 18 [25b], *20* [25b], 29 [29], 40 [29], *54* [29], 123 [36a], *127* [36a]
Bodenbenner, K. 3 [25], *4* [25], 6 [18], *10* [18], 18 [36], *21* [36], 24 [6a], 25 [6a], 27 [6a], 28 [6a], 29 [6a], 31 [6a], 40 [6a], 41 [6a], 42 [6a], 44 [6a], *54* [6a], 73 [29], 80 [29], 81 [29], 83 [29], 84 [29], 90 [29], *97* [29], 111 [21], 116 [21], *126* [21], 138 [31a], *170* [31a]

Author Index

Boeden, H.-F. 117 [27b], *126* [27b]
Böhme, H. 86 [87], 92 [87], *99* [87]
Boer, F. P. 5 [5], 6 [5], *10* [5]
Böttler, T. 48 [62], *56* [62]
Bohlmann, F. 158 [71b], *172* [71b]
Bollinger, J. M. 3 [27], *4* [27], 28 [22a], 29 [31], 30 [22a], 33 [22a], 40 [31], *54* [22a, 31], 111 [20], *126* [20], 176 [8], *179* [8]
Bolstad, A. N. 23 [1a], *53* [1a], 85 [82c], 92 [82c], *98* [82c]
Boos, H. 72 [47f], *97* [47f], 131 [14], 135 [14], *169* [14]
Borch, R. F. 70 [27c], 71 [27c, 40b], *97* [27c, 40b], 132 [16], 137 [30], *169* [16], *170* [30], 177 [11b], 178 [11b], *179* [11b]
Bormann, D. 135 [24a], *169* [24a]
Borner, P. 3 [25], *4* [25], 6 [17, 18], *10* [17, 18], 18 [34, 36], *21* [34, 36], 24 [6a], 25 [6a, 13], 27 [6a], 28 [6a, 13], 29 [6a], 31 [6a, 13], 34 [13], 40 [6a], 41 [6a], 42 [6a], 44 [6a], *54* [6a, 13], 71 [30], 72 [30], 73 [29, 30], 80 [29], 81 [29], 83 [29], 84 [29], 85 [30], 87 [30], 88 [30], 90 [29, 30], 93 [30], 94 [30], *97* [29, 30], 108 [16a], 111 [21], 116 [21], *126* [16a, 21], 128 [1], 138 [1, 31a], 140 [1], *168* [1], *170* [31a]
Bos, H. J. T. 52 [81], *56* [81]
Boschan, R. 112 [24b], *117* [24b], 119 [32], 120 [32], 121 [24b, 32], 124 [24b], *126* [24b, 32]
Bothner-By, A. A. 175 [7], *179* [7]
Bott, K. 71 [42], *97* [42]
Bouillon, C. 72 [38], *97* [38]
Bowers, A. 123 [35c], *126* [35c]
Bräuniger, G. 161 [81a], *172* [81a]
Brass, W. 1 [5], 2 [5], *3* [5], 6 [9], 10 [9], 62 [9a], *96* [9a]
Brechbühler, H. 133 [21a], *169* [21a]
Breckhoff, W. E. 151 [44a], 163 [44a], *171* [44a]
Bredereck, H. 133 [19a, b, 20b], 134 [19c], 136 [28], *169* [19a, b, c, 20b], *170* [28]
Breslow, R. 16 [14], *20* [14], 71 [36], *97* [36], 129 [7], 130 [7], *169* [7]
Brody, F. 158 [67], *171* [67]
Brookhart, M. 3 [32], *4* [32], 15 [11c], 16 [11c], *20* [11c], 60 [5c], 63 [5c], *96* [5c], 174 [6a], 175 [6a], 176 [6a], 177 [6a], *179* [6a]
Brouwer, D. M. 7 [22], *10* [22], 15 [10a, b], 19 [10a, b], *20* [10a, b], 174 [6c, d], 175 [6c, d], *179* [6c, d]
Brown, D. R. 71 [45], *97* [45]
Brown, R. A. 70 [24], *96* [24]

Brumlick, G. C. 72 [52], *97* [52]
Buckles, R. E. 100 [4], 110 [4], 111 [4, 19], 112 [19], 117 [19], 121 [19], *125* [4], *126* [19]
Büchi, H. 133 [21a], *169* [21a]
Bülow, C. 52 [85], *56* [85]
Bürgle, P. 129 [8], 130 [8], *169* [8]
Burger, B. 51 [73], *56* [73]
Burness, D. M. 72 [51b], *97* [51b]
Burwell, Jr., R. L. 77 [67a], *98* [67a]

C

Cahours, A. 1 [2, 3], *3* [2, 3]
Calin, M. 3 [31, 32, 33], *4* [31, 32, 33], 6 [11], *10* [11], 15 [11a, b], *20* [11a, b], 60 [3, 5a, b], 63 [3, 5a], 66 [15], 69 [5a], *96* [3, 5a, b, 15], 174 [5a, b, 6b], 175 [5a, b, 6b], *179* [5a, b, 6b]
Caple, G. 149 [39b], *170* [39b]
Capon, B. 24 [7b], *54* [7b], 101 [5b], *125* [5b]
Cargioli, J. D. 60 [10c], *96* [10c]
Cassey, H. N. 71 [31a], *97* [31a]
Černý, M. 117 [27a], *126* [27a]
Chassin, C. 32 [27d], *54* [27d]
Chassin, R. 32 [27d], *54* [27d]
Chekhuta, O. M. 142 [34c], *170* [34c]
Cherdron, H. 142 [36], *170* [36]
Chia, H. L. 149 [39a], *170* [39a]
Christmann, O. 161 [80], *172* [80]
Ciacin, J. 108 [14], *126* [14]
Ciba, Ltd. 71 [40d], *97* [40d], 133 [18], *169* [18]
Clemett, C. J. 176 [9e], *179* [9e]
Coffey, G. P. 70 [25], *96* [25]
Coffi-Nketsia, S. 77 [68], *98* [68]
Collie, J. N. 1 [9], *4* [9], 47 [51], *55* [51], 155 [60], *171* [60]
Comisarow, M. B. 5 [4], 6 [4], *9* [4], 66 [16], 67 [16], *96* [16]
Comte, V. 47 [56], *55* [56]
Cook, A. H. 47 [52a], *55* [52a], 158 [68d], *171* [68d]
Cook, D. 6 [4], *10* [4]
Coppen, M. J. 123 [36b], *127* [36b]
Cordua, R. 158 [71a], *172* [71a]
Corse, J. 112 [24b], 117 [24b], 121 [24b], 124 [24b], *126* [24b]
Cort, L. A. 82 [77b], *98* [77b]
Costisella, B. 25 [12b], 30 [12b], *54* [12b], 83 [79a], 91 [79a], *98* [79a]
Coxon, J. M. 42 [11b], *54* [11b]
Crawford, T. H. 153 [55b], *171* [55b]
Crompton, H. 85 [83], *98* [83]
Cross, A. D. 123 [35c], *126* [35c]

Curphey, T. J. 72 [50], 97 [50]
Curry, M. J. 25 [12a], 30 [12a], 54 [12a], 82 [78b], 98 [78b]

D

Danielisz, M. 16 [15], 20 [15], 71 [35a], 97 [35a], 129 [5a], 169 [5a]
Daniloff, S. 69 [18], 96 [18]
De Boer, T. J. 3 [35], 4 [35], 86 [79b], 98 [79b], 177 [11c], 179 [11c]
Decker, H. 1 [4], 2 [16], 4 [2, 16], 46 [48a], 52 [48a], 55 [48a]
Decoret, C. 151 [44b], 171 [44b]
Degani, I. 46 [44a], 55 [44a]
De Gottrau, H. 2 [19], 4 [19]
Delfs, D. 36 [38], 55 [38], 78 [69], 98 [69], 141 [34a], 142 [34a], 170 [34a]
DeMember, J. R. 18 [30a], 21 [30a], 34 [22b], 54 [22b]
Denney, D. B. 108 [14], 126 [14]
Deno, N. C. 5 [4], 10 [4], 19 [46], 21 [46], 66 [13], 96 [13]
Denot, E. 123 [35c], 126 [35c]
Dickoré, K. 154 [57b], 167 [57b], 171 [57b]
Diels, O. 158 [70b], 161 [70b], 171 [70b]
Dierichs, H. 158 [71a], 172 [71a]
Dilthey, W. 2 [16], 4 [16], 48 [62], 49 [66, 68a,b,c,d], 51 [73, 79a], 52 [66, 79a], 53 [68a,b,c, 90], 56 [62, 66, 68a,b,c,d, 73, 79a], 57 [90], 158 [68a, 70a, 71a], 164 [88], 171 [68a, 70a], 172 [71a, 88]
Dimroth, K. 6 [13], 10 [13], 18 [35a], 21 [35a], 25 [9], 26 [9], 45 [42], 46 [44b, 45a], 47 [42, 45a,b], 49 [42], 50 [72], 54 [9], 55 [42, 44b, 45a,b], 56 [72], 69 [19], 72 [46], 90 [46], 95 [19, 101, 102], 96 [19], 97 [46], 99 [101, 102], 131 [10], 147 [38b,c], 149 [40], 150 [41, 42, 43], 151 [40], 154 [10, 40, 57a, 59], 155 [10, 59, 61a], 158 [67, 69, 71b, 73], 159 [76b,c, 78], 161 [59, 81a,b, c, 82], 162 [42, 82, 84, 85], 163 [40, 42, 85, 86], 164 [76c], 165 [85, 89], 166 [86, 90], 167 [10, 57a, 59, 61a, 90], 168 [90], 169 [10], 170 [38b,c, 40, 41, 42], 171 [57a, 59, 61a, 67, 69], 172 [71b, 73, 76b,c, 78, 81a, b,c, 82, 84, 85, 86, 89, 90]
Döbling, W. 158 [71a], 172 [71a]
Dolby, L. J. 125 [38a,b], 127 [38a,b]
Domaschke, L. 82 [77c], 98 [77c]
Donamura, L. G. 70 [24], 96 [24]
Dorofeenko, G. N. 18 [25c], 20 [25c], 50 [70b], 51 [78], 52 [84b], 56 [70b, 78, 84b]
Dovey, W. C. 53 [87], 57 [87]
Dreux, J. 152 [47b], 171 [47b]

Duffin, G. F. 72 [49a], 97 [49a]
Dumas, J. 1 [7], 4 [7], 18 [22], 20 [22]
Dunitz, J. D. 72 [47f], 97 [47f], 131 [14], 135 [14], 169 [14]
Dusseau, C. H. V. 3 [35], 4 [35], 86 [79b], 98 [79b], 177 [11c], 179 [11c]
Duttenhöfer, A. 2 [19], 4 [19], 26 [18a], 47 [18a], 54 [18a]

E

Effenberger, F. 23 [1b], 53 [1b], 133 [19a,b, 20b], 134 [19c], 136 [28], 169 [19a,b,c, 20b], 170 [28]
Eicher, T. 16 [14], 20 [14], 71 [36], 97 [36], 129 [6a,b,c, 7], 130 [6a,b,c, 7], 169 [6a, b,c, 7]
Eichhorn, I. 143 [37a], 170 [37a]
Eiden, F. 47 [58], 55 [58]
Eisenmenger, U. 140 [33], 141 [33], 170 [33]
Elderfield, R. C. 51 [74a], 56 [74a]
Elsinger, F. 72 [47f], 97 [47f], 131 [14], 135 [14], 169 [14]
Emel'yanova, L. I. 18 [28], 20 [28]
Epel'baum, I. V. 142 [34c], 170 [34c]
Epstein, W. W. 108 [15], 126 [15]
Eschenmoser, A. 72 [47f], 97 [47f], 131 [14], 133 [21a], 135 [14, 24a,b], 169 [14, 21a, 24a,b]
Ewing, S. 14 [12], 20 [12], 24 [7a], 54 [7a]

F

Fabrycy, A. 2 [17c], 4 [17c], 24 [5c], 53 [5c]
Farcasiu, D. 51 [77b], 56 [77b]
Farcasiu, M. 51 [77d], 56 [77d]
Faworskii, A. 2 [17b], 4 [17b], 19 [41], 21 [41]
Faworsky, A. see Faworskii, A.
Fellion, Y. 152 [48], 171 [48]
Felner, I. 72 [47f], 97 [47f], 131 [14], 135 [14, 24a], 169 [14, 24a]
Fischer, E. 142 [36], 170 [36]
Fischer, G. 45 [43], 47 [43], 48 [43], 49 [64], 50 [43], 52 [82a,b, 84a], 55 [43], 56 [64, 82a,b, 84a], 155 [62], 171 [62]
Fischer, G. W. see Fischer, G.
Fischer, J. 51 [79a], 52 [79a], 56 [79a]
Fischli, A. 135 [24a], 169 [24a]
Fletcher, Jr., H. G. 117 [29, 30], 120 [29, 30], 121 [29, 30], 122 [29, 30], 126 [29, 30]
Floret, E. 49 [68d], 56 [68d]
Florian, W. 71 [39], 72 [39], 84 [39], 94 [39], 97 [39], 131 [12b], 132 [12b], 133 [12b], 134 [12b], 135 [12b], 136 [12b], 137 [12b], 169 [12b]

Author Index

Fochi, R. 46 [44a], 55 [44a]
Föhlisch, B. 129 [8], 130 [8], *169* [8]
Forbes, C. E. 71 [35b], 97 [35b], 129 [5b], *169* [5b]
Frangopol, P. T. 153 [54], *171* [54]
Freiberg, J. 25 [12b, 25], 30 [12b, 25], 54 [12b, 25], 83 [79a], 85 [85], 98 [79a, 85]
Friedel, C. 1 [6], *3* [6]
Friedmann, N. 19 [46], *21* [46]
Friedrich, E. C. 3 [26], *4* [26], 9 [27], *10* [27], 24 [6b], 25 [6b], 29 [6b], 30 [6b], 31 [6b], 40 [6b], 41 [6b], 54 [6b], 84 [80], 90 [80], 98 [80], 112 [24a], 113 [24a], 117 [24a], 120 [24a], 121 [24a], 124 [24a], *126* [24a]
Frenzel, G. 129 [6a], 130 [6a], *169* [6a]
Freudenberg, K. 83 [73], 98 [73]
Fuchs, O. 3 [25], *4* [25], 6 [17], *10* [17], 18 [34], *21* [34], 71 [30], 72 [30], 73 [30], 85 [30], 87 [30], 88 [30], 90 [30], 93 [30], 94 [30], 97 [30], 108 [16a], *126* [16a]
Fürst, A. 123 [35b], *126* [35b]
Fugnitto, R. 48 [61], 49 [61], 56 [61]

G

Galinos, A. 18 [20], 19 [20], *20* [20]
Galton, S. A. 29 [28d], 40 [28d], 54 [28d], 94 [100], 99 [100]
Gantmakher, A. R. 142 [34c], *170* [34c]
Garrido Espinosa, F. 29 [32], 40 [32], 41 [32], 54 [32], 83 [75], 98 [75], 117 [28a], 118 [28a, d], 121 [28a, d], 124 [28d], *126* [28a, d]
Gash, K. B. 112 [22b], *126* [22b]
Gassmann, P. G. 105 [11], *126* [11]
Gastaldi, C. 2 [16], *4* [16], 158 [68b], *171* [68b]
Gavat, M. 51 [77a, c], 56 [77a, c]
Geldern, L. 71 [37a], 97 [37a]
Gerhart, F. 71 [40c], 97 [40c], 131 [13], *169* [13]
Gerlach, A. 14 [9], 15 [9], *20* [9], 31 [20], 39 [20], 54 [20], 177 [11d], 178 [11d], *179* [11d]
Gildemeister, E. 6 [9], *10* [9]
Gillespie, R. J. 19 [45], *21* [45], 66 [12a], 67 [12a], 96 [12a], 176 [9d], *179* [9d]
Gipp, R. 71 [32b], 97 [32b]
Glick, R. 100 [1a], 101 [1a], 102 [1a], *125* [1a]
Gold, H. 2 [24], 3 [24], *4* [24], 18 [29], *20* [29], 24 [8], 27 [8], 33 [8], 34 [8], 36 [8], 37 [8], 38 [8], 54 [8], 70 [22], 71 [22], 73 [22], 96 [22], 131 [12a], *169* [12a]
Golding, B. 135 [24b], *169* [24b]

Goldstein, J. H. 175 [7], *179* [7]
Gompper, R. 161 [80], *172* [80]
Goodrich, R. A. 35 [37b], 55 [37b]
Gorin, P. A. J. 117 [31], 121 [31], *126* [31]
Gosink, T. A. 149 [39b], *170* [39b]
Greif, N. 72 [49b], 97 [49b], 147 [38a], 159 [76b], *170* [38a], *172* [76b]
Grib, A. V. 76 [62, 63], 98 [62, 63]
Gribi, H. P. 72 [47f], 97 [47f], 131 [14], 135 [14], *169* [14]
Griot, J.-P. 152 [47b], *171* [47b]
Gross, H. 25 [12b], 30 [12b], 30 [25], 54 [12b, 25], 82 [76], 83 [79a], 85 [85], 91 [79a], 93 [76], 98 [76, 79a, 85]
Grüne, A. 46 [50], 55 [50]
Grunwald, E. 111 [22a], 112 [22a], *126* [22a]
Gschwend, H. 72 [47f], 97 [47f], 131 [14], 135 [14], *169* [14]
Günther, H. 175 [7], *179* [7]
Gutweiler, K. 142 [36], *170* [36]

H

Hafner, K. 16 [15], *20* [15], 47 [52b], 51 [52b], 55 [52b], 71 [35a], 72 [53a], 97 [35a], 98 [53a], 129 [5a], 164 [87], *169* [5a], *172* [87]
Hamaide, N. 103 [7a], *126* [7a]
Hamann, K. 142 [35], *170* [35]
Hansen, A. M. 129 [6c], 130 [6c], *169* [6c]
Hantzsch, A. 1 [15], 2 [20], *4* [15, 20]
Hart, H. 3 [27], *4* [27], 13 [8], 14 [8, 12], *20* [8, 12], 29 [28a, c], 40 [28a, c], 41 [28a], 54 [28a, c]
Hartshorn, M. P. 18 [25b], *20* [25b], 29 [29], 40 [29], 42 [11b], 54 [11b, 29], 123 [36a, b], *127* [36a, b]
Hatz, E. 133 [21a], *169* [21a]
Hazen, J. R. 105 [11], *126* [11]
Heaton, B. G. 106 [12], *126* [12]
Heck, R. 100 [1a], 101 [1a], 102 [1a], *125* [1a]
Hederich, V. 13 [7], *20* [7], 23 [2, 3], 27 [2], 28 [2], 29 [2], 30 [2, 3], 31 [2, 3], 34 [2, 3], 40 [2], 41 [2, 3], 44 [2, 3], 53 [2, 3], 139 [31b, c], 148 [31c], 149 [31c], *170* [31b, c]
Hedgley, E. J. 117 [29, 30], 120 [29, 30], 121 [29, 30], 122 [29, 30], *126* [29, 30]
Hein, F. 1 [5], *3* [5], 18 [21], *20* [21]
Heinbach-Juhasz, S. 109 [17a], *126* [17a]
Heinrich, C. P. see Heinrich, P.
Heinrich, P. 6 [21], 7 [21], *10* [21], 18 [39], *21* [39], 25 [14], 28 [14], 31 [14], 33 [14], 34 [14], 54 [14], 90 [96], 95 [96, 101, 102], 99 [96, 101, 102], 147 [38b], *170* [38b]

Heiszwolf, G. J. 70 [23], 96 [23]
Helferich, B. 83 [74], 98 [74]
Helmkamp, G. K. 18 [32], 21 [32], 34 [33], 35 [33], 55 [33], 71 [31a,b], 97 [31a,b]
Henseleit, E. 133 [19b], 169 [19b]
Hermann, H. D. 142 [36], 170 [36]
Hess, N. V. 111 [19], 112 [19], 117 [19], 121 [19], 126 [19]
Hesse, G. 139 [32a], 170 [32a]
Heyns, K. 29 [32], 40 [32], 41 [32], 54 [32], 83 [75], 98 [75], 117 [28a,d], 118 [28a,d], 121 [28a,d], 124 [28d], 126 [28a,d]
Hill, D. W. 2 [21a], 4 [21a]
Hilbers, C. W. 3 [33], 4 [33], 176 [9a], 177 [9a], 179 [9a]
Hilgetag, G. 35 [37a], 38 [37a], 55 [37a]
Hilt, A. 142 [35], 170 [35]
Hinz, G. 2 [23], 3 [23], 4 [23], 6 [8], 10 [8], 18 [27], 20 [27], 37 [40a], 38 [40a], 47 [40a], 55 [40a], 70 [21], 71 [21], 72 [21], 73 [21], 96 [21]
Hodge, J. D. 19 [46], 21 [46]
Höft, E. 30 [25], 54 [25], 82 [76], 93 [76], 98 [76]
Hölzl, F. 19 [43], 21 [43]
Hoffmann, P. 159 [78], 172 [78]
Hoffmann, R. W. 85 [84], 91 [84], 98 [84]
Hofmann, K. A. 18 [25a], 20 [25a], 33 [18b], 46 [18b], 54 [18b]
Hofmann, P. 2 [23], 3 [23], 4 [23], 6 [8], 10 [8], 18 [27], 20 [27], 37 [40a], 38 [40a], 47 [40a], 55 [40a], 70 [21], 71 [21], 72 [21], 73 [21], 96 [21]
Hogeveen, H. 3 [33], 4 [33], 15 [11d], 19 [44b], 20 [11d], 21 [44b], 66 [14], 96 [14], 176 [9a,b,c], 177 [9a,c], 178 [9c], 179 [9a,b,c]
Holm, R. H. 71 [35b], 97 [35b], 129 [5b], 169 [5b]
Hoogewerff, S. 18 [24], 20 [24]
Hünig, S. 5 [3], 9 [3], 16 [16], 17 [16], 20 [16], 30 [24], 54 [24], 59 [71], 71 [33, 37a], 80 [71], 81 [71], 82 [71], 90 [71], 91 [71], 92 [71], 94 [71], 97 [33,37a], 98 [71]
Hütz, W. 11 [1], 20 [1]
Horner, L. 72 [43], 90 [43], 93 [43], 97 [43]
House, H. O. 70 [23], 96 [23]
Houser, J. J. 19 [46], 21 [46]
Huber-Emden, H. 34 [36a], 55 [36a]

I

Iffland, D. C. 58 [103], 99 [103]
Ingold, C. K. 67 [104], 99 [104]
Ingraham, L. L. 104 [9a,b], 111 [22a], 112 [22a], 126 [9a,b, 22a]
Irvine, F. M. 52 [86], 56 [86]
Isaeva, L. S. 76 [62], 98 [62]
IUPAC 9 [25], 10 [25]
Izzo, P. T. 130 [9], 169 [9]

J

Jacobi, E. 46 [50], 55 [50]
Jäger, H. 139 [32a], 170 [32a]
Jaffe, F. 142 [36], 170 [36]
James, G. H. 26 [18c], 54 [18c]
Jennen, J. J. 5 [1], 9 [1]
Jernow, J. L. 91 [92b], 99 [92b]
Jochinke, H. 83 [74], 98 [74]
Johnson, D. H. 3 [28], 4 [28], 34 [34], 55 [34], 105 [10], 126 [10], 173 [2], 179 [2]
Johnston, W. H. 46 [49], 55 [49]
Jones, F. R. 139 [32b], 170 [32b]
Jung, H. A. 18 [30b], 21 [30b], 27 [19b], 29 [19b], 34 [19b], 54 [19b], 173 [1], 179 [1]

K

Kabachnik, M. I. 70 [23, 26], 96 [23, 26]
Kabuss, S. 9 [26], 10 [26], 28 [27a,b], 29 [30], 30 [27b], 32 [27a], 36 [39], 39 [39], 40 [30], 42 [27a,b], 43 [27a], 44 [27a,b], 54 [27a,b, 30], 55 [39], 87 [88], 90 [88], 99 [88], 129 [3a,b], 137 [3a,b,c], 139 [3b,c], 142 [3a,b], 143 [3b,c], 144 [3b,c], 146 [3a,b], 147 [3a], 148 [3b], 149 [3b], 169 [3a,b,c]
Kachler, J. 1 [8], 4 [8], 18 [22], 20 [22]
Kaiser, H. 47 [52b], 51 [52b], 55 [52b], 164 [87], 172 [87]
Kalish, N. 71 [40a], 97 [40a], 132 [17], 169 [17]
Kalk, F. 161 [81b], 172 [81b]
Kakihana, T. 72 [47e], 97 [47e], 135 [25], 169 [25]
Karger, M. H. 77 [67c], 98 [67c]
Kaslow, C. E. 46 [49], 55 [49]
Kassatkin, F. 18 [22], 20 [22]
Kato, H. 155 [61b], 167 [61b], 171 [61b]
Katritzky, A. R. 49 [70a], 56 [70a]
Keese, R. 135 [24b], 169 [24b]
Kehrmann, F. 2 [19], 4 [19], 26 [18a], 47 [18a], 54 [18a]
Kelemen, J. 47 [58], 55 [58]
Kemp, D. S. 72 [51a], 97 [51a]
Kendall, J. 47 [51], 55 [51], 155 [60], 171 [60]
Kende, A. S. 130 [9], 169 [9]
Kent, R. E. 85 [82a], 98 [82a]
Kergomard, A. 77 [68], 98 [68]
Kersting, F. 72 [49a], 97 [49a]

Author Index

Kindler, W. 46 [48b], 55 [48b]
King, J. F. 18 [37], 21 [37], 27 [19a], 29 [19a], 34 [19a], 40 [19a], 54 [19a], 114 [25a, 26], 115 [26], 116 [26], 126 [25a, 26]
King, L. see King, L. C.
King, L. C. 47 [56], 55 [56], 167 [91], 172 [91]
King, T. P. 51 [74a], 56 [74a]
Kinzebach, W. 46 [44b], 55 [44b]
Kiprianov, A. I. 47 [53], 55 [53]
Kirk, D. N. 18 [25b], 20 [25b], 29 [29], 40 [29], 54 [29], 123 [36a, b], 127 [36a, b]
Kirmreuther, H. 18 [25a], 20 [25a]
Kirmse, W. 32 [27c], 54 [27c]
Kirrmann, A. 6 [10], 10 [10], 18 [31], 21 [31], 24 [4b], 29 [4b], 53 [4b], 103 [7a, b, c], 104 [7b, c, 8a], 126 [7a, b, c, 8a]
Klages, F. 6 [7, 11, 20], 10 [7, 11, 20], 18 [19a, b, 23, 26, 30b], 19 [19a, b, 23, 26], 20 [19a, b, 23, 26, 30b], 27 [19b], 28 [16], 28 [23], 29 [19b], 30 [23], 33 [23], 34 [16, 19b, 35a], 35 [35b], 36 [35a, b], 38 [16, 35a, b], 48 [60], 54 [16, 19b, 23, 35a, b], 56 [60], 60 [2b, 4, 6a], 62 [2b, 4, 6a], 77 [65], 78 [65], 96 [2b, 4, 6a], 98 [65], 158 [68c], 171 [68c], 173 [1], 179 [1]
Kloosterziel, H. 70 [23], 96 [23]
Knox, W. R. 46 [49], 55 [49]
Koch, A. 18 [22], 20 [22]
Kochetkov, N. K. 52 [83], 56 [83]
Kocourek, J. 117 [27a], 126 [27a]
Köbrich, G. 45 [41a], 55 [41a], 151 [44a], 163 [44a], 170 [44a], 171 [44a]
Koeppl, G. W. 173 [3], 179 [3]
Koller, F. 123 [35b], 126 [35b]
Konishiroku Photo Ind. 72 [48], 97 [48], 136 [27], 170 [27]
Kornblum, N. 58 [103], 70 [24, 25], 96 [24, 25], 99 [103]
Kosak, H. J. 72 [52], 97 [52]
Kovács, Ö. see Kovács, Ö. K. J.
Kovács, Ö. K. J. 24 [6c], 29 [6c], 30 [6c], 31 [6c], 54 [6c], 124 [37a, b], 125 [37a, b], 127 [37a, b]
Krafft, W. 47 [55], 55 [55], 154 [59], 161 [59], 167 [59], 171 [59]
Kraus, W. 32 [27d], 54 [27d]
Krebs, A. 16 [14], 20 [14], 71 [36], 97 [36], 129 [7], 130 [7], 169 [7]
Kress, A. 152 [51], 171 [51]
Krishnamurthy, G. S. 173 [3], 179 [3]
Krivun, S. V. see Kriwun, S. W.
Kriwun, S. W. 52 [84b], 56 [84b]
Kröhnke, F. 154 [57b], 167 [57b], 171 [57b]
Kroke, H. 149 [40], 151 [40], 154 [40, 58],
163 [58], 166 [90], 167 [58, 90], 168 [90], 170 [40], 171 [58], 172 [90]
Kroning, E. 2 [23], 3 [23], 4 [23], 6 [8], 10 [8], 18 [27], 20 [27], 37 [40a], 38 [40a], 47 [40a], 55 [40a], 70 [21], 71 [21], 72 [21], 73 [21], 96 [21]
Kugucheva, E. E. 70 [23], 96 [23]
Kuhn, S. J. 5 [4], 6 [4], 9 [4]
Kundiger, D. 85 [82b], 98 [82b]
Kunert, F. 3 [25], 4 [25], 6 [18], 10 [18], 18 [36], 21 [36], 24 [6a], 25 [6a], 27 [6a], 28 [6a, 26a], 29 [6a], 31 [6a], 32 [26a], 34 [26a], 40 [6a], 41 [6a, 26a], 42 [6a], 44 [6a, 26a], 54 [6a, 26a], 73 [29], 80 [29], 81 [29], 83 [29], 84 [29], 90 [29], 97 [29], 111 [21], 116 [21], 126 [21], 138 [31a] 170 [31a]
Kuznetzov, E. V. 50 [70b], 56 [70b]

L

Laasch, P. 70 [27a, b], 71 [27a, b], 96 [27a], 97 [27b], 137 [29], 170 [29]
Láng, L. K. 24 [6c], 29 [6c], 30 [6c], 31 [6c], 54 [6c], 112 [23], 124 [23, 37a, b], 125 [37a, b], 126 [23], 127 [37a, b]
Lang, W. 123 [35a, b], 126 [35a, b]
Lambert, J. B. 3 [28], 4 [28], 34 [34], 55 [34], 105 [10], 126 [10], 173 [2], 179 [2]
Lampe, F. W. 13 [5a, b], 20 [5a, b]
Larsen, J. W. 14 [12], 20 [12], 24 [7a], 54 [7a]
Laubert, G. 161 [82], 162 [82, 83], 172 [82, 83]
Lecher, H. 33 [18b], 46 [18b], 54 [18b]
Leisten, J. A. 66 [12a], 67 [12a], 96 [12a]
Leroux, Y. 106 [6c], 125 [6c]
Levy, G. C. 3 [32], 4 [32], 15 [11c], 16 [11c], 20 [11c], 60 [5c, 10c], 96 [5c, 10c], 174 [6a], 175 [6a], 176 [6a], 177 [6a], 179 [6a]
Lewis, A. J. 42 [11b], 54 [11b]
Libermann, G. S. 18 [28], 20 [28]
Lieb, F. 159 [75, 77], 160 [77], 172 [75, 77]
Lieske, C. N. 125 [38b], 127 [38b]
Lillya, C. P. 71 [32a], 97 [32a]
Lindegren, C. R. 104 [9b], 126 [9b]
Little, G. R. 42 [11b], 54 [11b]
Löliger, P. 135 [24b], 169 [24b]
Löschner, A. 129 [6b], 130 [6b], 169 [6b]
Loewenstein, A. 173 [4], 179 [4]
Lombard, R. 53 [87], 57 [87], 152 [51], 171 [51]
Losinger, S. 53 [88], 57 [88], 158 [68e], 171 [68e]
Lücke, E. 71 [37a], 97 [37a]
Lukaszyk, G. 78 [65], 98 [65]
Lwoski, W. 24 [7b], 54 [7b], 101 [5a], 125

[5a]
Lyudvig, E. B. 142 [34c], *170* [34c]

M

Mach, W. 6 [13], *10* [13], 18 [35a], *21* [35a], 25 [9], 26 [9], 50 [72], *54* [9], *56* [72], 69 [19], 95 [19], *96* [19], 159 [76c], 164 [76c], *172* [76c]
Mackor, E. L. 3 [33], *4* [33], 19 [44a,b], *21* [44a,b], 176 [9a], 177 [9a], *179* [9a]
MacLean, C. 3 [33], *4* [33], 19 [44a,b], *21* [44a,b], 176 [9a], 177 [9a], *179* [9a]
Märkl, G. 158 [65], 159 [75, 76a, 77], 160 [77], 163 [86], 165 [65], 166 [86], *171* [65], *172* [75, 76a, 77, 86]
Maier-Hüser, H. 47 [54], *55* [54], 77 [66], *98* [66]
Makarova, L. G. 12 [3], 17 [3], *20* [3], 28 [17], *54* [17], 75 [58b], *98* [58b]
Makletsova, N. V. 142 [34c], *170* [34c]
Mandell, L. 175 [7], *179* [7]
Mantell, G. J. 143 [37b], *170* [37b]
Marino, J. P. 91 [92b], *99* [92b]
Marshall, J. L. 105 [11], *126* [11]
Martin, D. 73 [55], *98* [55]
Martin, R. H. 13 [5a,b], *20* [5a,b]
Marvell, E. N. 149 [39b], *170* [39b]
Mastryukova, T. A. 70 [23, 26], *96* [23, 26]
Mateescu, G. 51 [77a], *56* [77a], 153 [54], *171* [54]
Matthiae, H. 140 [33], *141* [33]
Mazur, Y. 77 [67c], *98* [67c]
McElvain, S. M. 23 [1a], 25 [12a], 30 [12a], *53* [1a], *54* [12a], 82 [78a,b], 85 [82a,b,c], 92 [82c], *98* [78a,b, 82a,b,c]
McGeachin, S. G. 71 [41], *97* [41], 133 [20a], *169* [20a]
McGregor, P. T. 130 [9], *169* [9]
McKenna, J. 71 [45], *97* [45]
McKenna, J. M. 71 [45], *97* [45]
McManus, S. P. 24 [7a], *54* [7a]
Medvedev, S. S. 142 [34c], *170* [34c]
Meerwein, H. 2 [22, 23, 24], 3 [22, 23, 24, 25], *4* [22, 23, 24, 25], 5 [2], 6 [8, 9, 10, 12, 17, 18], 8 [2], 9 [2], *10* [8,9,10, 12, 17, 18], 12 [4], 13 [7], 18 [4, 27, 29, 31, 33, 34, 36], 19 [29], *20* [4, 7, 27, 29], *21* [31, 33, 34, 36], 23 [2, 3], 24 [4a, 6a, 8], 25 [6a, 11a, 13], 27 [2, 4a, 6a, 8, 11a], 28 [2, 6a, 11a, 13, 26a], 29 [2, 4a, 6a], 30 [2,3], 31 [2, 3, 6a, 11a, 13], 32 [26a], 33 [8, 11a], 34 [2, 3, 4a, 8, 13, 26a], 36 [4a, 8, 11a, 38], 37 [8, 40a, b], 38 [4a, 8, 40a], 39 [11a], 40 [2, 4a, 6a], 41 [2, 3, 4a, 6a, 11a, 26a], 44 [2,
3, 6a, 11a], 44 [4a, 6a, 11a, 26a], 45 [4a], 47 [4a, 40a, 54], *53* [2, 3, 4a], *54* [6a, 8, 11a, 13, 26a], *55* [38, 40a,b, 54], 65 [20b], 70 [20a, 21, 22, 27a,b,c], 71 [21, 22, 27a, b,c, 28, 30, 32b, 39], 72 [21, 30, 39], 73 [20a, 21, 22, 29, 30], 74 [22], 75 [21], 77 [66], 78 [69, 70], 80 [29], 81 [29], 83 [29], 84 [29, 39], 85 [28, 30], 87 [30, 89], 88 [30], 89 [92a], 90 [20a, 28, 29, 30], 93 [30, 97], 94 [30, 39], *96* [20a,b, 21, 22, 27a], *97* [27b, 28, 29, 30, 32b, 39], *98* [66, 69, 70, 89, 92a, 97], 104 [8b], 108 [16a], 111 [21], 116 [21], *126* [8b, 16a, 21], 128 [1, 2, 4], 131 [12a,b], 132 [12b], 133 [12b], 134 [12b], 135 [12b], 136 [12b, 28], 137 [12b, 29], 138 [1, 31a], 139 [31b,c], 140 [1, 2, 33], 141 [33, 34a], 142 [34a], 148 [31c], 149 [31c], *168* [1], *169* [2, 4, 12a,b], *170* [28, 29, 31a,b,c, 33, 34a]
Meier, D. P. 155 [61c], 167 [61c], *171* [61c]
Meier-Mohar, T. 19 [43], *21* [43]
Mersch, R. 70 [27a,b], 71 [27a,b], *96* [27a], *97* [27b], 137 [29], *170* [29]
Merz, A. 159 [75, 77], 160 [77], *172* [75, 77]
Metzler, A. 33 [18b], 46 [18b], *54* [18b]
Meuresch, H. 6 [7], *10* [7], 18 [19a,b], 19 [19a,b], *20* [19a,b], 28 [16], 34 [16, 35a], 35 [35b], 36 [35a], 38 [16, 35a,b], *54* [16], *55* [35a,b], 60 [2b], 62 [2b], *96* [2b]
Meuwsen, A. 34 [36b], *55* [36b]
Meyer, E. F. 72 [47f], *97* [47f], 131 [14], 135 [14], *169* [14]
Mezheritskaya, L. V. 18 [25c], *20* [25c]
Michel, W. 150 [43], *170* [43]
Mihai, G. 154 [56], *171* [56]
Miljkovic, D. 135 [24b], *169* [24b]
Miller, P. 71 [32a], *97* [32a]
Miller, S. I. 173 [3], *179* [3]
Mögling, H. 34 [36b], *55* [36b]
Möllenkamp, H. 161 [81a], *172* [81a]
Moffat, J. 47 [56], *55* [56]
Moffatt, M. E. 5 [4], 6 [4], *9* [4]
Morschel, H. 13 [7], *20* [7], 23 [3], 28 [26a], 30 [3], 31 [3], 32 [26a], 34 [3, 26a], 36 [38], 41 [3, 26a], 44 [3, 26a], *53* [3], *54* [26a], *55* [38], 78 [69], *98* [69], 139 [31c], 141 [34a], 142 [34a], 148 [31c], 149 [31c], *170* [31c, 34a]
Mühlbauer, E. 6 [11], *10* [11], 18 [23], 19 [23], *20* [23], 28 [23], 30 [23], 33 [23], *54* [23], 60 [4], 62 [4], 77 [65], 78 [65], *96* [4], *98* [65]
Müller, E. 34 [36a], *55* [36a]
Müller, K. 135 [24b], *169* [24b]

Müller, W. 153 [53], 158 [53], *171* [53]
Mumme, E. 18 [24], *20* [24], 62 [9c], *96* [9c]
Mutalapova, R. I. 72 [44b], *97* [44b]
Mutz, G. 158 [69], *171* [69]
Muxfeldt, H. 132 [15], *169* [15]

N

Naar-Colin, C. 175 [7], *179* [7]
Nagai, M. 72 [47d], *97* [47d], 135 [23], *169* [23]
Namanworth, E. 3 [29], *4* [29], 6 [6], *10* [6], 19 [11e], *20* [11e], 60 [1], 63 [1], *96* [1]
Neidlein, R. 86 [87], 92 [87], *99* [87]
Nenitzescu, C. D. 2 [21b], *4* [21b], 45 [41b], 48 [63], 51 [41b, 63, 77a,b,c], *55* [41b], *56* [63, 77a,b,c], 153 [54, 55a], 154 [56], 156 [63], 157 [64], 158 [63], 161 [63], 164 [64], *171* [54, 55a, 56, 63, 64]
Nentwig, J. 70 [27b], 71 [27b], *97* [27b]
Nesmeyanov, A. N. 12 [3], 17 [3], 18 [28], *20* [3, 28], 28 [17], 52 [83], *54* [17], *56* [83], 75 [58a,b, 60], 76 [58a, 60, 62, 63], 77 [64], *98* [58a,b, 60, 61, 62, 63, 64]
Nesterov, L. V. 72 [44b], *97* [44b]
Neubauer, G. 154 [57a], 161 [81a,b], 162 [85], 163 [85], 165 [85], 167 [57a], *171* [57a], *172* [81a,b, 85]
Neumer, J. F. 173 [4], *179* [4]
Nippe, B. 72 [43], 90 [43], 93 [43], 94 [43], *97* [43]
Novak, E. R. 101 [1d], 102 [1d], *125* [1d]
Noyce, D. S. 109 [18], *126* [18]
Nürrenbach, A. 72 [46], 90 [46], *97* [46], 147 [38c], *170* [38c]

O

Oae, S. 100 [2], 105 [2], 107 [2], *125* [2]
O'Brien, D. H. 3 [30, 31, 32, 34], *4* [30, 31, 32, 34], 6 [7, 11, 16], *10* [7, 11, 16], 15 [11a], *20* [11a], 60 [2a, 3, 5a, 7], 63 [2a, 3, 5a, 7, 105], 65 [2a], 66 [7], 68 [7], *96* [2a, 3, 5a, 7], *99* [105], 174 [5a,b], 175 [5a,b], 178 [12a], *179* [5a,b, 12a]
Ochiai, M. 72 [47d], *97* [47d], 135 [23], *169* [23]
Ohta, M. 155 [61b], 167 [61b], *171* [61b]
Oishi, T. 70 [53b], 72 [47d], *97* [47d], *98* [53b], 135 [23], *169* [23]
Olah, G. A. 3 [27, 29, 30, 31, 32, 33, 34], *4* [27, 29, 30, 31, 32, 33, 34], 5 [4], 6 [4, 6, 7, 11, 15, 16, 19], 7 [14, 19], *9* [4], *10* [6, 7, 11, 14, 15, 16, 19], 15 [11a,b], 18 [30a], 19 [11e], *20* [11a,b,e], *21* [30a], 28 [22a], 29 [31], 30 [22a], 33 [22a], 34 [22b], 40 [31], *54* [22a,b, 31], 60 [1, 2a, 3, 5a,b, 6b, 7, 8a], 63 [1, 2a, 3, 5a,b, 6b, 7, 8a, 11a, 105], 64 [11a], 65 [2a], 66 [6b, 7, 8a, 15, 16], 67 [16, 17], 68 [7, 8a], 69 [5a, 8a], *96* [1, 2a, 3, 5a,b, 6b, 7, 8a, 11a, 15, 16, 17], *99* [105], 111 [20], *126* [20], 174 [5a,b, 6b], 175 [5a,b, 6b, 12b], 176 [8, 10], 177 [10], 178 [12a,b, 13], *179* [5a,b, 6b, 8, 10, 12a, b, 13]
Oles, S. R. 71 [40a], *97* [40a], 132 [17], *169* [17]
Olofson, R. A. 72 [51a], 91 [92b], *97* [51a], *99* [92b]
Olsen, B. A. 71 [31a], *97* [31a]
Oosterloo, G. 161 [81a], *172* [81a]
Ozog, F. see Ozog, F. J.
Ozog, F. J. 47 [56], *55* [56], 167 [91], *172* [91]

P

Pacák, J. 117 [27a], *126* [27a]
Palchkov, V. A. 51 [78], *56* [78]
Paquette, L. A. 72 [47b,c,e], *97* [47b,c,e], 135 [22b, 25, 26a], *169* [22b, 25], *170* [26a]
Pasto, D. J. 100 [3], 101 [3], 105 [3], 107 [3], *125* [3]
Paulsen, H. 29 [32], 40 [32], 41 [32], *54* [32], 83 [75], *98* [75], 110 [39], 117 [28a, d], 118 [28a,d,e], 121 [28a,d,e], 123 [39], 124 [28d], *126* [28a,d,e], *127* [39]
Pawellek, F. 28 [26a], 32 [26a], 34 [26a], 41 [26a], 44 [26a], *54* [26a]
Pearson, R. G. 11 [2], *20* [2], 82 [77b], *98* 77b]
Pedersen, C. 120 [33], 121 [33], 122 [33], 123 [34a], *126* [33, 34a]
Peligot, E. 1 [7], *4* [7], 18 [22], *20* [22]
Perepelitsa, E. M. 51 [76], *56* [76]
Perrin, C. L. 145 [37d], *170* [37d]
Pesaro, M. 72 [47f], *97* [47f], 131 [14], 135 [14, 24a], *169* [14, 24a]
Petersen, S. 72 [47a], *97* [47a], 135 [22a], 136 [22a], *169* [22a]
Peterson, P. E. 139 [31d], *170* [31d]
Peterson, R. A. 16 [14], *20* [14], 71 [36], *97* [36], 129 [7], 130 [7], *169* [7]
Pettitt, D. J. 18 [32], *21* [32], 34 [33], 35 [33], *55* [33], 71 [31a,b], *97* [31a,b]
Peyretti, G. L. 158 [68b], *171* [68b]
Pfeiffer, P. 2 [17a], *4* [17a], 15 [13], 18 [13], *20* [13], 62 [9d], *96* [9d]
Pfeil, E. 2 [23, 24], 3 [23, 24], *4* [23, 24], 6 [8], *10* [8], 18 [27, 29], *20* [27, 29], 24 [8], 27 [8], 33 [8], 34 [8], 36 [8], 37 [8, 40a], 38 [8, 40a], 47 [40a], *54* [8, 40a], 70 [21, 22],

71 [21, 22], 72 [21], 73 [21, 22], 74 [22], 96 [21, 22], 131 [12a], *169* [12a]
Piccard, J. 46 [47], 48 [47], *55* [47], 151 [45], 158 [66b], 161 [66b], *171* [45, 66b]
Pitcher, R. 72 [52], *97* [52]
Pitman, M. E. 117 [31], 121 [31], *126* [31]
Pittman, Jr., C. U. 5 [4], *10* [4], 19 [46], *21* [46], 24 [7a], *54* [7a], 66 [13], *96* [13]
Plattner, P. A. 123 [35a, b], *126* [35a, b]
Plesch, P. H. 139 [32b], *170* [32b]
Plotnikov, W. A. 47 [51], *55* [51]
Pohl, G. 73 [54], *98* [54]
Popov, E. M. 70 [26], *96* [26]
Posner, J. 16 [14], *20* [14], 71 [36], *97* [36], 129 [7], 130 [7], *169* [7]
Post, H. W. 86 [86], *99* [86]
Praill, P. F. G. 51 [74b, 75], *56* [74b, 75]
Prelog, V. 109 [17a], *126* [17a]
Price, C. C. 160 [79a], *172* [79a]
Pritzkow, W. 73 [54], *98* [54]
Pütter, R. 164 [88], *172* [88]
Pummerer, R. 47 [51], *55* [51]

Q
Quint, F. 53 [90], *57* [90], 164 [88], *172* [88]

R
Radmacher, W. 158 [70a], *171* [70a]
Ramsey, B. G. 3 [27], *4* [27], 12 [6], 13 [6], 19 [6], *20* [6], 28 [21], 31 [21], *54* [21], 60 [8b], 68 [8b], 90 [95], *96* [8b], *99* [95], 177 [11a], 178 [11a], *179* [11a]
Rasburn, J. W. 106 [12], *126* [12]
Reddelien, G. 18 [22], *20* [22]
Reichardt, C. 71 [34], *97* [34], 158 [71b], *172* [71b]
Reichert, K. H. 142 [35], *170* [35]
Reischl, A. 48 [59a], *56* [59a]
Renk, H.-A. 49 [65], *56* [65]
Reynolds, G. A. 53 [89], *57* [89], 155 [61c], 167 [61c], *171* [61c]
Rhomberg, A. 72 [44a], *97* [44a]
Richey, Jr., H. G. 19 [46], *21* [46]
Rieche, A. 82 [77a], 84 [81], 93 [99], *98* [77a, 81], *99* [99], 143 [37c], 145 [37c], *170* [37c]
Riedel, H. W. 16 [15], *20* [15], 71 [35a], *97* [35a], 129 [5a], *169* [5a]
Riedel, W. 153 [53], 158 [53], *171* [53]
Rio, G. 152 [48], *171* [48]
Roberts, J. D. 173 [4], *179* [4]
Roberts, R. M. 112 [24b], 117 [24b], 121 [24b], 124 [24b], *126* [24b]
Robertson, A. 52 [86], *56* [86]
Robinson, R. 52 [86], 53 [87], *56* [86], *57* [87]
Roedig, A. 49 [65], *56* [65]
Rogalski, W. 132 [15], *169* [15]
Rondestvedt, Jr., C. S. 143 [37b], *170* [37b]
Rosencrantz, D. R. 125 [38b], *127* [38b]
Ross, A. 17 [18b], *20* [18b], 51 [79b], 52 [79b], *56* [79b]
Rothenwöhrer, W. 32 [27d], *54* [27d]
Rottmann, J. 52 [82b], *56* [82b]
Royer, J. 151 [44b], 152 [47b], *171* [44b, 47b]
Rozenberg, B. A. 142 [34c], *170* [34c]
Ruby, P. R. 158 [67], *171* [67]
Rüchardt, C. 32 [26b], 34 [26b], 44 [26b], *54* [26b]
Rundel, W. 18 [35c], *21* [35c], 69 [19], *96* [19]
Ryabinina, V. E. 50 [70b], *56* [70b]
Rybinskaya, M. J. 52 [83], *56* [83]

S
Saalfrank, R. 131 [11], *169* [11]
Sack, H. 51 [79c], 52 [79c], *56* [79c]
Sagatys, D. S. 173 [3], *179* [3]
Sasse, H. J. 3 [25], *4* [25], 6 [17], *10* [17], 18 [34], *21* [34], 25 [13], 28 [13], 31 [13], 34 [13], *54* [13], 71 [30], 72 [30], 73 [30], 85 [30], 87 [30], 88 [30, 90], 89 [90], 90 [30, 90], 93 [30], 94 [30], *97* [30], *99* [90], 108 [16a], *126* [16a], 128 [1], 138 [1], 140 [1], *168* [1]
Saville, B. 51 [75], *56* [75]
Sazanova, V. A. 18 [28], *20* [28]
Schaafsma, S. E. 3 [35], *4* [35], 86 [79b], *98* [79b], 177 [11c], *179* [11c]
Schade, W. 84 [81], *98* [81]
Schäfer, H. 106 [6d], *125* [6d]
Schäfer, K. 1 [5], *3* [5], 18 [21], *20* [21]
Scheeren, J. W. 86 [79c], 91 [79c], 95 [79c], *98* [79c]
Scheffold, R. 72 [47f], *97* [47f], 131 [14], 135 [14], *169* [14]
Schiess, P. 149 [39a], *170* [39a]
Schiffler, G. 158 [73], *172* [73]
Schlosberg, R. H. 18 [30a], *21* [30a], 34 [22b], *54* [22b]
Schlosser, M. 49 [65], *56* [65]
Schmid, K. H. 73 [54], *98* [54]
Schmidpeter, A. 71 [37b], *97* [37b]
Schmidt, J. 1 [10], *4* [10]
Schmidt, M. 18 [20], 19 [20], *20* [20]
Schmidt, R. 52 [80a], *56* [80a]
Schmitz, E. 82 [77a], 84 [81], 93 [99], *98* [77a, 81], *99* [99], 143 [37a, c], 145 [37c], 158 [74], *170* [37a, c], *172* [74]
Schneider, G. 24 [6c], 29 [6c], 30 [6c], 31

[6c], 54 [6c], 112 [23], 124 [23], 37a, b], 125 [37a, b], 126 [23], 127 [37a, b]
Schneider, J. 85 [84], 91 [84], 98 [84]
Schneider, S. 32 [27c], 54 [27c]
Schneider, W. 17 [18b], 20 [18b], 51 [79b,c], 52 [79b,c], 56 [79b,c], 153 [53], 158 [53], 158 [71a], 171 [53], 172 [71a]
Schöllkopf, U. 71 [40c], 97 [40c], 131 [13], 169 [13]
Schön, N. 71 [39], 72 [39], 84 [39], 94 [39], 97 [39], 131–137 [12b], 169 [12b]
Schönberg, A. 46 [46], 55 [46]
Scholz, H. 83 [73], 98 [73]
Schreiber, J. 133 [21a], 169 [21a]
Schrodt, H. 3 [25], 4 [25], 6 [17], 10 [17], 18 [34], 21 [34], 25 [13], 28 [13], 31 [13], 34 [13], 54 [13], 71–73 [30], 85 [30], 87 [30], 88 [30], 90 [30, 93], 93 [30], 94 [30], 97 [30], 99 [93], 128 [1], 138 [1], 140 [1], 168 [1]
Schromm, K. 95 [102], 99 [102]
Schroth, W. 45 [43], 47 [43], 48 [43], 49 [64], 50 [43], 52 [82a,b], 84a], 55 [43], 56 [64, 82a,b, 84], 155 [62], 171 [62]
Schütz, G. 46 [46], 55 [46]
Schulz, M. 117 [27b], 126 [27b]
Schwarz, M. J. 125 [38a,b], 127 [38a,b]
Schweig, A. 158 [71b], 172 [71b]
Searles, Jr., S. 60 [10b], 61 [10b], 77 [67b], 96 [10b], 98 [67b]
Seel, F. 8 [23], 10 [23]
Seitz, G. 47 [58], 55 [58], 129 [5a], 169 [5a]
Semple, B. M. 49 [70a], 56 [70a]
Serve, M. P. 100 [3], 101 [3], 105 [3], 107 [3], 125 [3]
Seymour, D. 112 [24b], 117 [24b], 121 [24b], 124 [24b], 126 [24b]
Shdanow, J. A. 51 [78], 52 [84b], 56 [78, 84b]
Sherman, Jr., P. D. 18 [35b], 21 [35b], 24 [5a], 29 [5a], 53 [5a], 90 [94a], 95 [94a], 99 [94a], 101 [6a], 105 [6a], 125 [6a]
Shipov, A. E. 70 [23, 26], 96 [23, 26]
Shiyan, Z. V. 52 [84b], 56 [84b]
Shriner, R. L. 46 [49], 55 [49]
Shungijetu, G. I. 51 [76], 52 [84b], 56 [76, 84b]
Siemiaticky, M. 48 [59b, 61], 49 [61], 56 [59b, 61]
Siepmann, T. 158 [71b], 172 [71b]
Simalty, M. 53 [91], 57 [91]
Simchen, G. 133 [19a], 169 [19a]
Sinnott, M. L. 26 [18c], 54 [18c]
Slama, F. J. 139 [31d], 170 [31d]
Smiley, R. 58 [103], 99 [103]

Snyder, J. P. 131 [11], 169 [11]
Sommer, J. 3 [29], 4 [29], 6 [6], 10 [6], 19 [11e], 20 [11e], 60 [1], 63 [1, 11a], 64 [11a], 96 [1, 11a]
Sonntag, A. C. 108 [15], 126 [15]
Soyka, M. 46 [44b], 55 [44b]
Spille, J. 3 [25], 4 [25], 6 [17], 10 [17], 18 [34], 21 [34], 25 [13], 28 [13], 31 [13], 34 [13], 54 [13], 70 [27a], 71 [27a, 30], 72 [30], 73 [30], 85 [30], 87 [30], 88 [30], 90 [30], 93 [30], 94 [30], 96 [27a], 97 [30], 108 [16a], 126 [16a], 128 [1], 137 [29], 138 [1], 140 [1], 168 [1], 170 [29]
Städe, W. 159 [76b], 172 [76b]
Staněk, J. 117 [27a], 126 [27a]
Stein, W. 141 [34d], 170 [34d]
Steinberg, H. 3 [35], 4 [35], 86 [79b], 98 [79b], 177 [11c], 179 [11c]
Stephan, A. 72 [53a], 98 [53a]
Stephan, J. P. 53 [87], 57 [87]
Steppich, W. 6 [7], 10 [7], 18 [19b], 19 [19b], 20 [19b], 34 [35a], 36 [35a], 38 [35a], 55 [35a], 60 [2b], 62 [2b], 96 [2b]
Stetter, H. 48 [59a], 56 [59a]
Steuber, F. W. 159 [76b], 172 [76b]
Stevens, C. L. 85 [82a], 98 [82a]
Stopp, G. 71 [39], 72 [39], 84 [39], 94 [39], 97 [39], 131–137 [12b], 169 [12b]
Straus, F. 18 [17], 20 [17], 93 [98], 99 [98]
Strzelecka, H. 53 [91], 57 [91]
Suld, G. 160 [79a], 172 [79a]
Susan, A. B. 158 [66c], 171 [66c]
Suter, C. 149 [39a], 170 [39a]
Syrkin, Y. K. 76 [59], 98 [59]

T

Taft, R. W. 3 [27], 4 [27], 12 [6], 13 [5a, b, 6], 20 [5a, b, 6], 28 [21], 31 [21], 54 [21], 60 [8b], 90 [95], 96 [8b], 99 [95], 177 [11a], 178 [11a], 179 [11a]
Tamres, M. 60 [10b], 61 [10b], 77 [67b], 96 [10b], 98 [67b]
Tarbell, D. S. 101 [1d], 102 [1d], 125 [1d]
Tautou, H. 77 [68], 98 [68]
Tavs, P. 72 [44a], 97 [44a]
Teichmann, H. 35 [37a], 38 [37a], 55 [37a]
Tickle, T. 1 [9], 4 [9], 47 [51], 55 [51], 155 [60], 171 [60]
Tietze, E. 72 [47a], 97 [47a], 135 [22a], 136 [22a], 169 [22a]
Tochtermann, W. 139 [32a], 170 [32a]
Tolgyesi, W. S. 5 [4], 6 [4], 9 [4]
Tolmachev, A. L. 47 [53], 55 [53]
Tolstaya, T. P. 11 [3], 17 [3], 20 [3], 28 [17],

54 [17], 75 [58a, b], 76 [58a, b, 62, 63], 77 [64], *98* [58a, b, 62, 63, 64]
Toma, C. 152 [49, 50], 158 [72], 161 [50], *171* [49, 50], *172* [72]
Tomalia, D. A. 3 [27], *4* [27], 13 [8], 14 [8, 12], *20* [8, 12], 29 [28a, c], 40 [28a, c], 41 [28a], *54* [28a, c]
Toyo Rayon Co 142 [36], *170* [36]
Träger, H. 6 [11], *10* [11], 18 [23], 19 [23], *20* [23], 28 [23], 30 [23], 33 [23], 48 [60], *54* [23], 56 [60], 60 [4], *96* [4], 158 [68c], *171* [68c]
Trautwein, W.-P. 29 [32], 40 [32], 41 [32], *54* [32], 83 [75], *98* [75], 117 [28a, d], 118 [28a, d], 121 [28a, d], 124 [28d], *126* [28a, d]
Traverso, G. 47 [56, 57], *55* [56, 57]
Traylor, T. G. 145 [37d], *170* [37d]
Treibs, A. 49 [71], *56* [71]
Treichel, P. M. 35 [37b], *55* [37b]
Trost, B. M. 70 [23], *96* [23]

U

Ulrich, P. 53 [88], *57* [88], 158 [68e], 160 [79b], *171* [68e], *172* [79b]
Umbach, W. 141 [34d], *170* [34d]
Uquisa, R. 123 [35c], *126* [35c]
Utermann, J. 71 [33], *97* [33]

V

Van Allan, J. A. 53 [89], *57* [89], 155 [61c], 167 [61c], *171* [61c]
Vanderstichele, P. L. 85 [83], *98* [83]
van Dorp, W. A. 18 [24], *20* [24]
Venus, E. 2 [17b], *4* [17b]
Venus-Danilova, E. 69 [18], *96* [18]
Vialle, J. 72 [38], *97* [38]
Viditz, F. 19 [43], *21* [43]
Villiger, V. 1 [11, 12], *4* [11], 19 [42], *21* [42], 62 [9b], *96* [9b]
Vincenzi, C. 46 [44a], *55* [44a]
Volland, R. 161 [81c], *172* [81c]
von Eicken, S. 158 [73], *172* [73]
von Fellenberg, T. 1 [4], 2 [16], *3* [4], *4* [16], 46 [48a], 52 [48a], *55* [48a]
von Oefele, A. 1 [1], *3* [1]
von Pieverling, L. 1 [2], *3* [2]
Vorbrüggen, H. 133 [21b], *169* [21b]
Vorländer, D. 18 [24], *20* [24], 62 [9c], *96* [9c]

W

Wache, H. 150 [42], 162 [42, 84], 163 [42], 165 [42], *170* [42], *172* [42, 84]
Wagner, H. 52 [85], *56* [85]

Walden, P. 1 [10], *4* [10]
Walinsky, S. W. 91 [92b], *99* [92b]
Wallach, O. 1 [5], 2 [5], *3* [5], 6 [9], *10* [9], 19 [40], *21* [40], 62 [9a], *96* [9a]
Ward, H. R. 18 [35b], *21* [35b], 24 [5a], 29 [5a], *53* [5a], 90 [94a], 95 [94a], *99* [94a], 101 [6a], 105 [6a], *125* [6a]
Wartski, L. 6 [10], *10* [10], 18 [31], *21* [31], 24 [4b], 29 [4b], *53* [4b], 103 [7b, c], 104 [7b, c, 8a], *126* [7b, c, 8a]
Weber, H. J. 93 [98], *99* [98]
Wedekind, E. 18 [22], *20* [22]
Wegener, P. 72 [47c], *97* [47c], 135 [26b], *170* [26b]
Wegler, R. 141 [34b], 142 [34b], *170* [34b]
Wehrli, P. 135 [24b], *169* [24b]
Weingarten, H. J. 109 [18], *126* [18]
Weinstock, J. 107 [13], *126* [13]
Weintraub, L. 71 [40a], *97* [40a], 132 [17], *169* [17]
Weise, A. 73 [55], *98* [55]
Weiss, K. 153 [53], 158 [53], *171* [53]
Weissermel, K. 142 [36], *170* [36]
Welks, J. D. 89 [91], *99* [91]
Werner, A. 1 [13], 2 [13], *4* [13], 47 [51], *55* [51]
Werner, R. 32 [26b], 34 [26b], 44 [26b], *54* [26b]
Whipple, E. B. 175 [7], *179* [7]
White, A. M. 3 [33, 34], *4* [33, 34], 6 [15, 16, 19], *10* [15, 16, 19], 60 [6b, 7, 8a], 63 [6b, 7, 8a, 105], 66 [6b, 7, 8a], 67 [17], 68 [7, 8a], 69 [8a], *96* [6b, 7, 8a, 17], *99* [105], 175 [12b], 176 [10], 177 [10], 178 [12a, b, 13], *179* [10, 12a, b, 13]
Whitear, A. L. 51 [74b], *56* [74b]
Whiting, M. C. 26 [18c], *54* [18c]
Wiberg, E. 18 [20], 19 [20], *20* [20]
Wiberg, N. 73 [54], *98* [54]
Wiberg, K. B. 109 [17b], *126* [17b]
Wick, A. 135 [24a], *169* [24a]
Wiersum, V. E. 25 [10], 26 [10], *54* [10]
Wiles, A. 66 [12b], *96* [12b]
Wiley, R. H. 153 [55b], *171* [55b]
Willfang, G. 2 [24], 3 [24], *4* [24], 18 [29], *20* [29], 24 [8], 27 [8], 33 [8], 34 [8], 36–38 [8], *54* [8], 70 [22], 71 [22], 73 [22], 74 [22], *96* [22], 131 [12a], *169* [12a]
Willstätter, R. 46 [48b], 47 [51], *55* [48b, 51]
Wilson, B. D. 72 [51b], *97* [51b]
Winnacker, E. L. 135 [24a], *169* [24a]
Winstein, S. 3 [26, 32], *4* [26, 32], 9 [27], *10* [27], 15 [11c], 16 [11c], *20* [11c], 24 [6b], 25 [6b], 29–31 [6b], 40 [6b], 41 [6b], *54*

Author Index   191

[6b], 60 [5c], 63 [5c], 74 [56, 57], 75 [56], 84 [80], 90 [80], 96 [5c], 98 [56, 57, 80], 100 [1a, 4], 101 [1a,b,c], 102 [1a,b], 103 [1c], 104 [9a,b], 110 [4], 111 [4, 19, 22a], 112 [19, 22a, 24a,b], 113 [24a], 117 [19, 24a,b], 119 [32], 120 [24a, 32], 121 [19, 24a,b, 32], 124 [24a,b], 125 [1a,b,c, 4], 126 [9a,b, 19, 22a, 24a,b, 32], 174–177 [6a], 179 [6a]
Wisotsky, M. J. 5 [4], 10 [4], 66 [13], 96 [13]
Wizinger, R. 46 [50], 47 [58], 53 [88], 55 [50, 58], 57 [88], 158 [68e], 160 [79b], 171 [68e], 172 [79b]
Wolf, K. H. 45 [42], 46 [45a], 47 [42, 45a,b], 49 [42, 67], 55 [42, 45a,b], 56 [67], 131 [10], 149 [40], 150 [41, 42, 43], 151 [40], 154 [10, 40, 59], 155 [10, 59, 61a], 158 [67, 69], 161 [59, 89], 162 [42], 163 [40, 42, 86], 165 [89], 166 [86, 90], 167 [10, 59, 61a, 90], 168 [90], 169 [10], 170 [40, 41, 42], 171 [59, 61a, 67, 69], 172 [81a, 86, 89, 90]
Woods, L. L. 47 [58], 55 [58]
Woodward, R. B. 72 [51a], 97 [51a]
Wright, G. J. 42 [11b], 54 [11b]
Wunder, D. 45 [41a], 55 [41a], 151 [44a], 163 [44a], 170 [44a]
Wunderlich, K. 3 [25], 4 [25], 6 [18], 10 [18], 13 [7], 18 [36, 38], 20 [7], 21 [36, 38], 23 [2, 3], 24 [6a], 25 [6a], 26 [15], 27 [2, 6a], 28 [2, 6a, 15, 26a], 29 [2, 6a, 15], 30 [2, 3], 31 [2, 3, 6a, 15], 32 [26a], 33 [15], 34 [2, 3, 26a], 40 [2, 6a, 15], 41 [2, 3, 6a, 15, 26a], 42 [6a], 44 [2, 3, 6a, 26a], 53 [2, 3], 54 [6a, 15, 26a], 73 [29], 79 [29], 80 [72], 81 [29], 83 [29], 84 [29], 87 [72], 90 [29, 72], 97 [29], 98 [72], 108 [16b], 111 [21], 116 [21], 126 [16b, 21] 138 [31a], 139 [31b,c], 148 [31c], 149 [31c], 170 [31a,b,c]
Wynberg, H. 25 [10], 26 [10], 54 [10]
Wynn, M. 14 [12], 20 [12], 24 [7a], 54 [7a]

Y
Yamada, Y. 135 [24b], 169 [24b]
Yarova, L. N. 51 [78], 56 [78]
Yuen, G. U. 112 [22b], 126 [22b]

Z
Zange, E. 6 [20], 10 [20], 18 [26], 19 [26], 20 [26], 60 [6a], 62 [6a], 96 [6a]
Zechmeister, L. 46 [48b], 55 [48b]
Zenner, K.-F. 71 [32b], 97 [32b]
Zeyfang, D. 133 [20b], 169 [20b]
Zhdanov, Y. A. see Shdanow, J. A.
Zhukov, A. 18 [22], 20 [22]
Zhungietu, G. I. see Shungijetu, G. I.
Zimmer, G. 149 [39b], 170 [39b]
Zvereva, T. M. 142 [34c], 170 [34c]

Subject Index

A

$A_{AC}1$ mechanism 67, 68
$A_{AL}1$ mechanism 67, 68
Acceptor-donor complex 11, 12
Acceptors for leaving groups 28–32
Acetaldehydium ion 174, 175
Acetalization 133
Acetals 30, 93, 138, 142
— cyclic 30, 95, 143
— cyclic, disproportionation 145
— halogenation 82, 85
— reaction with boron trifluoride or antimony pentachloride 33
— reaction with trialkyloxonium salts 73
Acetic anhydride 51
Acetoacetic ester 162
Acetolysis 105, 110, 112, 124
— of brosylates 102, 103
Acetonedicarboxylic ester 150, 163
Acetophenone 78, 146
1,2-Acetoxonium ions 9, 117, 118
1,3-Acetoxonium ions 9, 117, 123
Acetoxonium perchlorates 123
1,2-Acetoxonium salts 117
2-Acetoxy-3-bromobutane 110
β-Acetoxyethyl esters 41
3-Acetoxy-2-halo-2,3-dimethylbutanes 111
cis-1-Acetoxy-2-hydroxycyclohexane 112
trans-2-Acetoxymethylcyclohexyl brosylate 124
Acetylacetonate 106, 150
Acetylacetone 154, 162
10-Acetyl-1-α-bromo-trans-decalin 106
Acetyl cation
— hydride acceptor 49
Acetyl chloride 82, 86
Acetyl cyanide 86
Acetyl fluoride 78
α-Acetylglucosyl chloride 117
β-Acetylglucosyl chloride 117
β-Acetylglucosyl fluoride 118
O-Acetyltetrahydrofuranium tetrafluoroborate 78
β-Acetylxylosyl chloride 118
Acidity of acidic oxonium ions 62
Acidium ions, cis–trans isomerism 176–178
Acids
— hard 11, 12
— soft 11
Acylating agents 47
Acylation of C–C double bonds 51

Acyl cations 5, 8, 61, 68
Acyl chlorides 51, 77, 92
— aromatic 52
O-Acyl cleavage 67, 68
— acyldialkyloxonium ions 77, 78
Acyldialkyloxonium ions, cyclic intermediates 107–109
Acyldialkyloxonium salts, reactions 77, 78
2-Acylfuran 164
Acylium salts 142
Acyloxy groups, leaving groups 28, 40
4-Acyloxypyrylium salts 47
Alcohols
— aliphatic 60, 61
— alkylation 73
— protonation 1–3, 11, 19, 60, 61, 63, 65
— secondary 65
— tertiary 51, 65
Aldehydes
— aromatic, alkylation 129
— protonation 1, 3, 15, 60
— tertiary alkyl- 129
— vinyl- 129
Aldehydium ions 64, 173
— acidic 62, 66
— hydride acceptors 139
— cis–trans isomerism 174–176
Alkoxides 28
— acceptors 30, 31
— transfers 138, 148
Alkoxyalcohols, bisprotonation 66
ω-Alkoxyalkyl halides 100
Alkoxyallyl cations 7
Alkoxy(amino)carbonium ions 131
γ-Alkoxybutyric esters 140
γ-Alkoxybutyryl chloride 108
Alkoxycarbonium halides, intermediates 93
Alkoxycarbonium ions 5, 32, 66, 138, cf. carboxonium ions
Alkoxycarbonium salts 87, 142, cf. carboxonium salts
— acyclic 142
— carbeniate acceptors 42
Alkoxycarbonylalkylidene-triphenylphosphoranes 131
γ-Alkoxycarboxylic acid halides, rearrangement 109
Alkoxycyclopropenylium salts 129
Alkoxydihydroxycarbonium ions 66
2-Alkoxy-1,3-dioxolanes 41, 42
β-Alkoxyethyl acetate 81, 138

β-Alkoxyethyl esters 41
Alkoxy groups 28, 30, 40, 41
Alkoxyhydroxycarbonium ions 66
Alkoxylation of alcohols 141
Alkoxypyrylium ions 16
Alkoxypyrylium salts 131
2-Alkoxypyrylium salts 47
4-Alkoxypyrylium salts 47
Alkoxytrifluoroborate 37, 41
Alkoxytropylium salt 129
2-Alkyl-2-aryl-1,3-dioxolanes 42
Alkyl aryl ethers 61
Alkylating agents 23, 26–28, 34, 35, 37, 58
— cationic 27
— in disproportionation of cyclic ketals 43
Alkylation, intermolecular 104
Alkylation, intramolecular 23, 24, 27–29, 34, 40, 41, 44, 101, 104, 108, 117, 119
N-Alkylaziridines 173
2-Alkyl-1,3-benzodioxolium ion 8
Alkyl cations 26, 28, 30, 58, 61, 63, 66
O-Alkyl cleavage 63, 65, 67
— acyldialkyloxonium ions 77, 78
Alkyl cyanates 73
Alkylene oxides 141, cf. epoxides
Alkyl halides 1, 27, 30, 34, 77, 137
— alkylating agents 34, 39
— tertiary 65
Alkyl methyl ethers, ω-unsaturated 104
O-Alkyloxiranium ion 173
Alkyloxonium ions 62, 65, 66
n-Alkyloxonium ions 63
Alkylphosphonic acid diesters 72
Alkylphosphorothionates 70
4-Alkylpyrylium salts 155
4-(Alkylseleno)pyrylium salts 47
4-(Alkylthio)pyrylium salts 47
O-Alkyl-δ-valerolactonium ion 8
α-D-Altropyranose, dioxolenium ion 117
Ambident anions 58
Ambident cations 58, 79, 112
— different energies 91
— high energy 80, 91
— low energy 80, 93
Amide acetals 84, 94, 132–134, 148
Amidines 132
Amidinium ions 132
Amidinium salts 133
— vinylogous 133
Amines 71, 132
— carboxylation 69
— primary 133, 152, 161
— secondary 129, 133, 137, 152, 161
— tertiary 131

Amino pentadienones 152
Amine oxides 71
Ammonia 132, 152
1,5-Anhydro-D-arabinitol 122
1,5-Anhydro-D-galactitol 122
1,5-Anhydro-D-gulitol 122
N-Anilinopyridinium salts 158
Anion-acceptors 28–32, 138, 144, 146
Anions 3, 17, cf. Complex anions
Anisole 78
-a-Nomenclature 9
Appearance potentials 13
Arabinose 118
Aryl aldehydes 70, 142
Aryl aldehydium ions 142
Arylamines 88, 133
Aryl-bis(1,3-dioxolenium) ions 40
Aryldialkoxycarbonium ions 15
Aryldiazonium salts 28, 32
Aryldiazonium tetrafluoroborates 34, 137
Aryldiimine, intermediate formation 32
2-Aryl-1,3-dioxolane 82
2-Aryl-1,3-dioxolenium ions 14, 94
Aryl ketones 70, 142
Aryl ketonium ions 142
— alkylating agents 143
— anion-acceptors 144
Aryl methyl ketones 53
Aryl-tris(1,3-dioxolenium) ions 40
Azabullvalene, derivatives 135
Azacyclooctatetraene, derivatives 135
Azides 73
Azulene derivatives 72, 164–166

B
Back-side participations 119, 120
Bases
— hard 11, 12
— soft 11, 12
Basicities of oxygen bases 60, 61
Beilstein nomenclature 9
Benzaldehyde 146
Benzaldehydium ion 8
Benzamides 165
Benzenediazonium tetrafluoroborate 75
Benzopyrylium ions 7
Benzoxonium ions 9
trans-2-Benzoyloxycyclohexyl tosylate 112
Benzyl bromide 82
4-Benzyl-4H-pyrans 166, 167
Benzylideneacetophenone 15
Benzylmagnesium chloride 151, 154
Bicycloheptanone derivatives 105
o,o'-Biphenylylenephenyloxonium ion 77

Subject Index

Bis(β-bromoethylester) 40
Biscinnamylideneacetone 15
Bis(dialkylcarbamoyl)mercury 131
Bis-1,3-dioxolenium ions 40
Bis(dihydroxycarbonium) ions 66
Bis(2H-pyran) 166
Bis(imidic ester) 131
Bisketene acetals 91
Bisoxonium ions 64
— disecondary 64
Boron ylide 139
p-Bromobenzenesulfonate group 100, cf. brosylates
cis-2-Bromocyclohexyl acetate 111
trans-2-Bromocyclohexyl acetate 111
Bromohydrin esters 114
Brosylates 100, 102, 105, 106, 111, 124
cis-2-Brosyloxymethylcyclohexyl acetate 124
2-Butanol 65
Butenolide, derivative 108
tert-Butoxide 154, 161, 162
n-Butyl cation 146
tert-Butyl cation 49, cf. trimethylcarbonium ion
sec-Butyl chloride 146
tert-Butyl chloride 49
2-Butyl methyl ether 65
Butyric esters 88, 108
Butyrolactone 88, 140
Butyrolactonium ion 88, 140

C

Camphor 93, 129
— diethyl ketal 129
— oxonium ion 129, 145
Carbenes, intermediates 89
Carbeniate ions 28, 32
Carbonates
— cyclic, O-alkylation 25
— protonation 68
Carbon double bonds
— alkylation 24
— protonation 24
Carbonic acid
— derivatives, protonation 60
— protonation 69, 178
Carbonic esters 6, cf. carbonates
Carbonium ions 63, 64, 73
— heteroatom-stabilized 7
— tertiary 68
Carbonium-oxonium ions 22
Carbonium salts
— alkoxide acceptors 41

— hydride acceptors 41
Carbonyl compounds 62, 71, 128
— O-alkylation 22
— O-alkyl cations 3
— reactions with nucleophiles 68
Carbosulfonium ions 128
Carboxonium ions 5–8, cf. oxonium ions, alkoxycarbonium ions
— acidic 66, 69, 79
— acyclic 6, 24
— alkoxide acceptors 30
— cyclic 7, 9, 59, 60
— cyclic, formation 24
— cyclic, intermediates 105
— formation 22, 31
— high energy 91, 92
— low energy 92, 93
— reaction routes 59
— reaction with nucleophiles 58
— secondary 6, 7
— steric effects 94, 95
— tertiary 6, 7, 79
— tertiary, formation 25
Carboxonium salts 62
— formation 30, 33
Carboxylic acids 6, 51, 61, 62, 133
— aliphatic 66
— alkylation 73
— aromatic 66
— protonation 3, 60, 61, 68, 176
— unsaturated 66
Carboxylic amides 71, 87, 131–133
— from nitrilium salts 137
— primary 131
— secondary 131
Carboxylic anhydrides 51, 77
— acyclic 67
— bisprotonation 67
— cyclic 67
Carboxylic esters 6, 42, 61, 77, 80, 132, 146
— alkylation 39, 70, 87
— hydrolysis 67
— O-alkyl cations 3
— protonation 1, 3, 60, 67, 68, 178
Catechol 95
— monobenzoate 95
Chalcones 52, 53, 69
Charge delocalization 12, 15, 16
Chlorobenzene 78
Chlorobutyrophenone 100, 105
2-Chlorocyclohexanol 119
2-Chloro-2-dichloromethyl-1,3-dioxolane 83
β-Chloroethyl acetate 81

β-Chloroethyl formate 82
ω-Chloro ketones 100
Chlorosulfonyl isocyanate 135
2-Chlorotropone 129
5-Chlorovalerophenone 100
α-Chlorovinyl ethers 85, 89
β-Chlorovinyl ketones 52
Cholestane derivatives 114, 123
Chromanones 46
Complex anions 3, 17, 18, 27, 32
— formation 33
— hydrogen halide dissociation pressures 19, 62
— preformed 33, 34
Copper(I) cyanide 106
Corrins, syntheses 131, 135
Coumarins 46, 47
Cryoscopic measurements 66, 67
Cyanide 28, 41, 80, 153
Cyanine dyes 134, 136
Cyanoacetic ester 162
2-Cyano-1,3-dioxolanes 41, 81
2-Cyanotetrahydrofuran derivatives 106
2-Cyanotetrahydropyran derivatives 106
Cyclitol esters 120
Cycloalkanone ketals 32
Cyclohexane derivatives, conformations 113, 114
trans-Cyclohexane derivatives 112
trans-1,2-Cyclohexanediol 119
cis-Cyclohexene-1,2-acetoxonium ion 84
Cyclopentadien 164
Cyclopropanonium ion 69

D

trans-Decalin derivatives 114
Dehydropyrans 150
4-Dehydropyrans 47, 155
cis-1,2-Diacetoxycyclohexane 119, 120
Dialkoxyacetic acid, bromination 85
Dialkoxy carbenes, intermediates 90
Dialkoxycarbonium hexachloroantimonates 34, 137
Dialkoxycarbonium ions 30, 89, 90, 138, 177, cf. oxonium ions
— acyclic 85
— acyclic, anion-acceptors 146
— alkylating agents 36
— cyanide acceptors 41
— cyclic 107
— intermediate formation 32
— cis–trans isomerism 177, 178
Dialkoxycarbonium salts 30
— acyclic, alkylating agents 145, 146, 147

— acyclic, formation 39
— alkylating agents 36
— cyclic 39
— dealkylation 86
Dialkoxycarbonium tetrafluoroborates 34
Dialkoxyhalomethane derivatives 30, 31
Dialkoxyhydroxycarbonium ions 66
Dialkoxymethyl chloride 86
2,2-Dialkoxytetrahydrofuran 140
4-Dialkylaminopyrylium salts 167
N,N-Dialkylaniline 46
O,O′-Dialkyl carbonate acidium ion 62
Dialkylcarbonates 71
2,2-Dialkyl-1,3-dioxolanes 42
Dialkyl ethers 1, 12, 35, 60, 61, 77
— adducts with antimony pentafluoride 38
— adducts with phosphorus pentafluoride 35
Dialkyloxonium ions 34, 62, 65, 66
Dialkyloxonium salts 19, 62, 78
— alkylation with diazoacetic ester 35, 36
Dialkylphosphinic acid esters 72
Dialkylphophorochloridates 35
Dialkylphosphorochloridothioates 35
Dialkyl sulfates 26, 47, 136
Dialkyl sulfides 12
Diarylcarbonium salts 148
Diarylethylenes 46
Diaryloxycarbonium ions 91
Diazoacetic esters 35, 38, 71
Diazoalkanes 28, 34, 38, 62
Diazomethane 35, 62
Diazonium salts 27, 41
Dibenzylideneacetone 15
Dibenzylideneacetonium halides 17
Di-n-butoxycarbonium salts 146
Di-tert-butyl carbonate 68
Di-n-butyl ether 62
Di-tert-butyl ketone 175
Di-n-butyloxonium salt 62
1,5-Dicarbonyl compounds 48
β-Dicarbonyl compounds 52
Dicarboxylic acids 66
Dichloroacetic acid 60
Diethoxycarbonium hexachloroantimonate 39
Diethoxyhydroxycarbonium ion 9
Diethyl carbonate 68
O,O′-Diethyl carbonate acidium ion 8
Diethyl ether 62, 142
Diethylmethyloxonium salts 38
Diethyloxonium ion 8
Diethyloxonium salts 62
Diethyl-n-propyloxonium salts 38

2,3-Dihydrofuran derivatives 26
Dihydrofurylium salts 2
Dihydroxycarbonium ions 9, 66, 67
Dihydroxycarbonium-oxocarbonium ions 66
Di-(p-iodophenyl) ether 76
Diisopropyl carbonate 68
Diisopropyl ether 88
Diisopropylethylamine 90
1,5-Diketones 49, 50, 166
Dimethoxy(phenyl)carbonium ion 95
2-(2′,6′-Dimethoxyphenyl)-1,3-dioxolenium ion 94
Dimethoxy(2,4,6-trimethylphenyl)carbonium ion 95
Dimethylacyloxonium ion 8
Dimethylaniline 129
N,N,-Dimethylbenzimidazolone 137
Dimethylcarbonate 68
Dimethylcarbonium ion 68
Dimethyl ether 26, 34, 62, 139
Dimethyloxonium salt 62
2,6-Dimethyl-γ-pyrone 1
— boron trifluoride acceptor 42
— hydrogen chloride complex 1
Dimethyl sulfate 131, 133
Dineopentyloxymethyl chloride 95
3,5-Dinitrobenzoic acid 73
1,3-Dioxacycloalkenium ions 107
1,3-Dioxane 9, 32
1,3-Dioxanylium ions 9
1,3-Dioxenium ions 9, 123, 124
Dioxocarbonium ions 66, 67
1,3-Dioxolane 9, 30, 42, 139
— derivatives 32, 41, 44
1,3-Dioxolanylium ions 9
1,3-Dioxolenium halides 82
1,3-Dioxolenium ions 30, 32
— bicyclic 119
— carbohydrate chemistry 117
— equilibrium 121
— formation 40, 41
— intermediates 110–112, 116, 119
— reaction with water 115
1,3-Dioxolenium salts 30, 41, 42, 138
— preparation 39
1,3-Dioxolenium tetrafluoroborate 81
1,3-Dioxolium ions 7
Diphenylcarbonium salts, carbeniate acceptors 42
Diphenylcyclopropenone 16, 129
Diphenyl ether 28, 75
2,6-Diphenylpyrylium salts 154
Diphenyl sulfoxide 87

Di-n-propylcyclopropenone 130
Disproportionations
— alkoxy halogenocomplex anions 33
— cyclic acetals 30, 139
— cyclic ketals 43
— dialkyl ether—phosphorus pentafluoride adduct 35
Dissociation constants of acidic oxonium ions 60, cf. pK_a values
Distribution equilibria 60
Disulfides 71
Divinyloxycarbonium ions 91

E

Electrocyclic reactions 164, 165
Electrophiles, reaction with triaryloxonium salts 76
Enamines 129, 165
Enamino carbonyl compounds 71, 133
Enolates 58, 70
Epichlorohydrin 37, 38, 140, 141
Epimerization
— cyclitol esters 120
— polyhydroxytetrahydropyran esters 120
Epoxides 42, 140, cf. alkylene oxides
Esterification 67
Etherates
— of dialkyloxonium salts 19, 35
Ethers 62, 71, 138
— alkylation 22
— alkylation with dialkoxycarbonium salts 87, 146
— alkylation with trialkyloxonium salts 36
— cleavages 65, 77
— cyclic 24, 36, 61, 103
— cyclic, polymerizations 141
— halogenation 93
— protonation 1–3, 11, 60, 61, 65
Ethoxide 80
2-Ethoxy-1,3-dioxolane 82
Ethoxymethylphenylcarbonium ion 9
O-Ethylacetophenonium ion 8
O-Ethylacetophenonium salts 143
O-Ethylbenzophenonium salts 142
Ethyl bromide 23
Ethyl γ-bromobutyrate 107
Ethyl ω-bromocarboxylic esters 107
O-Ethylbutyrolactonium ion 25
Ethyl γ-chlorobutyrate 87
Ethyldimethyloxonium ion 75
Ethyldimethyloxonium salts 38
Ethyl-N,N-dimethylurethane 136
Ethylene oxide 105
Ethyl esters 68

Ethyl formate, protonation 178
O-Ethylphthalidium ion 88
O-Ethyltetrahydrofuranium tetrafluoroborate 75

F
Flavylium salts 52
Formate acidium ion 8
Formylcholine iodide 84
Front-side participations 119–121
Furan derivatives 164

G
α-D-Glucopyranose 117
α-D-Glucose 117
Glycine 158
γ-Glycol monoethyl ether 75
Glycols 42, 133
— bisprotonation 66
— dialkyl ethers 140, 141
— diethylene- 163
— protonation in FSO_3H–SbF_5–SO_2 64
cis-Glycols 116
Glutacondialdehyde 48
Glutaric acid 49
Grignard compounds 154, 161, 168
— aryl- 151
Guaiacol 95

H
Halide acceptors 27, 30, 34, 104
Halide ions 80, 82, 85
— leaving groups 40
δ-Halobutyl methyl ether 104
γ-Halobutyric esters 108
4-Halobutyrophenones 105
γ-Halocarboxylic esters 109
— thermal cleavage 107
2-Halo-1,3-dioxolanes 82
α-Halo ethers 30, 33, 93
ω-Halo ethers 104
β-Haloethyl ester 40, 41
Halogen complexes 61
γ-Halo ketones 106
δ-Halo ketones 106
Halonium ions 111
5-Halo-2-pentanones 106
5-Halo-valerophenones 105
Hammett σ-constants 14
Heptafulvene derivatives 129
N-Heterocycles, alkylation 72
Heterocycles from pyrylium salts 158
Hexachloroantimonate(V) ion 17
Hexacyanocobaltate(III) 19

Hexacyanoferrate(II) 19
Hexacyanoferrate(III) 19, 46, 167
Hexoses, acylated 117
Hexyl cations 68
Homoallyl resonance 16
Hydride acceptors 30, 32, 46, 52, 53, 139
Hydride ion 28, 41, 45, 49, 53
Hydride transfers 139, 143, 148
Hydrogen bonds 19
— IR measurements 60
— protonated carboxylic acids 176
Hydrogen halides, dissociation pressures of oxonium salts 19, 62
Hydroxonium ion 63
Hydroxyallyl cations 7, 53, 69
— hydride acceptors 49
4-Hydroxybenzopyrylium salts 46
4-Hydroxybutyrophenone 105
Hydroxycarbonium ions 66, 69
Hydroxy-dihydrofuran 2
β-Hydroxyethyl esters 41
2-Hydroxy-2H-pyran 152
1-Hydroxyisochromane 145
β-Hydroxy ketals 51
Hydroxylamine 153
Hydroxymethoxymethylcarbonium ion 9
2-Hydroxy-5-methoxypentane 102
Hydroxy(phenyl)carbonium ion 9
N-(Hydroxyphenyl)pyridinium betaines 158
4-Hydroxypyrylium ions 155
Hydroxypyrylium salts 1
4-Hydroxypyrylium salts 47
5-Hydroxyvalerophenone 105

I
α-D-Idopyranose, 4,6-acetoxonium salt 117
α-D-Idose, acetoxonium ion 117
Imidic esters 131, 137
— cations 88, 94
— cyclic 135
Imidochlorides 137
Imines 128
Immonium ions 120
Iodide 70, 93
β-Iodoethyl formate 84
2-Iodo-2′-phenoxybiphenyl 77
Iron(III) halides 103
IR spectroscopy, hydrogen bonds 176
Isobutyraldehyde 104
Isocoumarins 46
Isomesityl oxide 51
Isopropyl alcohol 87, 88
Isopropyl formate, protonation 178
Isopropyl methyl ketonium ion 69

Isoxazoline 153, 156

K
Ketals 30
— cyclic 42, 145
— preparation 129
— reaction with antimony pentachloride or boron trifluoride 33
— reaction with trialkyloxonium salts 73
Ketene acetals 23, 59
— cyclic 25, 89
— hydrogen chloride addition 85
— protonation 26
Ketene O,N-acetals 134
Ketenes
— divinyl acetals 92
— intermediate 78
Ketones 6
— activation 129–131
— *tert* alkyl 129
— alkylation 12, 129–131
— aryl- 61, 129
— conjugated 61
— functional derivatives 129
— non conjugated, aliphatic 61
— protonation 1–3, 15, 60, 61
— β,γ-unsaturated 51
— vinyl- 129
Ketonium ions 173, 175
— acidic 62, 66
— cyclic 106
— intermediates 42, 105–106
— *cis-trans* isomerism 175, 176
Ketonium salts 12
Kinetically controlled reactions 79, 80, 85–87, 91, 93–95, 115, 133, 143, 148

L
Labeling experiments 107, 108, 112
Lactam acetals 136
Lactams 72, 131, 135, 136
Lactones 6, 25, 72, 88
Lactonium ions 59, 87, 88, 107, 108
Lactonium salts 108
Leaving groups 23, 28–30, 40
Lewis acid-base complexes 27, 33, 62
Lewis acids 28, 51, 77, 104, 140, 142
Lithium aryl compounds 151
Lyxose 118

M
Magic acid 63
Malodinitrile 164
Malonic ester 162, 164

— anion 70
α-D-Mannopyranose, dioxolenium ion 117
Mass spectroscopy 13
Mercaptals 71
Mercury-carbene complex 131
Mesityl oxide 51
Metal halides 27, 33, 34
— etherates 38
Methanolysis 112
Methosulfates 133
p-Methoxyacetophenone 78
ω-Methoxy-*n*-alkyl-*p*-bromobenzenesulfonates 100
4-Methoxybutyl brosylate 101
ω-Methoxybutyl halides 103
cis-3-Methoxycyclohexanecarboxylic acid 109
2-Methoxy-1,3-dioxolane 83
5-Methoxypentyl brosylate 103
4-Methoxypentyl chloride 102
ω-Methoxypentyl halides 103
O-Methylacetate acidium ion 8
Methyl bromide 65
Methyl brosylate 103
Methyl *trans*-3-chlorocyclohexanecarboxylate 109
2-Methyl-1,3-dioxacyclohex-1-enium ion 9
2-Methyl-1,3-dioxenium ion 9
2-Methyl-1,3-dioxolenium ion 80
2-Methyl-1,3-dioxolenium tetrafluoroborate 43, 44
Methylene compounds, reactive 129, 131, 133, 136, 137, 154
Methylene-2,5-dihydrofuran 26
Methylenepyrans 150
2-Methylenepyrans 155
4-Methylenepyrans 47, 155, 167
Methylenetriphenylphosphoranes 163, 166
Methyl ethers 104
Methylfluoride 34
Methylfluorsulfonate 26
Methyl halides 65, 104
Methyl hydrogen carbonate 68
Methyl ketones 53
Methyl orthocarbonate 13
Methyloxonium hexachloroantimonate 62
Methyloxonium ion 65, 68
N-Methyl-piperidine 89
2-Methyltetrahydrofuran 102-104
1-Methyltetrahydrofuranium ion 8
O-Methyltetrahydrofuranium ion 74
O-Methyltetrahydropyranium ion 74
2-Methyl-*cis*-4,5-tetramethylene-1,3-dioxacyclopent-1-enium ion 9

Subject Index

2-Methyl-*cis*-4,5-tetramethylene-1,3-dioxenium tetrafluoroborate 124
2-Methyl-*cis*-4,5-tetramethylene-1,3-dioxolenium ion 9
3-Methyl-2,4,6-triphenylpyrylium salts 152
Monoalkoxycarbonium ions 28, 30, 89, 90, 93
Monoalkyloxonium salts 62
Monohydroxycarbonium ions 66

N

Naphthalene derivatives 166, 167
Naphthols 52
4-(1-Naphthylmethyl)-4H-pyran 168
Neighboring-group participations 24, 100–125
— acetoxy group 111
— benzoyl group 101
— carboxylic ester groups 106–108, 114, 120, 125
— carboxylic ester groups, carbohydrates 116
— ether groups 102–105
— halogens 111
— keto carbonyl groups 105
— methoxy groups 102, *cf.* -ether groups
Nitriles 71, 128, 137
Nitrilium ions 128
Nitrilium salts 87, 137
Nitroalkane anions 58, 70
Nitrobenzene derivatives 161
Nitrobenzenesulfonic esters 47
Nitromethane 154, 161
— condensation 161
p-Nitrophenylhydrazine 133
Nitrosamines 71
NMR spectroscopy
— 1,2-acetoxonium ions of carbohydrates 117, 118
— aldehydium ions 174
— allylic coupling constants 175
— bromination of ortho carboxylic esters 86
— dialkoxycarbonium ions 177
— dialkoxycarbonium salts 14
— dimethoxymethylcarbonium ion 177
— 1,3-dioxolenium ions 14, 111, 120
— front-side participation 120
— in strongly acidic solutions 13, 15, 19, 63–69, 174–178
— inversion of trialkyloxonium ions 173
— ketonium ions 174–176
— low temperature 3, 14, 19, 63, 83, 111, 173–178
— oxonium salts 3
— protonation of carbonyl compounds 174
— protonation of dicarboxylic acids 66
— protonation of esters 66
— triethoxycarbonium ion 178
— vinyl coupling constants 175
Nonmetal halides 27, 33, 34
— etherates 38
Norbornanone 16
Norbonenone 16
Norcamphor, oxonium ion 145
Nucleophiles 65
— addition to ambident cations 79–91, 128
— addition to carboxonium ions 59
— addition to hydroxycarbonium ions 69
— anionic 65, 70, 80–91
— energy of addition products with carbonium ions 81
— neighboring-group participation 100, 101
— reaction with pyrylium salts 149, 150
— reaction with trialkyloxonium salts 70–72, 74–76
— steric requirements 89
— strong 80–82, 91, 94, 95, 146
— uncharged molecules 70, 83
— weak 70, 80–82, 86, 91, 95, 146
Nylon 133

O

Orthocarbonic esters 30, 36, 73, 93
Orthocarboxylic esters 30, 36, 39, 138, 142, 146, 148
— acyclic 42, 116
— bicyclic 112
— bromination 86
— cyclic 81, 108, 120, 121
— cyclic, halides 83
— cyclic, partial hydrolysis 116
— halides 85
— idose 117
— intermediates 84
— reaction with acetyl chloride 36
— reaction with antimony pentachloride or boron trifluoride 33
Orthodiacetates 112, 117, 124
Orthoformates 116, 129
Oxalic acid diester 85
Oxenium-carbonium ions 59
Oxenium ions 5
Oxidation
— anodic 106
— pyrylium salts 157
Oxinium ions 5

Oxocarbonium ions 5, 66, 67
3-Oxo-1,2-dithioles 72
Oxonium ions, cf. carboxonium, dialkoxycarbonium, trialkyloxonium ions
— acidic 22, 58, 63
— acyclic 6, 8
— bicyclic 173
— classification 5
— cyclic 6, 23, 24, 29
— primary 6, 8, 22
— saturated 6, 24, 60
— secondary 6, 8, 18, 22
— tertiary 6, 8, 18, 22, 58, 100–125
Oxonium salts
— acidic 60–62
— ambident tertiary, reactions 78–95
— reactions 58–95
— secondary 2, 17, 28
— tertiary 17, 27
— tertiary, intermediate formation 37
Oxygen
— atmospheric 49
— bases, basicities 61
— coordination number 8
Oxygen heterocycles
— preformed 23, 24, 45
— syntheses 23

P

Pentaacetylglucose 117
Pentadienones 45, 49, 149
Pentaphenylpyrylium salts 151, 154
Pentatriafulvenes 130
Pent-2-en-4-yn-1-one 48
Pentoses, acylated 117
Peptides 133
Permanganate 46
Phenanthrene derivatives 168
Phenolic ethers 49
Phenols 52, 61, 133
— alkylation 73
— from pyrylium salts 161
Phenylacetylene 52
2-Phenylbenzodioxolium ion 95
2-Phenyl-1,3-dioxolane 139
— derivatives 84
Phenylhydrazine 152, 158
Phenylnitromethane 162
Phenyloxonium ion 8, 62
Phenyloxonium salts 62
Phosphabenzene 159
Phosphine 159
Phosphine oxides 72
Phosphines 72

Phosphonium iodide 159
Phosphoric acid triesters 72
Phosphorins 159
Phosphorous acid triesters 72, 87
Phosphoryl fluoride 35
Photochemical rearrangements 154, 166
Pivalaldehydium ion 69
pK_a values
— acidic oxonium ions 60, 61
— protonated ethers 61
— protonated oxygen bases 61
Polyamides 133
Polyamidines 133
Polyethers 142
Polyhydroxytetrahydropyran esters 120
Polymerization, tetrahydrofuran 142
Polyphosphoric acid 49
Propenes, diacylation 51
Pyranols 46, 151, 152
Pyranolphenylhydrazide 153
Pyrans 45
2H-Pyrans 46, 149, 150, 154, 156, 160, 163, 166
4H-Pyrans 45, 154, 155, 163, 166, 167
Pyrazoline derivatives 153, 156
Pyridine derivatives 152, 158
Pyridine N-oxides 158
Pyridinium-N-carboxylic acid 158
Pyridinium salts 158
α-Pyridone 70
Pyrones 131
— alkylation 16
— reaction with electrophiles 46, 47
— reaction with nucleophiles 46
α-Pyrones 46, 47, 150
γ-Pyrones 46, 131, 150, 155
— alkylation 26, 47
— condensation with reactive methylene compounds 47
— protonation 47
— reaction with Grignard compounds 46
Pyrylium halides 17
Pyrylium ions 7, 9
Pyrylium salts 2, 149
— $C_1C_2C_2$ syntheses 47, 50–52
— $C_1C_3C_1$ syntheses 47, 50, 51
— C_1C_4 syntheses 47, 50–52
— $C_2C_1C_2$ syntheses 47, 50, 53
— C_2C_3 syntheses 47, 50, 52, 53
— C_5 syntheses 47–49
— formation 43, 45–53
— fused rings 51, 52
— reactions 149–168

Q
Quinuclidine 173

R
Ring opening
— *cis*- of 1,3-dioxolenium ions 84, 117, 121
— diaxial 114
— diequatorial 114
— 1,3-dioxolenium ions 82–84, 113–116
— 2H-pyrans 151
— 4H-pyrans 166
— pyrylium ions 155
— stereospecificity 114–116

S
Salicylaldehyde 52
Schiff bases 71
4-Selenopyrone 47
Semicarbazide 133, 136
Silver
— acetate 110, 111
— hexafluoroantimonate 27, 29, 34, 40, 114
— perchlorate 100
— tetrafluoroborate 23, 27, 29–31, 34, 40, 105, 108, 113, 124
S_N1 reactions 74, 146
S_N2 reactions 59, 65, 101, 102, 111, 146
— intramolecular 24
— trialkyloxonium salts 73–75
Solvatochromy 158
Solvolysis 100, 107, 110, 139, *cf.* acetolysis, methanolysis
— brosylates 105
— mechanisms 113
— rates 100, 101, 105, 111, 125
sp^2 hybridization, oxygen in ketonium ions 174
Stability
— dialkyloxonium salts 11
— dialkylsulfonium salts 11
— trialkyloxonium salts 11, 12
— trialkylsulfonium salts 11, 12
— triphenyloxonium salts 12, 17
Stabilization
— carbonium ions 12, 13
— effect of alkyl groups 15
— energy 13
Stereochemistry 102, 103, 114–116
Sulfides 71
Sulfonium ions 11
Sulfoxides 71
Sugars
— derivatives, acetohalo- 83, 117
— esters 123

T
2,3,4,6-Tetra-O-acetyl-1,5-anhydro-D-glucitol 122
Tetraacetylidoses 117
Tetraarylnitrobenzene derivatives 162
Tetraarylphenols 162
Tetracycline derivatives 132
Tetrafluoroborate ion 17, 19, 34
Tetrahydroborate
— lithium 145
— sodium 137, 154
Tetrahydrofuran
— derivatives 104
— polymerization 78, 141, 142
Tetrahydrofuranium ions 64, 102, 103, 141, 142
Tetrahydropyran derivatives 104
Tetrahydropyranium ions 103, 173
Tetramethoxyethylene 85, 91
Tetramethylammonium bromide 114
3,3,4,4-Tetramethyl-1,3-dioxolenium salt 91
Tetramethylurea 136
2,3,4,6-Tetraphenylpyrylium salts 152
2,3,5,6-Tetraphenylpyrylium salts 151
Thermodynamically controlled reactions 79, 80, 82–84, 86, 87, 91–95, 115, 133, 149
Thioalcohols 11
Thiocarbonyl compounds 128
Thioethers 11
Thiolactams 72, 136
Thiolactones 72
Thiophene 72
4-Thiopyrone 47
Thiopyrylium salts 160
3-Thioxo-1,2-dithioles 72
Titration, oxygen bases with perchloric acid 60
Toluene, alkylation 146
p-Toluolsulfonic acid 121
Tosylates 105, 111, 124, 125
Tosyl chloride 123
Trialkoxycarbonium ions 30, 89, 90, 93
Trialkoxyhalomethane derivatives 30
Trialkyloxonium halides, existence 1, 3
Trialkyloxonium hexachloroantimonates 34–38
Trialkyloxonium hexafluorophosphate 35
Trialkyloxonium ions, *cf.* oxonium ions
— cyclic 101
— cyclic, intermediates 102–104
— cyclic, four-membered ring 105
— cyclic, three-membered ring 34, 104
Trialkyloxonium salts 2, 3, 30, 36, 62, 87, 149

— acyclic 34–39
— alkoxide acceptors 30, 138, 139
— alkylation of carboxylic esters 39
— alkylation of pyrones 47
— alkylation with- 70–73, 128–142
— catalysts 140
— formation 34–39
— mixed- 35
— reactions 70–75
Trialkyloxonium tetrachloroaluminate 35, 73
Trialkyloxonium tetrachloroferrate 38, 73
Trialkyloxonium tetrafluoroborates 34, 36–39, 73
— alkoxide acceptors 41
— exchange of anions 34
Trialkyl phosphates 35
Trialkylphosphorothionates 35, 38
2,4,6-Trialkylpyrylium salts 153
Trialkylselenonium salts 1, 3
Trialkylsulfonium salts 1, 3, 11
Trialkyltelluronium salts 1, 3
Triarylcarbonium salts 148
Triaryloxonium salts, reactions 75–77
Triarylphosphorins 159
Tri-n-butyloxonium salts 38, 39
Tri-$tert$-butyloxonium salts, existence 39
Tri-$tert$-butylphosphorins 159
Triethylamine 161
Triethoxycarbonium ion 23, 178
Triethyl orthoformate 36
Triethyloxonium salts 36, 38
Triethylphosphine 93, 94
Trifluoroacetic acid 105, 106
Trihydroxycarbonium ion 68
Triisopropyloxonium salts, existence 39
Tri(neopentyloxy)methane 86
Trimethoxycarbonium ion 13
Tri(p-methoxyphenyl)methane 148
Trimethylamine 84
4,6,8–Trimethylazulene 164
Trimethylcarbonium ion 39, 63, 65, 68, cf. $tert$-butyl cation
Trimethyloxonium hexafluoroantimonate 34
Trimethyloxonium ion 68
Trimethyloxonium salts 36, 38
2,4,6-Trimethylpyrylium perchlorate 164
Trimethylpyrylium salts 51, 151

2,4,6-Trinitrobenzene sulfonate 18
2,4,6-Trinitrobenzenesulfonic acid 35
Trioxane 142
Triphenylcarbonium perchlorate 53
Triphenylcarbonium salts 23, 30
— alkoxide acceptors 30, 39
— cyanide acceptors 41
— hydride acceptors 46, 49
Triphenylcarbonium tetrafluoroborate 13, 34
3,5,7-Triphenyl-4H-1,2-diazepine 153
2,4,6-Triphenylnitrobenzene 161
Triphenyl orthoformate 92
Triphenyloxonium halides 17, 75
2,3,5-Triphenylphenol 161
2,4,6-Triphenylphenol 161
Triphenylphosphine oxide 163
2,4,6-Triphenylpyrylium perchlorate 152
2,4,6-Triphenylpyrylium salts 150, 152–154, 162, 166
Tri-n-propyloxonium salts 38
Tris-p-aminophenyloxonium ion 76
Tris(hydroxymethyl)phosphine 159
Tris-p-iodophenyloxonium ion 76
Tris-p-nitrophenyloxonium ion 76
Tris(trimethylsilyl)phosphine 160
Tropone 16, 129

U
Urea acetals 136, 137
Ureas 71, 131, 136
Urethanes 71, 131, 136

V
Vinyl ethers 23, 59, 93, 129
— cyclic 25, 105
4-Vinyl-1,3-dioxolenium tetrafluoraborate 40
Vinyl phosphonium ions 131

W
Walden inversion 120

X
Xanthene 148
Xanthylium ion 7
Xanthylium salts 148
α-D-Xylopyranose, 1,3-dioxolenium ion 118
Xylose 118